South Africa and the Global Hydrogen Economy

The Strategic Role of Platinum Group Metals

SOUTH AFRICA
AND THE
GLOBAL HYDROGEN
ECONOMY

THE STRATEGIC ROLE
OF
PLATINUM GROUP METALS

MAPUNGUBWE
INSTITUTE FOR STRATEGIC REFLECTION (MISTRA)

Mapungubwe Institute for Strategic Reflection (MISTRA)
First floor, Cypress Place North
Woodmead Business Park
142 Western Service Road
Woodmead 2191
Johannesburg

First published September 2013

© MISTRA 2013

ISBN 978-1-920655-68-6

Published by Real African Publishers
on behalf of the Mapungubwe Institute for Strategic Reflection
(MISTRA)

First floor, The Mills
66 Carr Street
Newtown, Johannesburg 2001

Copy editor: Angela McClelland
Indexer: Jackie Kalley

Printed and bound in South Africa

MAPUNGUBWE INSTITUTE (MISTRA)
[A NON-PROFIT COMPANY][104-474-NPO]
REGISTRATION NUMBER 2010/002262/08
["THE INSTITUTE"]

The PGM Report is dedicated to,

and in memory of, the late

Thabang Makubire

CONTENTS

FOREWORD

South Africa's endowment of mineral resources is legend. Valued at some US$2.5 trillion (about R250 trillion), these reserves constitute a national asset from which the nation can collectively draw immeasurable benefit. While mining, refining and export are in themselves important, the significance of these endowments lies in building a mature industrial cluster that combines extraction, manufacturing of machinery and value-added products, and development of engineering services – all of which can be used domestically and for export.

Platinum group metals (PGM), of which the country possesses over three-quarters of known global reserves, are one sub-sector with such possibilities, given their utility which includes jewellery, electronic goods, catalytic convertors and hydrogen fuel cells. It is on the latter that this research initiative, *South Africa and the Global Hydrogen Economy: The Strategic Role of Platinum Group Metals* has chosen to focus. The platinum catalyst is a core component of proton exchange membrane fuel cells which are emerging as a dominant hydrogen fuel cell type, capable of powering automobiles and acting as stationary devices to provide electricity.

Herein lies the confluence of factors that have inspired the Mapungubwe Institute (MISTRA) to interrogate this issue: the extensive reserves of a unique class of minerals, the utility of such natural endowments in a nascent energy value chain, and the fact that possibilities exist for such energy to be generated with as little destruction to the environment as possible.

In its long-term vision, government has set its sights on South Africa supplying about 25% of global platinum-based fuel cells by the turn of this decade. This may be a tad ambitious: but it does focus the mind on the variety of interventions required to plan appropriately for the tide, and ride the crest of a wave.

This report examines the challenges that attach to this ambition. To what extent are PGM relevant to the emergent hydrogen economy and is this economy truly on the rise? What is the state of global research on hydrogen fuel cell technology? What are the lessons that can be learnt from experience on the emergence of a 'disruptive technology'? Is the country's knowledge base, and are its capabilities suited to, and being mobilised for, the changes that are required?

At core is the issue of a knowledge-based network, already manifest in the Hydrogen South Africa (HySA) initiative in which the Department of Science and Technology has invested some resources. If – or indeed, when – the hydrogen economy takes off in earnest, the demand for PGM and fuel cells will be significantly impacted. Yet this will escalate along with capabilities to recycle PGM, which are largely non-perishable. Related to all these probabilities is the potent mix of geo-political issues relating to global security of supply, PGM trading arrangements that minimise disruptive price volatility, and social stability within the mining communities and South Africa at large. Similarly, the extent to which key role-players in the PGM-mining sector – mining corporations, government, workers and communities – are able to forge a compact informed by mutual strategic interest, will be a crucial part of the equation.

The ability to ride the crest of a wave also depends on the courage to take the plunge. This implies, among others, starting today to invest in highly skilled human capital across technologies and segments of the hydrogen fuel cell value chain. It also requires the courage to expand massively, the creation of fuel cell demand within South Africa itself.

A few countries across the globe, including the USA, Germany, Canada, Japan, China and South Korea are undertaking extensive research on hydrogen fuel cell technologies. Needless to say, among these, and between them and South Africa, can be expected some level of subliminal or even open competition. But, as the researchers note in this report, no single country can excel in this field without partnering with others. Each country will have to determine its own balance between pursuit of relative self-sufficiency, and mutually-beneficial global networking across the public and private spheres.

Besides literature reviews and interviews with experts in this field, the researchers have also attempted a novel approach, to apply the optimisation model of the South African energy system to fuel cell-based road transport.

The fledgling nature of the hydrogen economy means that many questions are as yet unanswered. MISTRA's modest aim is to contribute to the continuing exploration of, and discourse on, this subject.

The researchers, peer reviewers and other partners in this project deserve our profound gratitude. And so do the funders who have made it possible for us to undertake this work.

Joel Netshitenzhe
Executive Director

Authors

Ayender Makhuvela
Assistant Researcher, Mapungubwe Institute for Strategic Reflection (MISTRA), South Africa.

Velaphi Msimang (Ph.D.)
Head of Knowledge Economy & Scientific Advancement (KESA) Faculty, Mapungubwe Institute for Strategic Reflection (MISTRA), South Africa.

Radhika Perrot
Senior Researcher, Mapungubwe Institute for Strategic Reflection (MISTRA), South Africa

Fátima Ferraz (Ph.D.)
Environmental Consultant, FADO CONSULT, South Africa.

Vítor Ferreira (Ph.D.)
Assistant Professor, Polytechnic Institute of Leiria, Portugal.

Adrian Stone
Lecturer, Faculty of Engineering & the Built Environment, University of Cape Town, South Africa.

Bruno Merven
Lecturer, Faculty of Engineering & the Built Environment, University of Cape Town, South Africa.

Mamahloko Senatla
Lecturer, Faculty of Engineering & the Built Environment, University of Cape Town, South Africa.

ACKNOWLEDGEMENTS

MISTRA would like to take this opportunity to thank the various partners who contributed to this research outcome which has been codified into a publication. For the past two years numerous conversations across various sectors from the public and private sectors, academia and civil society helped to enrich the insights contained in this publication.

We would like to extend our gratitude to the following:

Chapter Reviewers:
Dr Bonakele Mehlomakulu (South African Bureau of Standards), Mr Paseka Leeuw (University of the Witwatersrand), Prof. Phuti Ngoepe (University of Limpopo), Prof. Harold Annegarn (Sustainable Energy Technology and Research Centre), Vatsal Bhatt (Brookhaven National Laboratory), Dr Jesika Singh (University of Limpopo)

Core Project Team:
Ms Joyce Lesia, Ms Ayender Makhuvela, Mr Wilson Manganyi, Dr Mosibudi Mangena, Mr Lufuno Marwala, Mr Tebogo Matsimela, Dr Velaphi Msiman, Dr Peter Mukoma, Ms Nosipho Mzamo, Mr Joel Netshitenzhe, Ms Radhika Perrot, Mr Rets'elisitsoe Taole

Hydrogen South Africa (HySA) strategic and technical advisors:
Dr Cordellia Sita (MISTRA Fellow), Prof. Vladimir Linkov (MISTRA Fellow), Dr Dmitri Bessarov (MISTRA Fellow), Ms Mandy Mtyelwa (HySA), Dr Cosmas Chiteme (HySA)

Contributors through scheduled interviews:
Dr Iraj Abedian (Pan African Capital Holdings), Dr Sakib Khan (Enerleq), Magatho Mello (Maraswi Consulting Services), Feizel Matthews (Alfene Project Management cc), Ms Anthea Bath (Terracotta, formerly with Anglo Platinum), Ms Ingrid Davids (Ballard), Karim Kassam (Ballard)

Editorial Team:
Mr Joel Netshitenzhe, Dr Velaphi Msimang, Mr Wilson Manganyi, Ms Rachel Browne, Mr Nkoe Montja

MISTRA Staff:
Ms Hope Prince and Ms Thabang Moerane for providing logistical support
Mr Loyiso Ntshikila for making sure that the project was within budget

Project Funders:
Department of Science and Technology
Xstrata

MISTRA Funders and Donors:

- Adcorp
- Ahanang Hardware and Construction
- Anglo Platinum
- Aveng
- Baswa
- Batho Batho Trust
- Brimstone
- Chancellor House Holdings
- Cyril Ramaphosa
- Darene Foundation
- De Beers Consolidated Mines Limited
- First Rand Foundation
- Ford Foundation
- Lincoln Mali
- Liphosa Matodzi
- Matemeku
- Mathews Phosa
- MTN Group
- Mvelaphanda Management Services
- Nedbank
- Ogilvy
- Roger Jardine
- Safika Holdings
- Sexwale Family Foundation
- Shanduka Group
- Simeka Group
- South African Breweries
- Standard Bank
- Transnet Foundation
- Yellowwoods

ABBREVIATIONS

3M	Minnesota Mining and Manufacturing Company
ALT	alternative scenario
ADM	advanced micro devices
BASF	Badische Anilin- und Soda-Fabrik
BBBEE	Broad-Based Black Economic Empowerment
BBL	barrel (of crude oil)
BEV	battery electric vehicle
BRT	bus rapid transport
CARB	California Air Resources Board
CHP	combined heat power
CND	Canadian dollars
CO_2	carbon dioxide
COG	coal gasification
CRM	critical raw material
CSP	concentrated solar power
CTL	coal to liquid
DOE	Department of Energy (US)
DST	Department of Science and Technology
EIA	Energy Information Administration (US)
EITI	Extractive Industries Transparency Initiative
ELE	water electrolysis hydrogen production
EPA	Environmental Protection Agency (US)
ERC	Energy Research Centre (CT, RSA)
ESOP	Employee share ownership programmes
FCEV	fuel cell electric vehicle
FCV	fuel cell vehicle
FIT	feed in tariffs
GH_2	gaseous hydrogen
GJ	gigajoule
GTL	gas to liquid
H_2	hydrogen
HCV	heavy commercial vehicles (> 8500 kg GVM)
HEV	hybrid electric vehicle
HFCT	hydrogen and fuel cell technology
HySA	Hydrogen South Africa
ICE	internal combustion engine
ICMM	International Council on Mining and Metals
IFCI	Institute for Fuel Cell Innovation
ILC	industry life cycle
IP	intellectual property
IPC	international patent classification

IPP	independent power producers
Ir	Iridium
IRP	integrated resource plan
LCR	local content requirements
LCV	light commercial vehicle
MBT	minibus-taxi
MCV	medium commercial vehicle
MEAs	membrane electrode assemblies
Mt	megaton (million metric tons)
MYPD3	multi-year price determination year three
NAAMSA	National Association of Automobile Manufacturers of South Africa
Natis/eNatis	South African National Traffic Information System
NOCON	unconstrained scenario
NRC	National Research Council
NREL	National Renewable Energy Laboratories (US)
O&M	operation & maintenance
OEM	original equipment manufacturer
Os	osmium
PAFC	phosphoric acid fuel cells
Pd	palladium
PEM	proton exchange membrane
PEM	proton exchange membrane / polymer electrolyte membrane
PEMFC	proton exchange membrane fuel cell
PGM	platinum group metals
pkm	passenger.kilometres
PRO	public research organisation
Pt	platinum
PV	photovoltaics
Rd	rhodium
RD&D	research, development and demonstration
RDI	research, development and innovation
REACH	registration, evaluation, authorisation and restriction of chemicals
REF	reference scenario
REIPP	Renewable Energy Independent Power Producer
REIPPP	Renewable Energy Independent Power Producer & Procurement Programme
RIS	regional innovation systems
RPS	renewable portfolio standard
Ru	ruthenium
S&T	science & technology
SAPIA	South African Petroleum Industry Association
SATIM	The South African Times Model, a TIMES-MARKAL linear optimisation model of the South African energy system

SFU	Simon Fraser University
SMR	steam methane reforming
SOFC	solid oxide fuel cells
SSI	sectoral systems innovation
STS	science and technology studies
SUV	sport utility vehicle
t.km	ton.kilometres
TIFC	Toshiba fuel cell power system
TIS	technological innovation system
TPES	total primary energy supply
UBC	University of British Columbia
UG2	Upper Group 2
UNCSD	United Nations Conference on Sustainable Development
UOP LLC	Universal Oil Products Limited Liability Company
UTC	United Technologies Corporation
WBCSD	World Business Council for Sustainable Development
ZEV	zero emissions vehicles

LIST OF FIGURES AND TABLES

CHAPTER 6:

CHAPTER 7:

CHAPTER 8:

CHAPTER 1

PGM AND OTHER STRATEGIC MINERALS IN THE HYDROGEN ECONOMY: SOUTH AFRICA'S BOON?

Ayender Makhuvela and Velaphi Msimang

1 INTRODUCTION

Cited in the country's National Development Plan (NDP) Diagnostic document (National Planning Commission, 2011), a recent Citibank report ranks South Africa as arguably the richest country in the world in terms of mineral resources. At US\$2.5 trillion, more than 90 per cent of these endowments are attributed to the country's Platinum Group Metal (PGM) reserves. Consisting of platinum, palladium, rhodium, iridium, ruthenium, and osmium, PGM are preferred because of their unique and often inimitable[1] physical and chemical properties in numerous industries, including – in order of size of current demand – catalytic converters, jewellery, and industrial applications.

Contrasting this strategic reality are the recent and on-going media reports on platinum mines, which depict an image of a South African PGM industry marked by strife, and apparently in terminal decline (Harvey, 2013). While the ebbs and flows of labour relations and global business cycles are critical in determining the strategic importance of mineral endowments, attention to these issues should not distract from the long-term opportunities which, in turn, can help address the short-term challenges.

This chapter aims to highlight the significant role that PGM play – and can

---

1. Their unique chemical and physical properties, including durability, resistance to corrosion and oxidation, high melting points, electrical stability, and catalytic abilities, make them indispensable, with a notable reduction in performance being observed in those applications where non-precious metals are substituted for platinum (Yang, 2009).

play – in society and the importance for South Africa to fulfil a responsible role in the stewardship of these resources for the world. Specifically, it looks back at the history of the role of PGM, explores current uses of these precious metals, recent trends in their demand and applications, and assesses possible applications for the near and longer term future.

Done at greater length in other chapters, much attention is paid to the prospect of PGM uses in the prospective hydrogen economy and the transition towards a low carbon global energy system. Particularly with the advent of intermittent renewable energy, the electrolytic and 'load levelling' capabilities of a combination between the renewable energy conversion technology and the fuel cells suggest a potentially significant role for PGM (Aabakken, 2006). Other mineral resources of which the country has significant reserves, including titanium and manganese, also have a role in this envisaged low carbon energy system.

The chapter is informed in part by interviews with key stakeholders in the industry, and also by a review of the existing literature.

FROM PLATINA TO PRECIOUS METALS[2]

Going as far back as 100 BC, the Inca civilisations of the South Americas are some of the earliest recorded users of platinum, which they would alloy with copper for use as ornaments (Platinum Guild International, 2012). Earlier uses going back to 7 BC by Egyptians have also been recorded (McDonald & Hunt, 1982).

When they first encountered it during the 16th century, Spanish Conquistadors named it platina (for 'little silver') because they regarded platinum nuggets – at purity levels averaging between 77 and 86 per cent – as a nuisance in their search for alluvial gold. Only in the 18th century was it realised that platina was in fact a group of elements rather than a single metal.

First noted for its unique catalytic properties by Sir Humphrey David in his 1817 lecture to the Royal Society of London (Brenan, 2008), industrial uses of platinum began in the late 19th century, with the use of platinum for boilers used in sulphuric acid manufacturing. These and other milestones in the applications of PGM are illustrated in Figure 1, which also reflects their impact on the price of these precious metals.

A more detailed description of the trends in applications of PGM and possibilities for the future is provided further on.

---

2. Unless otherwise specified, this section borrows from Cramer (Cramer, 2000).

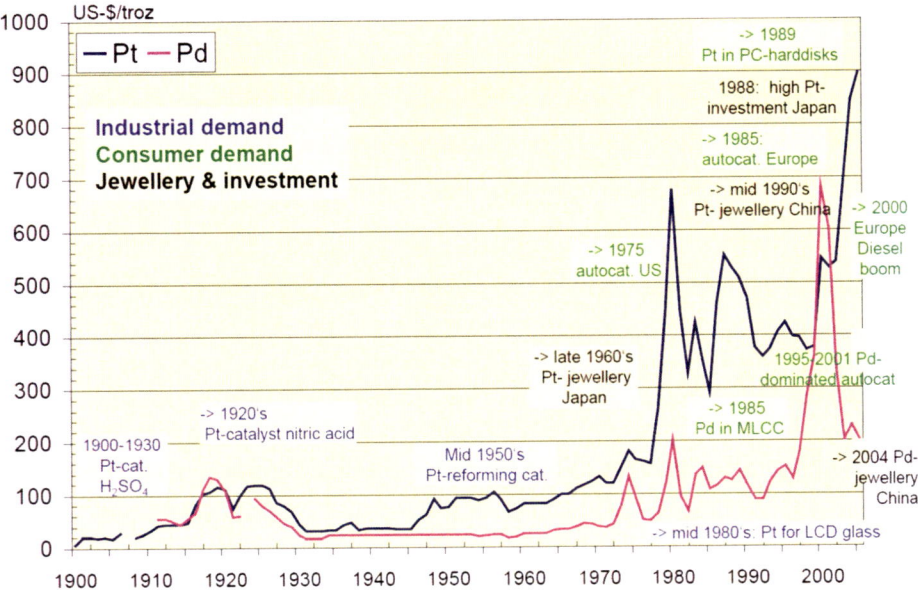

Figure 1: Long-term price trajectories of platinum and palladium and milestones in their applications. Source: (Hageluken, 2006)

2 WORLD RESERVES OF PLATINUM GROUP METALS

South Africa is endowed with an estimated 85 per cent of the world PGM known reserves. This, it is postulated, resulted from high temperature magmatic processes that formed the Bushveld Igneous Complex, which extends over 300 kilometres, with the eastern limb being in the north-east part of the country and the western limb in the south-west (Cawthorn, 2010). In South Africa, PGM are mined from three different ore bodies, including the Merensky Reef, the Upper Group 2 (UG2), and the Platreef.

As will be appreciated following a description of their contemporary uses, PGM have been argued to be 'too precious for fuel cells' (Garland, *et al.*, 2002), with some even proposing military interventions to assure security of PGM supplies (Burgess, 2010).

3 CONTEMPORARY USES OF PLATINUM GROUP METALS

PGM have assumed an important role in relation to numerous applications, 'including vehicle catalysts for controlling vehicle pollution, chemical catalysts and coatings, dental alloys, electronic components and computer hard discs, fuel cells for power generation, glassmaking equipment, investment coinage, jewelry [sic], medicines, and petroleum catalysts for gasoline refining' (Blewis & Wilburn, 2010 pg. 7). According to the International Platinum Group Metals Association (IPA), 'one in four of the goods manufactured today either contains PGMs or had PGMs play a key role in their manufacture' (IPA, n.d.).

Bearing testimony to growing prospects that further industrial applications of PGM are in the offing, examinations of worldwide literature (Gavin, 2010) and patents (Seymour, 2008) going as far back as the early 1980s[3] until 2008 reveal an increasing trend in the focus of research and commercialisation of PGM-based technologies. In the family of PGM elements, interest in platinum and ruthenium (in publications) and platinum and palladium (patents) leads that in the other platinum group elements (in the order stated). As testimony to the uniqueness of their physical and chemical properties, these same studies reveal that interest in PGM outpaces that in other mineral resources.

As is further described in the chapter on the geopolitics of PGM in a hydrogen economy, all developed and emerging economies have placed high priority on the importance of access to PGM, and have instituted a number of policy measures as a result.

4 TRENDS IN PGM DEMAND AND APPLICATIONS

Particularly in Europe, where diesel engine-powered cars are preferred, a major platinum application is in catalytic converters, in which the metal catalyses the oxidation and reduction of pollutant gases that would otherwise lead to photochemical smog. The use of diesel as a fuel source and the stringent environmental regulations within the EU – and increasing stringency in developing and emerging economies – underpin the high demand for platinum-based catalytic converters.

In populous emerging economies such as China and India, industrialisation and growing concerns about air pollution and climate change are also pointing towards a growing role for PGM, with PGM-based original equipment manufacturers (OEM) opening major facilities in these countries (Chen, 2001), or announcing the intent to do so (Umicore, 2011).

Driven by US and European environmental legislation to stem smog-causing exhaust emissions from their car industries in the 1970s, and enabled

---

3. Predominantly due to the growth of catalytic converter and jewellery markets for them, PGM mined since the early 1980s account for more than 80% of all PGM ever mined (Hageluken, 2012).

by investment in research and development (R&D), demand for PGM in catalytic converters increased rapidly to overtake all other previous applications and boost overall demand for these minerals (Whitburn, 2012).

In Figure 2 global demand for platinum in catalytic converters (over the 10-year period from 2002) is shown, with the EU reflecting the largest uptake as a result of the popularity of diesel-fuelled cars.

Other than in transportation, platinum has numerous uses in other

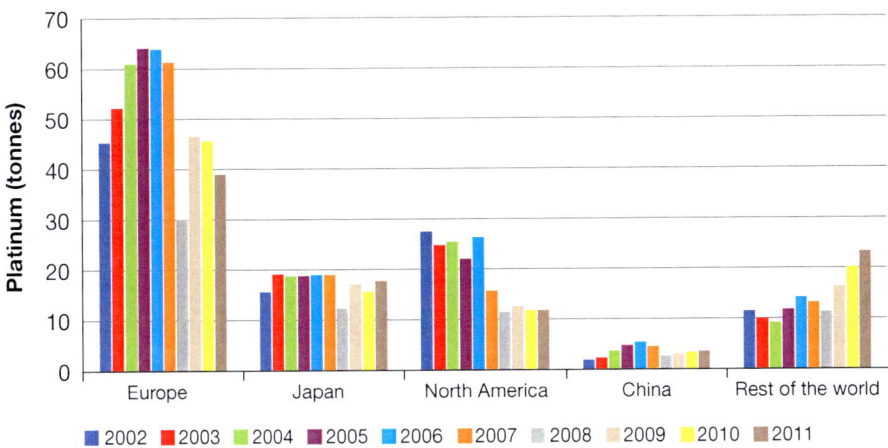

Figure 2: Platinum auto catalyst demand in the world (Johnson Matthey Plc, 2013)

sectors as shown in Figure 3 and briefly described earlier. Particularly in the populous developing and emerging economies, the improving standard of living is reflected in a marked increase in the demand for PGM-based goods from the sectors identified.

In most countries that manufacture jewellery (the second biggest market for platinum), the mineral is alloyed with other PGM such as palladium, ruthenium and iridium and alongside other metals such as copper and cobalt to optimise its working characteristics and wear properties. In this mix, platinum normally constitutes 85 per cent of fabricated jewellery. This improves its strength and resistance to tarnishing over time, and enables it to permanently retain its shape (Johnson Matthey Plc, 2013).

Industries that use platinum include the chemical and electronics sector, glass and glass fibre manufacturing, the medical sector, and petroleum refining.

Figure 4 shows demand for platinum jewellery. In China, the jewellery

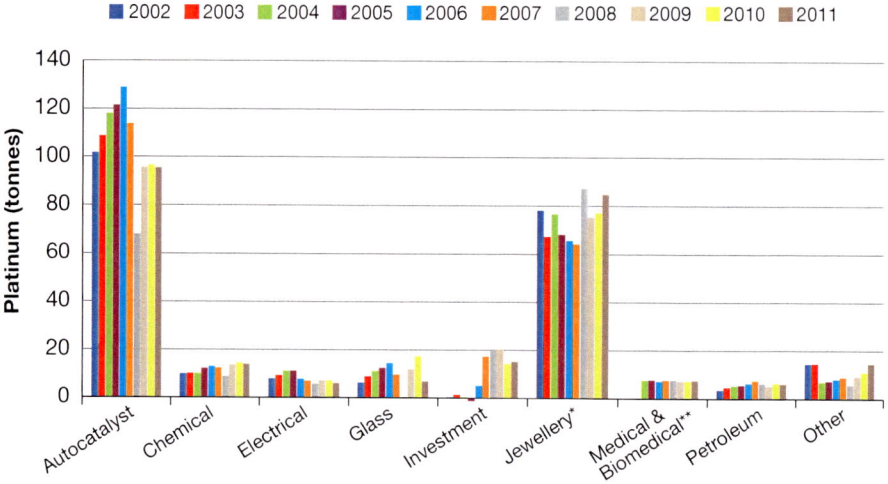

Figure 3: Platinum demand by sector (Johnson Matthey Plc, 2013)

market for the mineral exceeds that for all other markets by far (Johnson Matthey Plc, 2013). At more than 60 tonnes, demand was at an all-time high in 2009 and 2012.

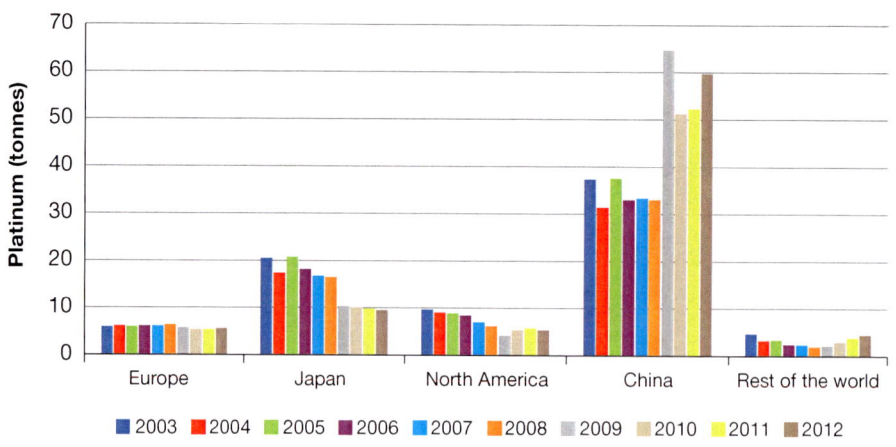

Figure 4: Platinum jewellery demand (Johnson Matthey Plc, 2013)

5 Prospects for future PGM applications

While research and patenting[4] trends are not definitive indicators of prospects for technologies or industries to emerge, they do signal the scope of possibilities (Griliches, 1998). Carefully done, a regional focus of these also indicates the prospective geography of markets for the emergent products and services. To this extent, the analyses by Seymour (Seymour, 2008) and Gavin (Gavin, 2010), which were briefly discussed earlier, provides useful leads in the probe for possible applications that may emerge in the short- to medium-term (using patents), and in the longer term (research publications).

5.1 Short- to medium-term prospects

Apparent in the patterns reflected in Figure 5, the unique catalytic and electrical properties of PGM are likely to see them continue to play a significant role in markets for both homogeneous and heterogeneous catalysts, semiconductors of electronics, and fuel cells and batteries.

Based on analysis of more than 13 000 documents, and further described by Seymour (Seymour, 2008), a sampled analysis of PGM applications (limited to palladium and platinum, and between 2003 and 2007) is illustrated in Figure 6. Evident in it are applications where platinum (represented by red dots) is likely to remain dominant (e.g. 'fuel cell

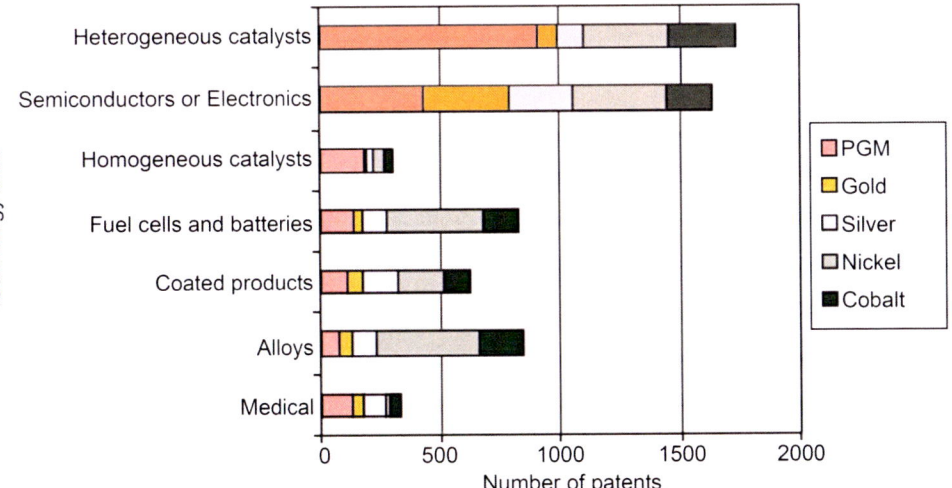

Figure 5: Patents of various metals in selected fields of application. Source: (Seymour, 2008)

---

4. A patent is a document that describes an invention and grants the inventor proprietary rights over the use and/or production of the described invention.

electrodes' and 'silicone rubber polysiloxane'), where palladium will dominate (e.g. 'plating deposited substrate', which represents electronic applications), or where both metals will remain important (e.g. 'exhaust engine oxide', representing catalytic converter applications).

Indeed patenting activity in auto catalysts suggests that any presumption that innovation around the internal combustion engine has ground to a halt would be mistaken. In any case, PGM are used in both electric (battery or fuel cell), and internal combustion engine-powered cars. As stated before, this same analysis (Seymour, 2008) showed that patenting of PGM-related technologies is trending in an upward trajectory.

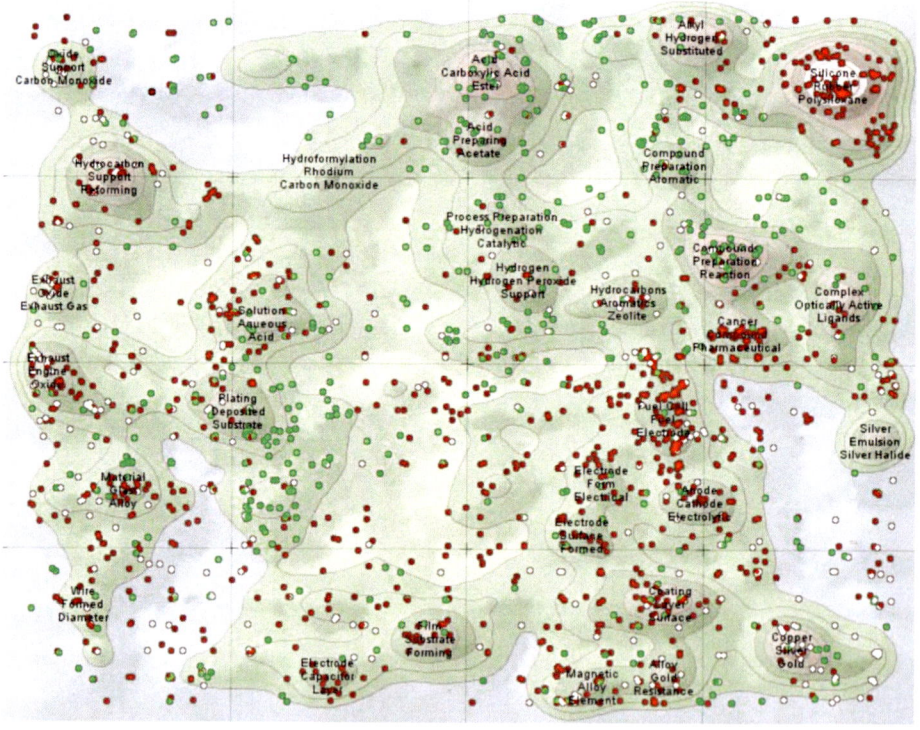

Figure 6: Fields of application of platinum patents (red dots), palladium patents (green dots), and patents covering both of them (white dots) for the period 2003–2007. Source: (Seymour, 2008)

Ranked by the number of patents they registered in the US, and shown in Table 1, most of the top 10 assignees are multinationals headquartered in the US, Europe and Asia.

A phenomenon of interest is the strong correlation between platinum

Rank	'Platinum' in patent title (1,611 patents)	'Platinum' in patent title, abstract or claims (8,878 patents)	'Platinum' in patent full-text but not title, abstract or claims (35 663 patents)
1	Micron Technology	Micron Technology	Micron Technology
2	General Electric	General Electric	Semiconductor Energy laboratory
3	Shin-Etsu Chemical	IBM	Fuji Photo Film
4	UOP LLC	Samsung Electronics	Eastman Kodak
5	Engelhard Corporation[5]	AMD	Canon KK
6	Dow Corning	Matsushita	Matsushita
7	Matsushita	Shin-Etsu Chemical	General Electric
8	Texas Instruments	Intel Corporation	3M
9	Dow Corning	Infineon Technologies	IBM
10	IBM	Hitachi	NGK Insulators

Table 1: Top 10 assignees granted US patents (between 2001–2007). Adapted from Seymour (Seymour, 2008)

production and the number of platinum patents (Figure 7). In this particular case, there is reason to believe that the role of air pollution legislation in North America, which is often credited for the development and diffusion of catalytic converters (Tao, *et al.*, 2011), played an instrumental role in driving increased demand for PGM.

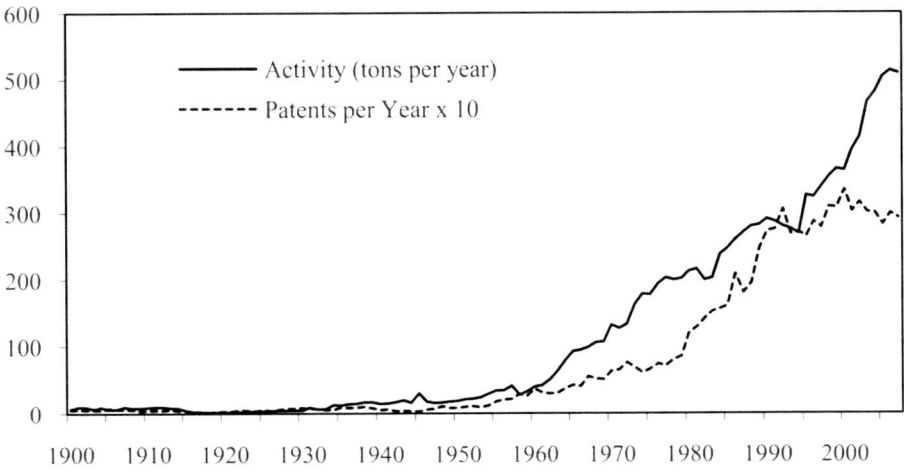

Figure 7[6]: The relationship between platinum production and platinum patents. Source: (Connelly, 2010)

---

5. Now BASF Catalysts.
6. The units used in the graph are in USA tons. Conversion to British / SA units: 1 tonne equivalent to 1.12 ton.

5.2. MEDIUM- TO LONGER TERM PROSPECTS

Normally assumed to be representative of basic research, and thus still relatively far from commercialisation, the non-patent literature provides an indication of medium- to long-term prospects for technologies or industries that may emerge.

Gavin's analysis of more than 5,000 non-patent publications published over the 10-year period running from 1998, indicates that platinum was the largest contributor to the 73 per cent increase in PGM research observed (Gavin, 2010). Based on the location of the primary researcher, trends in the geographic distribution of these publications are shown in Figure 8. The increase in PGM research from Asia clearly outpaced that in Europe and

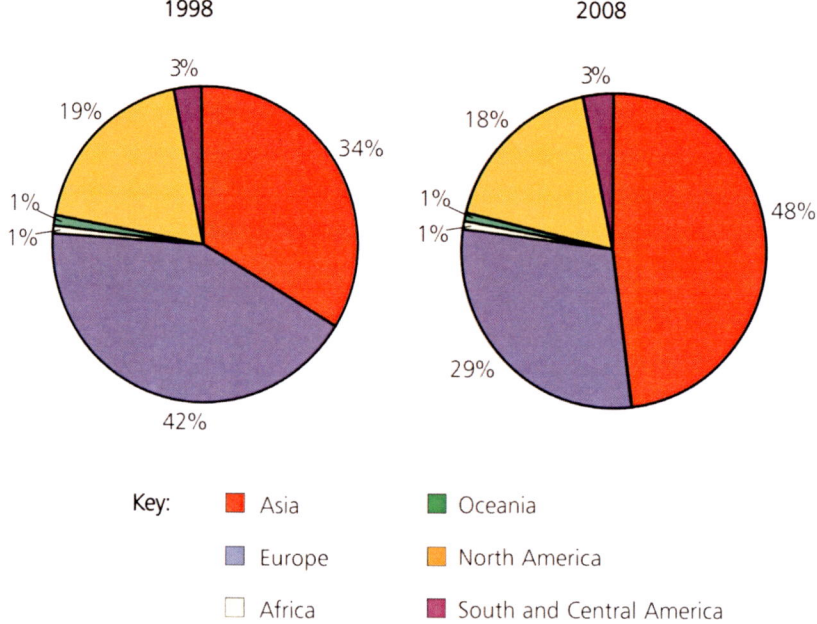

Figure 8: Comparison of the relative geographical distribution of papers on PGM research published in 1998 and 2008. Source: (Gavin, 2010)

North America. Done by the same author, a further breakdown of Asia reveals that, followed by South Korea, India, Taiwan, and Japan, China accounts for much of the acceleration of PGM research in the region[7].

Coupled with parallel efforts by China to lock in PGM supplies from South Africa (Lapper, 2010), elements of a scenario where raw PGM are shipped to China for downstream manufacturing begin to loom large.

---

7. Over the same period of analysis, China quadrupled their PGM research outputs. South Korea's and India's outputs grew by a factor of 3 and 2, respectively.

6 The PGM Supply Chain

Schematised in Figure 9, the PGM supply chain begins (upstream) with exploration and mining of ore bodies, primary mineral processing, and smelting to separate the PGM from the base metals that they are associated with. This is followed by a hydrometallurgical process of leaching the PGM concentrate in aqua-regia solutions, which is further sent for refining to separate and extract each of the PGM. A number of processes involving other stakeholders then follow towards the development of value-added products, their use, disposal and recycling.

This supply chain, which is further described in the geopolitics chapter

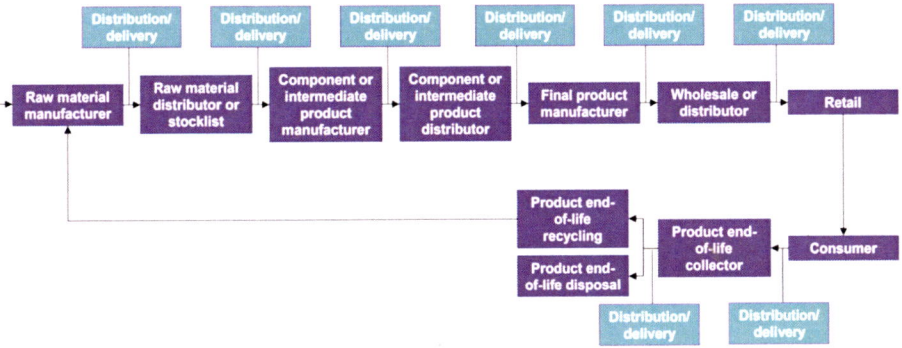

Figure 9: Illustration of the PGM upstream and downstream supply chain. Source: (Alonso, *et al.*, 2007)

and elsewhere (Steinweg, 2008), is dominated by a few major companies that are members of the International Platinum Association. These include (upstream) Anglo Platinum (South Africa), Norilsk Nickel (Russia), Impala (South Africa), and (downstream) Umicore (Belgium), Heraeus (Germany), Johnson Matthey Plc (United Kingdom), Tanaka (Japan), and BASF (Germany).

6.1 PGM Recycling

The physical and chemical properties of PGM make them non-perishable and non-consumable, which implies restorability to their pure form. Together with the prospect of substitution, the recycling of PGM is regarded as one of the most strategic and quickest means to escape dependency on South Africa and mitigate the impact of price volatility characteristic of commodity markets. Indeed (particularly in the case of platinum), as illustrated in Figure 10, recycling trends seem to track price trajectories of these precious metals – though account should also be taken of factors such as, for instance: higher demand for vehicles with catalytic converters, and the higher number of vehicles with such converters now reaching the end of their life cycle.

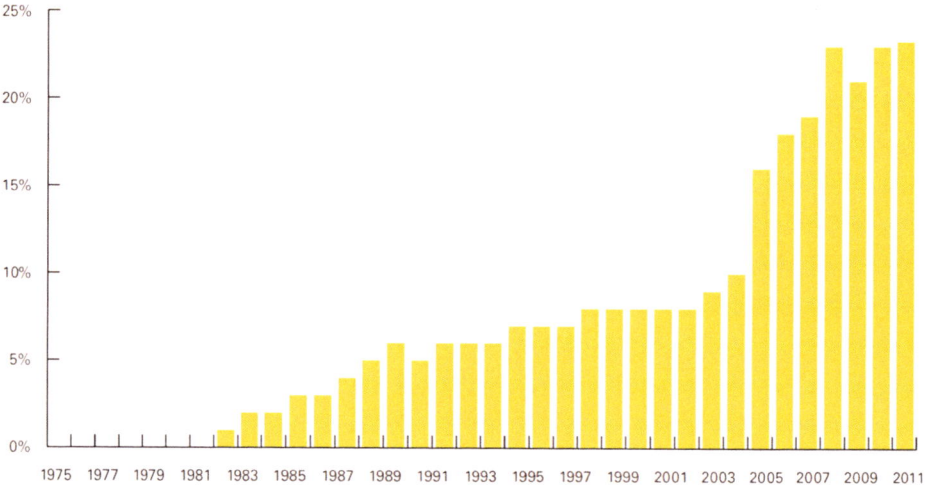

Figure 10: Platinum recycled as a percentage of total demand. Source: (Whitburn, 2012)

However, as has been modelled and is demonstrated by the results of such modelling in Figure 11, recycling also has a dampening effect on the fluctuation of platinum prices. This poses a challenge for miners of virgin ores as the profitability of their operations is constrained by their high input costs (certainly relative to those of recyclers).

The recycling rate of platinum is highest in industry applications (80–90%), then the catalytic converter industry (50–55%), and lowest in the electronics sector (0–5%). (International Resource Panel, 2011.) Recovery rates (of platinum) of more than 95% are possible (Hageluken, 2012). However, the end-use devices that contain PGM first have to reach the gate of the recycling facility. Of the PGM derived from devices at end of use, catalytic converters contribute most of the recycled platinum, followed by the jewellery industry.

Signalling the potential recognised in PGM recycling, Figure 12 shows an escalation in the number of innovations in recycling technologies. Indeed, it also tracks the PGM price trends and other factors noted earlier.

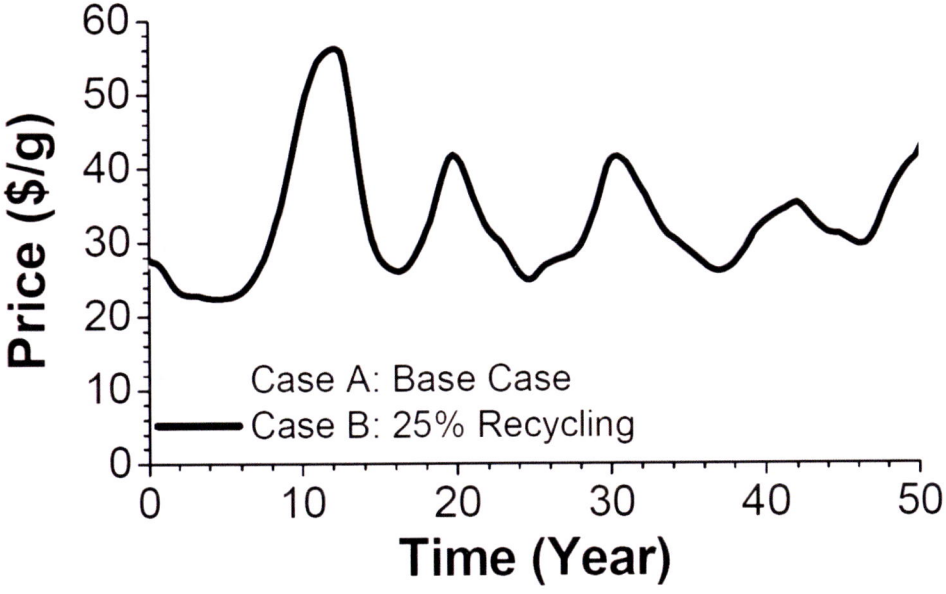

Figure 11: 50-year price profile for platinum market models for base case ('Case A') and 25% of base case recycling rate ('Case B'). Source: (Alonso, *et al.*, 2009)

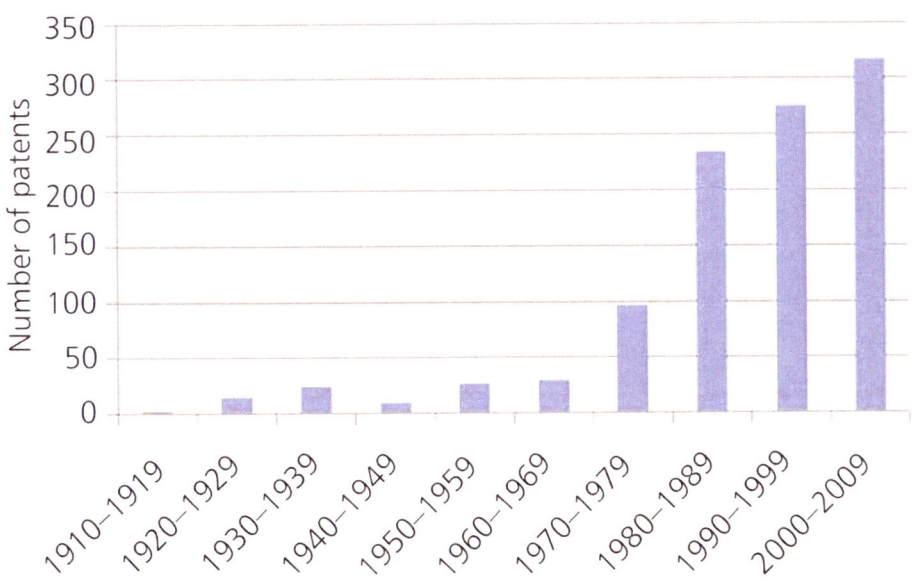

Figure 12: Number of PGM recycling patents since 1910. Source: (Hageluken, 2012)

6.2   PALLADIUM AND OTHER PGMs

Like platinum, palladium has similar applications within the auto catalyst, electrical and jewellery sectors (Johnson Matthey Plc, 2013). Russia leads the world in terms of supply of palladium, as shown in Figure 13.

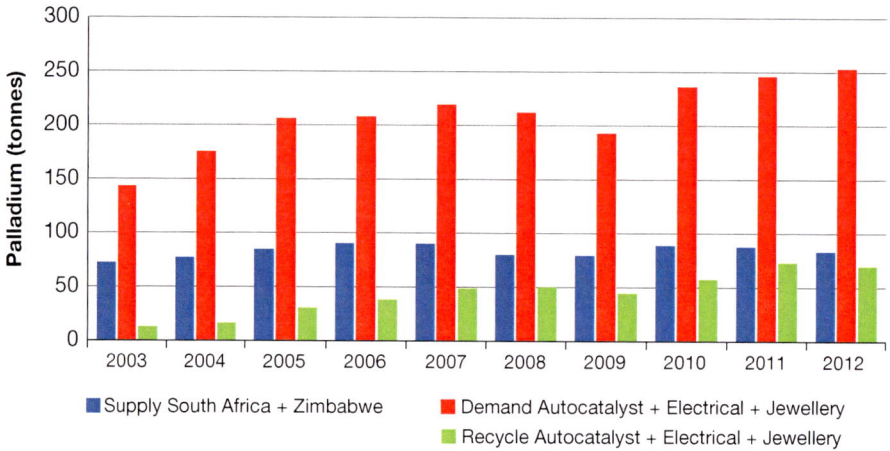

Figure 13: Palladium demand, supply and recycling (Johnson Matthey Plc)

In Russia, PGM mining is linked to nickel mining, with PGM recovered from the production process as by-products. The chemical properties of palladium are similar to those of platinum with the advantage of its lower price; hence it is used as a substitute for platinum in most applications other than fuel cells and diesel catalytic convertors (Forrest & Clarke, 2006). However, due to its susceptibility to poisoning by trace metals found in diesel fuel, its utility is limited to petrol-powered vehicles.

As noted earlier, the observed increase in recycling figures of auto-catalysts is partly as a result of the increased purchases of new vehicles and the scrapping of old ones that incorporate this technology. Recycling trends also tend to track prices of these precious metals (Johnson Matthey, 2012).

The enforcement of stricter emissions control standards in the past decade-and-a-half has led to vehicle manufacturers complying by increasing catalyst loading while gradually substituting the use of platinum with palladium in petrol-fuelled vehicles.

Rhodium is the other significant PGM largely produced in South Africa and Zimbabwe, followed by Russia. Like the other PGM, rhodium also finds application mainly in auto-catalysts.

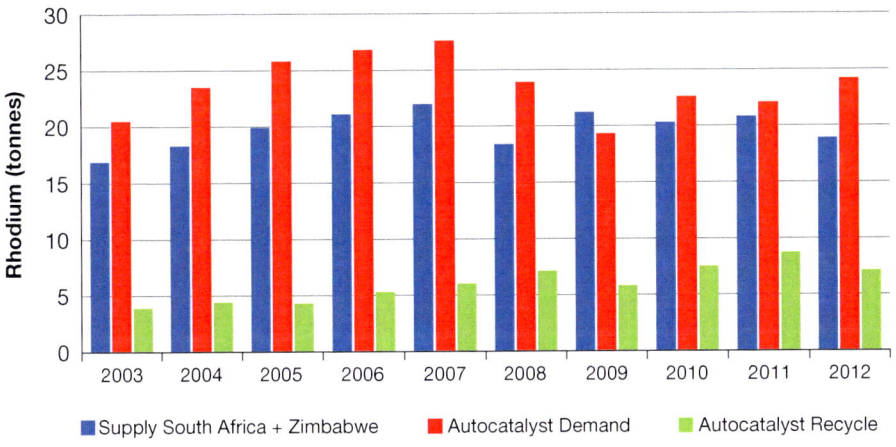

Figure 14: Rhodium supply, demand and recycle (Johnson Matthey Plc)

7 NEW ENERGY PARADIGM

As described in the chapter on the state of the global hydrogen economy, governments and private sector stakeholders in most developed economies are investing resources towards the development of a hydrogen economy, through which they hope to escape the stranglehold of a few oil-producing countries on their transport sectors, and diversify their energy systems away from fossil fuels. However, a recent multi-year research report launched at the 2012 Rio+20 United Nations Conference on Sustainable Development (UNCSD), and developed by more than 500 international experts in fields related to energy, raised a number of serious concerns around an apparent lack of appreciation for the time, risk and effort required to transition energy systems to new sources, and the mismatch between the required and actual efforts necessary to realise such transformations. Among these are the multi-decadal timescales required for new energy sources to displace incumbent ones (Figure 15)[8], the mistaken focus of global efforts on energy supply-side technology solutions and neglect of near term energy demand side options, the need for well-co-ordinated supply and demand side policy measures (exemplified by the successful Brazilian ethanol industry), and the inadequate investment into cultivation of the 'absorptive capacity'[9] necessary for

8. Among noticeable trends are the lag times involved in the penetration of new sources (see the pink line for renewable energy), and (since 1975), the fact that the decline in the role of coal and biomass has slowed down, and the gradual ascendance of gas. Recent discoveries of new oil reserves and breakthroughs in hydraulic fracturing technology also suggest the likely persistence of (shale) oil well into the future.

9. Defined and discussed at length elsewhere (Cohen & Levinthal, 1990), absorptive capacity is the ability to recognise, assimilate, and take advantage of (commercial) opportunities provided by new knowledge sourced externally.

developing countries to successfully acquire and indigenise – and thus derive maximum benefit from – capabilities in new low carbon energy solutions.

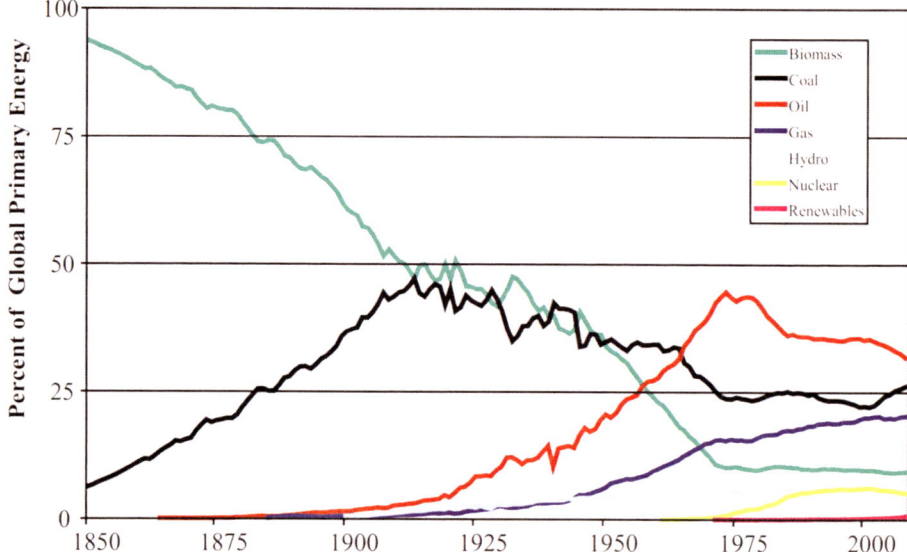

Figure 15: Evolution of primary energy shown as shares of different energy sources. Source: (Johansson, *et al.*, 2012)

Though their efficiencies are higher than those of the incumbent technologies, the lower overall system efficiency of fuel cells challenges their potential to play a significant role in the envisaged low carbon energy system, and thus raises questions about the merits of the hydrogen economy in a world of increasingly costly energy resources. Nevertheless, to the extent that markets exist for these technologies, the opportunity should not be missed to leverage them and thus maximise value from South Africa's PGM endowment.

While the dominance of the platinum-based fuel cell in cars and PGM availability largely restricted to South Africa also raises concerns about security of supplies, the non-perishability and non-consumability of PGM imply that the global stock of platinum will remain largely constant, and thus recoverable from the various end-use applications that include these precious minerals. It can also be argued that the pick-up in demand as uses of PGM are expanded – as in fuel cell technology – will imply greater demand for mined PGM; while, at the same time, recycling will reduce the pace at which global PGM reserves are depleted.

South Africa views hydrogen and fuel cell technology as a platform to secure the country's future energy needs and advance its scientific and economic interests (The South African Agency for Science and Technology Advancement, 2012). Hydrogen is an abundant element, although it is found in nature bound to other elements. If the unbound hydrogen molecule could be utilised as a fuel carrier, it would play a critical role in reducing the world's dependence on fossil fuels. The strategic posture of the European Union (EU), which sources 80 per cent of its transportation energy from outside of the region, is to encourage its members to develop approaches that reduce such dependence (Pollet, 2012). To that end, hydrogen and fuel cell technology affords countries the ability to leverage locally available fuel sources for use in their transport sectors.

Described in further detail elsewhere (including in other chapters herein), the importance of fuel cell technology lies in its utilisation of a catalysed electrochemical reaction to convert chemical energy into electricity, with water and heat as its by-products (Pollet, 2012)[10].

When the fuel cell is used with hydrogen from renewable energy, the environmental-friendliness of the process is enhanced manyfold.

8 STRATEGIC ROLE OF PGM AND OTHER STRATEGIC MINERALS IN THE HYDROGEN ECONOMY

In his presentation at the PreCOP17 conference hosted by MISTRA, Dr Vladimir Linkov proposed a hydrogen energy system life cycle, in which the

Components	Type of PGM used
Platinum to precursors	platinum and palladium
Precursor to catalysts	platinum, palladium, ruthenium, iridium, osmium and rhodium
Catalyst to Membrane Electrode Assemblies (MEAs)	platinum, palladium, ruthenium, iridium and osmium
High purity hydrogen generation	iridium and ruthenium
Membranes for hydrogen separation	palladium
Metal hydrides	palladium
Super capacitors and batteries	ruthenium and iridium

Table 2: PGM applicability in a hydrogen energy system life cycle. (Linkov, see further illustration in Figure 16)

10. Pollet, B. G., Staffell, I. & Shang, J. L. (2012). Current status of hybrid, battery and fuel cell electric vehicles: From electrochemistry to market prospects, Electrochimica Acta, manuscript submitted for publication p. 6.

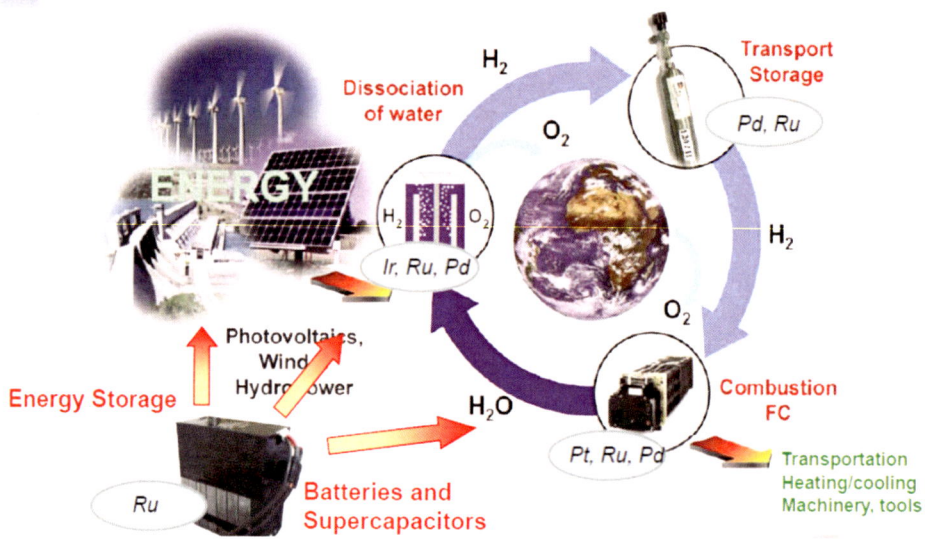

Figure 16: PGM value chain in hydrogen economy. (Linkov, 2011)

whole of the PGM play a significant role in the hydrogen economy (Linkov, 2010). South Africa seeks to play a leading role in the development of the fuel cell catalysts, using organic precursors, which are inputs in the development of the fuel cells.

8.1. PGM FOR HYDROGEN PRODUCTION AND STORAGE

Figure 16 illustrates the value chain of the PGM elements, and how these elements contribute to the production of hydrogen and energy storage. The proposed pathways of utilising electric energy sourced from renewable sources can also be used to electrolyse water as the cleanest pathway to produce hydrogen, providing a zero carbon footprint process. An electrolyser produces hydrogen from clean water by splitting the latter into constituent elements, i.e. hydrogen and oxygen. This process results in hydrogen embrittlement of the material of construction caused by the interaction between the hydrogen and the ferrous lattice, which is a hindrance in this process. (Nelson, *et al.*, 1971). It is therefore essential to use an alloy of iridium, ruthenium and palladium in hydrogen production through electrolysis (Linkov, 2011; Bishop & Stern, 1963).

Other PGM catalysts employed in the dissociation of water are ruthenium and palladium, where ruthenium accelerates the splitting of the hydrogen atom (Linkov, 2011) while palladium readily absorbs hydrogen up to 900 times its own volume (LANL, n.d.). Palladium is thus used in industrial applications for hydrogenation and dehydrogenation.

As discussed in the chapter on the state of the hydrogen economy, most hydrogen produced for commercial use is derived from conventional fossil fuels. The role of PGM in this process has been confirmed by others (Kikuchi, *et al.*, 2006).

8.2 PGM AND HYDROGEN STORAGE

Hydrogen can be stored either as a gas or a liquid. Yet its storage poses a huge technical challenge because of its low density. Typically, a gas cylinder weighing 66.6 kg is required to store 0.74 kg of hydrogen gas (Afrox, 2013).

Material-based storage devices can be used to store hydrogen in small volumes. Metal hydrates, which are one variety of these, efficiently absorb hydrogen when the surface is coated with PGM, allowing for faster kinetics making the hydrate resistant to poisoning by carbon monoxide (Linkov, 2011). 'Metal hydrates can be either pure metal such as palladium, titanium, or zirconium. They can also be intermetallic compounds or alloys made up of two or more metals such as iron titanium and/or lanthanum-nickel' (Motyka, 2011).

Titanium in combination with PGM is used as a catalyst in the storage process of hydrogen in metal hydrides (Linkov, 2011). Metal hydrates of palladium, aluminium, and iron (in combination with titanium as a catalyst) are the practically preferred storage for hydrogen because of their safety and efficiency (Chandhuri & and Muckerman, 2005). The amount of hydrogen stored per litre is increased when titanium is used as a catalyst. Even in small amounts, it increases the absorption rate of hydrogen by metal hydrates. South Africa hosts the second largest reserves of titanium in the world and this gives the country a comparative advantage of using titanium in the hydrogen economy complex (Department of Mineral Resources, 2011).

Over the years research has been conducted to reduce platinum loading in fuel cells while retaining the optimum efficiency of the catalyst. The US Department of Energy has managed to reduce the platinum content by a factor of five and is currently at less than 0.2 g/kilowatt (Breakthrough Technologies Institute, 2012). Reducing platinum loading of the fuel cells would mitigate for the price of platinum as a major impeding factor in the overall cost of the catalyst (Johnson Matthey Plc, 2013). The activity of a catalyst can be kept high with reduced platinum loading by keeping the core of the catalysts non-platinum; for example, using cobalt metal and coating the surface with just a single ethyl layer of platinum (Linkov, 2011).

PGM in their totality are therefore of great utility to the hydrogen economy value chain: platinum is important for platinum-based fuel cells; palladium is important in the storage of hydrogen; iridium and ruthenium are useful in the generation of pure hydrogen, and battery storage for energy is supported by iridium and ruthenium. In brief, the functional capabilities of PGM enable smooth linkages in a renewable energy value chain.

9 OTHER STRATEGIC MINERALS

As can be seen in Figure 17, South Africa ranks high in world reserves of a number of other minerals identified as critical or important to several developed and emerging economies. Further described and discussed elsewhere (defra, 2012), such minerals include manganese, titanium, fluorspar, vanadium, chromium, and nickel. The mineral utility of PGM has found congruence with other strategic minerals like titanium and manganese. Top economic sectors dependent on these and others include the high-tech materials, energy, logistics, life-sciences, chemical, food and water industries.

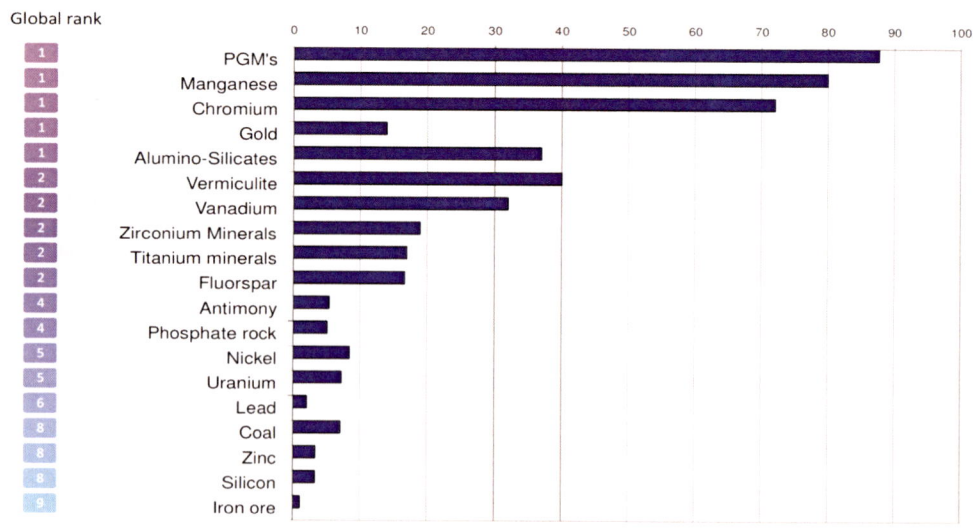

Figure 17: South African reserves for key minerals (based on 2008 data). x-axis shows % of global. Source: (Baxter, 2011)

10 Conclusion

From being considered a nuisance by Spanish Conquistadors, the significance of PGM grew around the middle of the 20th century, when demand for them and markets for their application increased at a tremendous rate. Volume drivers of this demand were largely the catalytic converter and jewellery markets, with Europe and China leading these, respectively. The unique physical and chemical properties of these metals have made them strategic to numerous important industries; and most developed and emerging economies are taking a number of steps to address concerns around access to them.

Growth of the world population and the global economy implies growth in demand for PGM, with the patent and research publications literature indicating the potential for further diversification of PGM markets, and Asia assuming a leading role in the longer term. Demand for these minerals is therefore bound to increase, and with South Africa hosting the overwhelming proportion of global reserves for these, their careful stewardship is imperative if the country is to derive benefit from their growing role in the global economy.

As these precious minerals are extracted, processed, and value-added into products, they will gradually become more available to the rest of the world, where capabilities are being developed and legislation being put in place to enhance their recovery from host devices at the end of their life cycle. Though the uses of PGM are expected to expand, the demand for mined PGM, and thus the country's significance as a supplier of PGM, may somewhat gradually decline. This probability – and the opportunity to maximise benefits from the PGM resources by enhancing capabilities up- and downstream of their supply chain – demands urgent investment in the development of the associated skills and capabilities.

Especially because its development pathway and exact architecture are still in the experimental phase[11] – and thus the game is still open for new players – the hydrogen economy offers significant opportunities whereby to focus the development of the country's technological capabilities. The know-how that will result from this investment can only lead to beneficial consequences for the country. The material conditions diagnosed in the South African NDP can thus be transformed fundamentally, and the country's developmental objectives realised.

---

11.   This is further elucidated in the chapter on the state of the hydrogen economy.

Bibliography

Aabakken, J. (2006). *Power Technologies Energy Data Book*, Golden, Colorado: US Department of Energy.

Afrox, T. D. (2013). [Interview] (10 April 2013).

Alonso, E., Field, F., Gregory, J. & Kirchain, R. (2007). Material availability and the supply chain: Risks, effects, and responses. *Environmental Science & Technology*, 41(19), 6649–6656.

Alonso, E., Field, F. R., Roth, R. & Kirchain, R. E. (2009). *Strategies to address risks of platinum scarcity for supply chain downstream firms*. s.l., Institute for Electrical and Electronics Engineers, 1–6.

Anon. (n.d.).

Baxter, R. (2011). *Brief history and overview of the mining sector*, s.l.: Chamber of Mines of South Africa.

Bishop, C. R. & Stern, M. (1963). *Means of preventing embrittlement in metals exposed to aqueous electrolytes*. New York/United States of America, Patent No. 3 109 734.

Blewis, D. I. & Wilburn, D. R. (2010). Platinum group metals – the world supply and demand. *U.S. Geological Survey open file report*, p. 60.

Bonesteel, E. (2008). *Hydride Storage of Hydrogen*. [Online] Available at: www.slowmoving water.com/hydride2.htm [Accessed January 2013].

Breakthrough Technologies Institute. (2012). *2011 fuel cell technologies market report*, Washington, DC: US Department of Energy, Energy Efficiency & Renewable Eenergy.

Brenan, J. M. (2008). The Platinum-Group Elements: 'Admirably Adapted' for Science and Industry. *Elements*, Volume 4, 227–232.

Burgess, S. (2010). *The Effect of China's Scramble for Resources and African Resource Nationalism on the Supply of Strategic Southern African minerals. What can the United States do?* Colorado Springs, Colorado: Air Force Academy.

Cawthorn, G. R. (2010). The Platinum Group Elements Deposits of the Bushveld Complex in South Africa. *Platinum Metals Review*, 205–215.

Chandhuri, S. & Muckerman, J. (2005). *The Role of Titanium in Hydrogen Storage*. Washington DC, www.bnl.gov/isd/documents/31287/sh_feature_6.pdf 2–14.

Chen, J. (2001). New autocatalyst plant opens in Shanghai – Johnson Matthey plant to help in effort to improve air quality. *Platinum Metals Review*, 45(4), p. 175.

Cohen, W. M. & Levinthal, D. A. (1990). Absorptive capacity: A new perspective on learning and innovation. *Administrative Science Quarterly*, 35(Special Issue), 128–152.

Connelly, M. (2010). *An Analysis of Innovation in Materials and Energy*. s.l.: University of Cincinatti.

Cramer, L. A. (2000). Presidential Address: Platinum Perspectives. *The Journal of the South African Institute of Mining and Metallurgy*, 273–280.

Defra. (2012). *A Review of National Resource Strategies and Research*, London: Department for Environment Food and Rural Affairs.

Department of Mineral Resources. (2011). *The Beneficiation Strategy for the Minerals Industry*. s.l., s.n., p. 12.

European Commission Enterprise and Industry. (2010). *Critical raw material for the EU: report for the working group on defining critical raw materials*, s.l.: European commision.

Forrest, S. A. & Clarke, B. (2006). *End-users, Recyclers and Producers: Shaping Tomorrow's PGM Market and Metal Prices*. Oxford, UK, The South African Institute of Mining and Metallurgy, 307–316.

Garland, N. L., Milliken, J., Carlson, E. & Wagner, F. (2002). Platinum: Too precious for fuel cell vehicles? *Society for Automotive Engineers Technical Papers*, 1(1896).

Gavin, H. (2010). Platinum Group Metals Research from a Global Perspective. *Platinum Metals Review,* 54(3), 166–171.

Griliches, Z. (1998). Patent Statistics as Economic Indicators: A Survey. In: Z. Griliches, ed. *R&D and Productivity: The Econometric Evidence*. s.l.: University of Chicago Press, 287–343.

Hageluken, C. (2006). Markets for the catalyst metals platinum, palladium and rhodium. *Metal*, 60(1–2), 31–42.

Hageluken, C. (2012). Recycling the Platinum Group Metals: A European Perspective. *Platinum Metals Review*, January, 29–35.

Hageluken, C. (2012). *Towards Bridging the Materials Loop – How Producers and Recyclers can Work Together*. Brussels: Umicore.

Harvey, J. (2013). *SA's platinum miners face new rival for investment flows*. [Online] Available at: http://www.bdlive.co.za/business/mining/2013/07/26/sas-platinum-miners-face-new-rival-for-investment-flows [Accessed 4 August 2013].

International Partnership for Hydrogen and Fuel cell in the economy. (2013). [Online] Available at: www.iphe.net

International Resource Panel. (2011). *Recycling Rates of Metals – a Status Report*, Paris: United Nations Environment Programme.

IPA. (n.d.). *25 Prominent and Promising Applications Using Platinum Group Metals*. Munich, Germany: IPA.

Johansson, T. B., Nakicenovic, N., Patwardhan, A. & Gomez-Echeverri, L. (2012). *Global Energy Assessment*, Luxembourg: International Institute for Applied Systems Analysis.

Johnson Matthey Plc. (2012). *Media Recycling Johnson Matthey Plc*. [Online] Available at: http://www.platinum.matthey.com/media/1393539/recycling.pdf [Accessed 20 November 2012].

Johnson Matthey Plc. (2012.) *Publications: Platinum Today, The world's leading authority on platinum group metals*. [Online] Available at: http://www.platinum.matthey.com/publications [Accessed 25 February 2013].

Johnson Matthey Plc. (2013). *Platinum 2013*, Hertfordshire, England: Johnson Matthey Plc.

Kikuchi, E., Uemiya, S. & Matsuda, T. (2006). Hydrogen production from methane steam reforming assisted by use of membrane reactor. *Natural Gas Conversion*, Volume 61, 509–515.

Koek, M., *et al.* (2010). A review of the PGM industry, deposit models and exploration practices: implications for Australia's potential. *Resource Policy*, 20–35.

Kwang, S. R., K, K. M. & Nam-Gyu Park, S. H. C. (2002). Symmetric redox supercapacitors with conducting polyaniline electrodes. *Journal of Power Sources*, 305–309.

LANL. (n.d.). PERIODIC TABLE OF ELEMENTS: *Los Alamos National Laboratory*. [Online] Available at: http://periodic.lanl.gov/46.shtml [Accessed 4 August 2013].

Lapper, R. (2010). *China seals African platinum deal.* [Online] Available at: http://www.ft.com/intl/cms/s/0/6c44a70e-67f6-11df-af6c-00144feab49a. html#axzz2QpKQaIJn [Accessed 1 August 2013].

Linkov, V. (2010). *Hydrogen Economy: the Role of PGMs.* Johannesburg, s.n. 1–30.

Linkov, V. (2011). *PGM in the fuel cell value chain and other related uses.* Pretoria, s.n., 2–13.

Long, J. T. (2013). *Recycling, Not Mining, is the Future for Securing Immediate Platinum Group Metal Supply.* [Online] Available at: http://www.theaureport.com/pub/na/15068 [Accessed 18 July 2013].

McDonald, D. & Hunt, L. B. (1982). *A History of Platinum and its Allied Metals.* London: Johnson Matthey.

Motyka, T. (2011). Hydrides for Processing and Storing Tritium. *WSRC-MS-2000-00061,* 187–197.

Mudd, G. (2011). Key trends in the resource sustainability of platinum group elements. *Ore Geology Reviews,* 106–117.

National Planning Commission. (2011). *Material conditions diagnostic.* [Online] Available at: http://www.npconline.co.za/MediaLib/Downloads/Home/Tabs/Diagnostic/ Diagnostic_Material_Conditions.pdf [Accessed 15 July 2013].

Nelson, H. G., Williams, D. P. & Tetelman, A. S. (1971). *Embrittlement of a Ferrous Alloy in a Partially Dissociated Hydrogen Environment,* s.l.: Metallurgical transitions.

Petersen, H. L. a. L. S. (2011). *Risø Energy Report 10, Energy for smart cities in an urbanised world,* Denmark: Scanprint.

PGM Database. (n.d.). *Browse PGM.* [Online] Available at: http://www.pgmdatabase.com/ jmpgm/index.jsp [Accessed 18 November 2012].

PGM Database. (2012). *Browse PGM.* [Online] [Accessed 18 November 2012].

PGM Database. (n.d.). *Browse PGM.* [Online] Available at: http://www.pgmdatabase.com/ jmpgm/index.jsp [Accessed 18 November 2012].

Platinum Guild International. (2012). History of Platinum. [Online] Available at: http://platinumguild.com/en-US/sale-support/about-platinum/history-platinum [Accessed 3 August 2013].

Pollet, B. S. S. J. L. (2012). Current status of hybrid, battery and fuel cell electric vehicles: from electrochemistry to market prospects. *Electrochemistry Acta,* manuscript submitted for publication.

Seymour, R. (2008). Platinum Group Metals Patent Analysis and Mapping. *Platinum Metals Review,* 52(4), 231–240.

Speirs, J., Gross, B., Gross, R. & Houari, Y. (2012). *Energy Materials Availability Handbook,* London: UK Energy Research Centre.

Steinweg, T. (2008). *A Sputtering Process, An Overview of the Platinum Group Metals Supply Chain,* Amsterdam: Stichting Onderzoek Multinationale Ondernemingen Centre for Research on Multinational Corporations.

Tao, L., Garnsey, E., Probert, D. & Ridgman, T. (2011). Innovation as response to emissions legislation: revisiting the automotive catalytic converter at Johnson Matthey Plc. *R&D Management,* 40(2), 154–168.

The South African Agency for Science and Technology Advancement. (2012). [Online] Available at: http://saasta.ac.za/images/stories/HydrogenF.pdf

Umicore. (2011). *Umicore to build new production line for automotive catalysts in China*, s.l.: Umicore.

United Nations ESCAP. (2011). *United Nations Economic and Social Commission for Asia and the Pacific.* [Online] Available at: http://www.unescap.org/ttdw/MCT2011/EGM/EGM1-8E.pdf [Accessed 25 January 2013].

US Department of Energy. (2009). *Platinum Availability and Economics for PEMFC Commercialisation*, s.l.: TIAX .

US Department of Energy, Energy Efficiency and Renewable Energy. (2011). *Fuel Cell Technologies Program*, s.l.: s.n.

US Geological Survey. (2012). *Mineral Commodity Summaries 2012*, s.l.: US Geological Survey.

Volfkovich, Y. M., *et al.* (n.d.). *Studies of super capacitor carbon eletrodes with high capacitance.* A. N. Frunkin Institute of Physical and Electrochemistry, Russian Academy of Sciences, Moscow, 159–163.

Whitburn, P. (2012). The goose that laid the platinum egg. *RE:VIEW*, 10–17.

Woodford, C. (2012). *Explainthatstuff.* [Online] Available at: http//www.explainthatstuff.com/how-supercapacitors-work.html [Accessed 18 January 2013].

Yang, C. (2009). An impending crisis and its implications for the future of the automobiles. *Energy Policy*, 1802–1808.

Yang, P., *et al.* (2013). *Hydrogenated ZnO Core Shell Nanocables for Flexible Supercapacitors and Self-Powered Systems.* American Chemical society, 1–10.

CHAPTER 2

THE GLOBAL HYDROGEN ECONOMY AND FUEL CELL TECHNOLOGY

Velaphi Msimang

1 INTRODUCTION

The vision of a hydrogen economy is more than 130 years old. In his novel, Jules Verne's characters imagine hydrogen produced via electrolysis as the energy option after the exhaustion of coal mines (Verne, 1874). In modern times, however, the concept is attributed to Bockris (Bockris, 2002), who proposed the idea as a response to rising concerns about global warming and air pollution.

Albeit with different sources in mind, to this day, this vision inspires many as momentum for intermittent sources of renewable energy, energy security and distributed energy generation continues to increase. As a portable and versatile carrier of energy, hydrogen it is argued – and illustrated in the top half of Figure 1 below – is an enabler of a stable, versatile and clean supply of a variety of renewable and other low-carbon sources[1] of energy.

While valid,[2] thermodynamic arguments about the overall higher inefficiencies of a circuitous approach, (using energy to raise hydrogen from which energy is generated) overlook the value proposition of portability and versatility, which the current electricity system cannot provide.

The attraction of the almost two-hundred-year-old fuel cells concept for application in a hydrogen economy arises from a number of their attributes,

---

1. The Japanese hydrogen economy 'Manifesto' envisioned waste heat from Japan's high temperature nuclear reactors (HTR) as a potential source of hydrogen. At the time, HTR work was being done by the Japanese and the Germans (Marchetti, 1973), the latter on what was later to be continued in South Africa as the Pebble Bed Modular Reactor (PBMR), whose researchers shared the vision (van Ravenswaay, et al., 2009).

2. In a world where the cost of energy is trending upwards, the notion of using energy to generate hydrogen from which energy can then be derived sounds irrational if portability and versatility are not important.

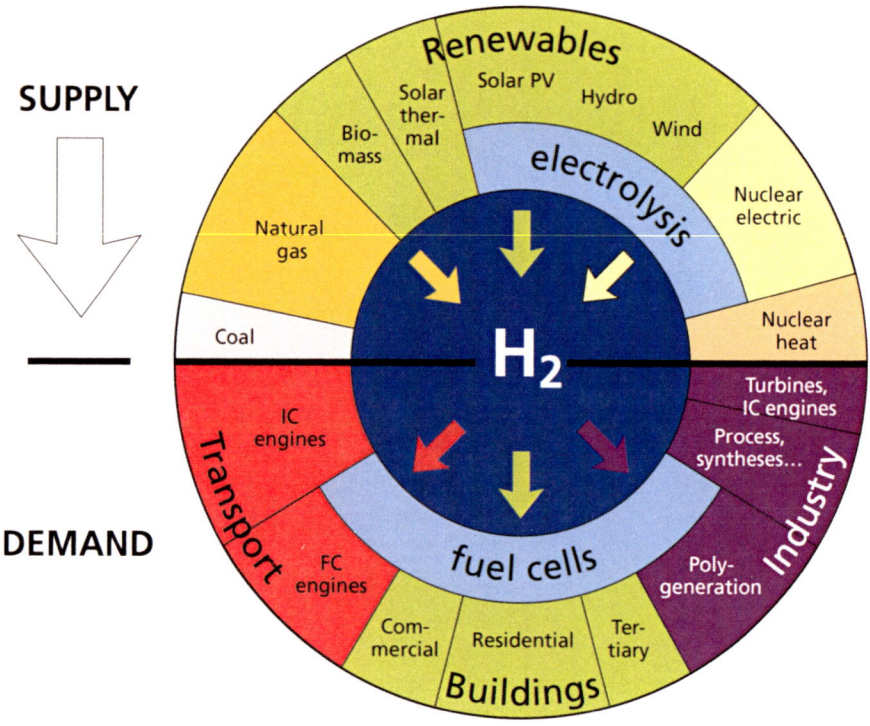

Figure 1: Primary energy sources, conversion technologies, and end use applications of hydrogen (European Commission, 2003)

including reliability and the high efficiencies with which they can convert chemical to electrical and/or thermal energy. Such a conversion, which occurs through a chemical reaction between hydrogen and oxygen, and catalysed by Platinum Group Metals (PGM) in some types of fuel cells – gives off water and heat as by-products. If the resultant heat can be channelled for useful purposes, the overall efficiency can be higher than 85%. For comparison, roughly 20% of the energy fed into a car powered by the internal combustion engine (ICE) is used to move it (Chu & Majumdar, 2012).

Fuel cell types are typically differentiated by the electrolyte employed in the design, and their operating temperatures. The five main types introduced here include:

- Alkaline fuel cells (AFC), which use potassium hydroxide in water (electrolyte) and operate at low temperatures (using nickel or platinum as a catalyst);

- Phosphoric acid fuel cells (PAFC), which use phosphoric acid (electrolyte) and a platinum catalyst on porous carbon electrodes;
- Proton exchange membrane fuel cells (PEMFC)[3], which typically operate at low temperatures and use a solid polymer (electrolyte) and platinum catalyst on porous carbon electrodes;
- Molten carbonate fuel cells (MCFC), which operate at high temperatures and use a molten carbonate-salt electrolyte in an inert, porous ceramic matrix; and
- Solid oxide fuel cells (SOFC), which operate at high temperatures and use a non-porous ceramic electrolyte.

Though their patterns of dominance do change[4], and their level of market penetration remains insignificant, the different types of fuel cells tend to dominate different market segments (Table 1), with the PEMFC and its variation – the direct methanol fuel cell (DMFC) – prevailing in portable devices, telecoms back-up power systems, residential combined heat and power (CHP) markets, and mobile applications. In order of sequence, the high temperature MCFC and SOFC dominate in multi- megawatts power utility and cogeneration applications.

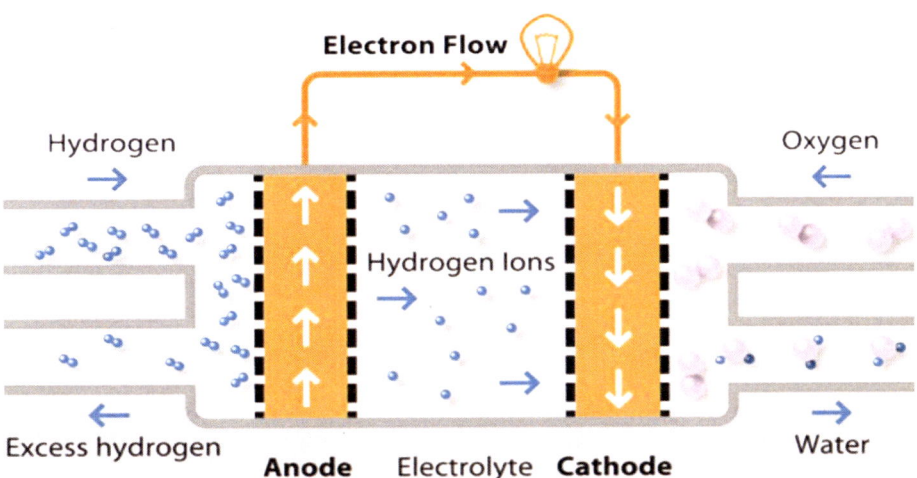

Figure 2: Schematic showing how a PEMFC works. Source: (Fuel Cell Today, 2013)

---

3. Figure 2 is a schematic representation of the structure and main processes of a PEMFC.
4. Before 1990, PEMFCs did not even feature on the radar of fuel cell options considered to be approaching commercialisation (Fueki, 1990) (Appleby, 1988). Conversely, the PAFC, which was dominant in the 90s (Figure 3(a)), has been overtaken by the MCFC and SOFC (Figure 7), then considered to be 20 and 30 years behind it, respectively (Cropper, et al., 2004).

Fuel Cell type	Operating temperature	System Output	Efficiency	Applications
Alkaline (AFC)	90–100°C	10kW–100kW	60–70% electric	• Military • Space
Phosphoric Acid (PAFC)	150–200°C	50kW–1MW (250kW module typical)	80–85% overall with CHP (36–42% electric)	• Distributed generation
Proton Exchange Membrane (PEMFC)	50–100°C	<250kW	50–60% electric	• Back-up power • Portable power • Small distributed generation • Transportation
Molten Carbonate (MCFC)	600°–700°C	<1MW (250kW module typical)	85% overall with CHP (60% electric)	• Electric utility • Large distributed generation
Solid Oxide (SOFC)	650°–1000°C	5kW–3MW	85% overall with CHP (60% electric)	• Auxiliary power • Electric utility • Large distributed generation

Table 1: Specification of each fuel cell type. Source: (Nordin, 2010)

Reflected in the patterns of patent applications that can be seen in Figure 4[5], and, more recently, fuel cell market behaviour (Figure 5), the world has repeatedly witnessed what could be labelled 'the rise, fall and rise of the hydrogen economy' over the last couple of decades as hydrogen fuel cells have been 'duelling it out' among themselves, i.e. the different types of fuel cells, and with other alternative energy technologies (AET)[6], including battery and hybrid electric cars in the transport sector, in the broader AET onslaught against well entrenched, continually advancing, and mainly fossil-fuel based incumbents.

As will become apparent further on, notable 'black swan'[7] events associated with these 'boom-bust' cycles include (boom) the US space programmes of the 1960s, (boom) the oil crisis of the 1970s, (boom) California's Zero Emission Vehicle (ZEV)[8] legislation from 1990, (boom) the US 2003 FreedomCAR

---

5. Borrowed from (Mock & Schmid, 2009).

6. The case of competition between the fuel cell car and the electric car and how attention to each of them alternated over a 20 year period, is discussed elsewhere (Bakker, *et al.*, 2012).

7. 'Black swan' events are those that are highly improbable, but whose occurrence result in significant impact (Taleb, 2010).

8. This legislation requires that car companies sell a certain percentage of zero emission cars in their fleet or forfeit the right to sell cars in the state. Many other states in the US, including New York, New Jersey and Maryland, have followed California by enacting the same legislation. Accounting for about 10% of car sales in the US, California is considered an important market by all major car companies (Ramsey & Rogers, 2012).

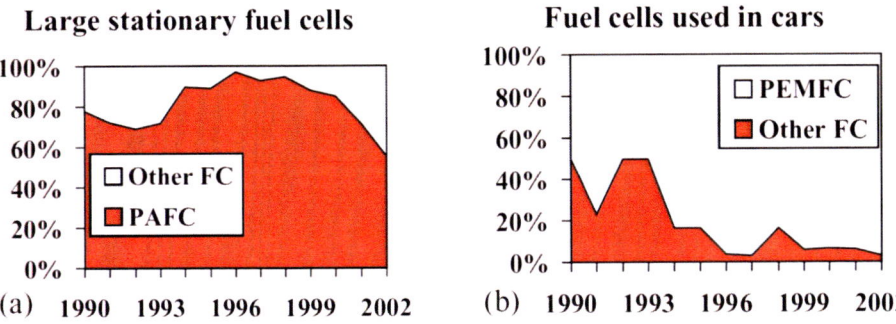

Figure 3: Competition in market share between (a) PAFC and other fuel cells (FCs) in large stationary power applications, and (b) PEMFC and other FCs in cars. Source: (Cropper, *et al.*, 2004)

programme, and (bust) stimulus packages following the 2008 global financial meltdown. Be that as it may, (boom) progress in laboratories and (bust) unfulfilled promises also played a crucial role.

The practical use of the fuel cell occurred more than a century after its invention. Following its first use for military and spacecraft applications – which are markets not typically driven by cost considerations[9] – at General Electric in the 1960s, research in fuel cell technologies, particularly for transport applications was intensified as a result of the 1970s oil crisis and environmental legislation in the US.

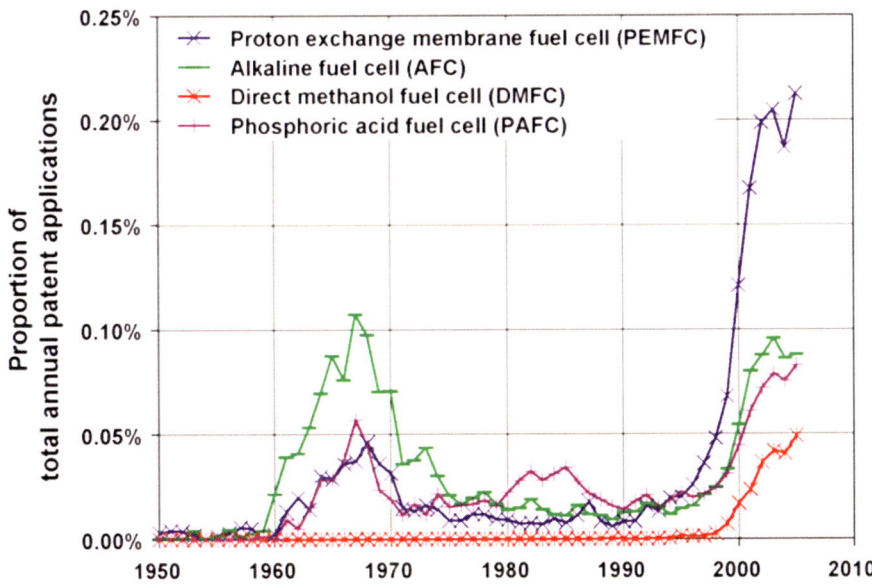

Figure 4: Relative (%) patent applications for different types of fuel cells (Mock & Schmid, 2009)

9. (Steinemann, 1999)

Despite persistent hopes that nuclear power 'too cheap to meter' would enable the generation of cheap hydrogen (Bockris, 1976), formidable challenges posed by the inefficiency of fuel cells and lack of charging stations caused the dream of a hydrogen economy to fade by the mid-2000s. Electric cars powered by batteries then took the mantle from fuel cells, and governments made significant investments into battery electric cars and their variations as part of their stimulus packages following the global financial

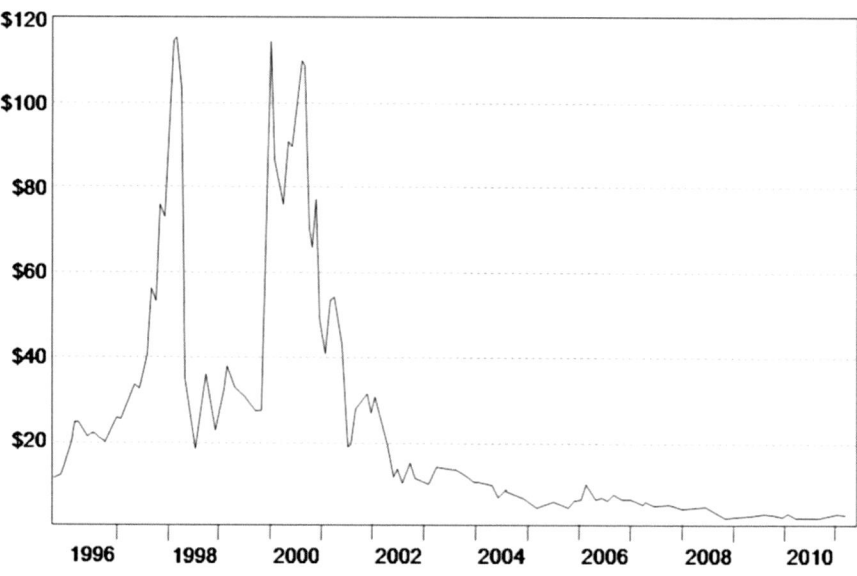

Figure 5: Stocks of fuel cell companies (Barbir, 2013)

meltdown of 2008 (Keane, 2012). In contrast to these dynamics, however, patent trends (see Figure 6) over the 10 year period from 2002 reflected a sustained dominance of fuel cells in the number of patents registered (Farley & Mesiti, 2012). Recent announcements by car manufacturers also suggest the focus is shifting away from the electric car back to the fuel cell power train (Ko, 2012).

Arguably reflected by the fluctuating stock price of fuel cell companies (Figure 5), the management of signals and expectations around the offerings and availability of fuel cell transportation options have also not been helpful (Bakker, 2010), with a long history of broken promises (*The Economist*, 2008)[10], questions about the sincerity of car manufacturers (van den Hoed, 2007), or regarding readiness of the technology (Bullis, 2009) (Bullis, 2009), (Nelson & Soda, 2013) (Nelson and Soda, 2013).

---

10. Many others recall this history (Cato, 2009), (Walsh & Moores, n.d.).

Particularly for developing countries with numerous socio-economic imperatives to address (Jun, *et al.*, 2013), such uncertainty as has been reflected in the apparently confusing signals sent out about the different technology options (typical of the fluid stage or 'era of ferment' in technological transitions (Utterback, 1994)), can induce policy paralysis because of the significant investments required by each of the options, all of which seem to be competing but could be complementary.

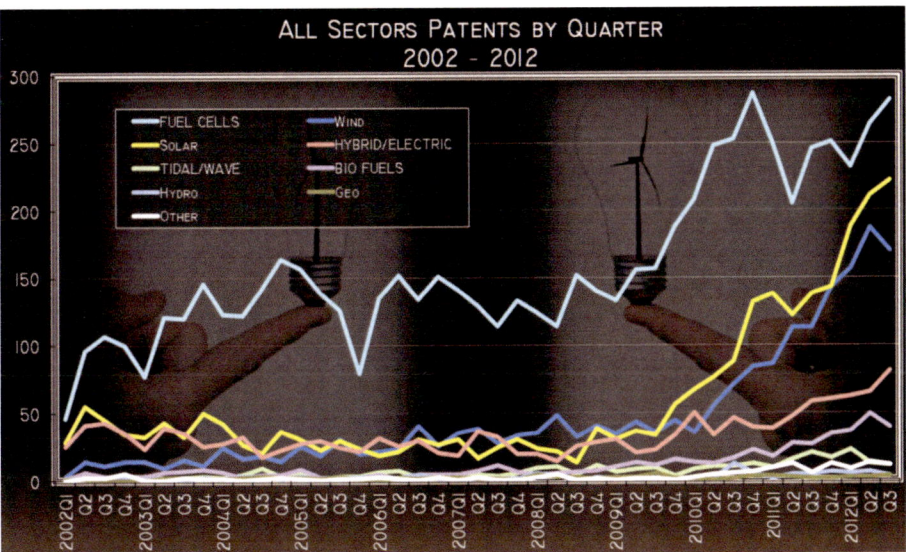

Figure 6: Trends in patents granted by the United States Patent Office for different types of clean energy technologies.

It is partly as a result of these uncertainties that the route of learning[11] from mature industries and approaching the innovation frontier as (economic) development gains traction has been recommended for catching up countries (Kim, 1980). Nevertheless, the advent 'new wave technologies' that are patent-intensive, science-based, and systems-embedded[12], suggests that 'the locus of knowledge creation and the forms through which knowledge is created will increasingly shape opportunities for learning, for innovation, and thus for growth and development' (Mytelka, 2003). While chances and the timing of their entry into mainstream industry are unknowable, fuel cells are one such technology.

In 2007, the government of South Africa approved a long-term strategy on

11. The three-pronged approach whereby this can be achieved involves 'emphasizing manpower development at various levels, accelerated introduction of foreign technologies, and the stimulation of domestic research and development (R&D) activities' (Choi, 1986).

12. In other words, these technologies require appropriate infrastructure for their deployment.

Hydrogen and Fuel Cells Technologies Research, Development and Innovation (HFCT RDI), whose main objective was 'to build sufficient capacity in research, development, and innovation to supply 25%[13] of the global hydrogen and fuel cell market's platinum group metals catalyst demand by 2020.' The strategic intent of the South African government was to enable technology development through its system of innovation to support its energy security thrust (Mange, 2010).

Early signs of its success include the country's entry into global fuel cell supply chains, the announcement of a number of fuel cell powered prototypes, graduation of Masters and Ph.D. students, and the unveiling of commercialisation initiatives (PMG, 2012)[14]. However, these gains are being made in a global fuel cell industry that is still struggling to penetrate the mainstream of the energy and transportation sectors (Behling, 2013), and whose prospects are subject to speculation. Indeed, a high level of uncertainty is reflected in estimations of prospective markets, which range between US$3 billion by 2030 (Warshay, 2012) and US$ 2.6 trillion by 2021 (PricewaterhouseCoopers, 2002).

In view of the fact that transportation applications offer the most significant potential markets for fuel cells, the somewhat cool reception

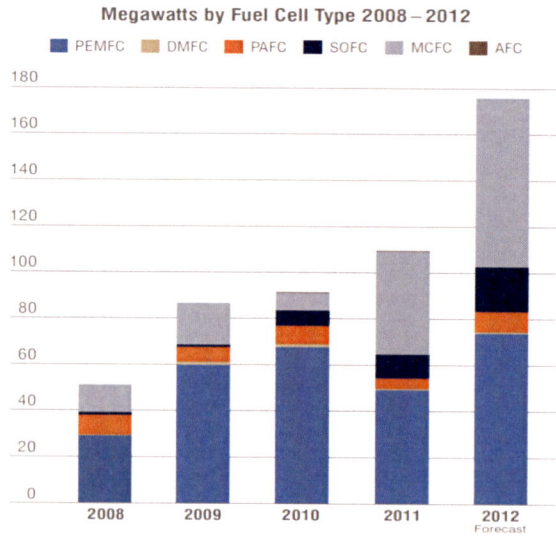

Figure 7: Shipments (measured in Megawatts) of different fuel cell types between 2008 and 2012. Source: (Fuel Cell Today, 2012)

13. The huge range of estimates of the value of markets for fuel cells in the future, and their dependence on the (evolving) responses of the different segments of the market, reflect the highly speculative value of these projections.
14. More accounts of these are available elsewhere (Wild, 2013), (FuelCellWorks, 2013).

(Smil, 2012) demonstrated towards electric cars[15] – which are significantly cheaper than their fuel cell counterparts – raises questions about prospective adoption of fuel cells.

Unless the adoption trend apparent in Figure 7 gets a significant boost, the market for platinum-bearing fuel cells is not likely to exceed 200 MW by 2030. From the power sector alone, global demand for more than 9000 GW (by 2035) has been predicted (International Energy Agency, 2012). In view of the 20 000 ounces of Platinum sold to fuel cell markets in 2010, as opposed to over 3 million ounces of the metal supplied to the catalytic converter market in the same year (Butler, 2011), and lag times associated with the diffusion of new technology, indications are that if it happens at all, it will take a long time for markets for fuel cells to become truly significant, even if the technical hurdles confronted by the technology were overcome today (AAAS, 2011), (Wilson & Grubler, 2011). Meanwhile, numerous on-going efforts to replace platinum from the fuel cell have been reported (Halper, 2013)[16].

SOME CHALLENGES AND WINDOWS OF OPPORTUNITY FOR SOUTH AFRICA

A number of options, challenges and opportunities arise as a result of the hypothetical transition to a hydrogen economy. Among these is the disruptive nature of fuel cells and how these can lead to the emergence of new dominant players and a resultant reconfiguration of the industry as new capabilities and firms displace old ones (Christensen, 2006).

Illustrated in other fields by Xing and Detert, 2010 and Kraemer, et al., 2011 (Kraemer, et al., 2011)[17], however, is that the globalisation of supply chains also raises questions about the appropriability[18] of the socio-economic benefits associated with developing fuel cell capabilities. Much hoped for jobs may thus be difficult to realise (Duhigg & Bradsher, 2012), and the country's problems of poverty, unemployment and inequality (PUI) may persist.

Further, among issues that could pose more challenges towards the ability of South Africa to translate its comparative advantage in access to the raw

---

15. Out of a global passenger car fleet of more than 900 000 000, only 180 000 are electric cars. At less than 0.02%, this means less than 2 of every 100 000 cars on the road are electric cars (Trigg & Telleen, 2013).

16. Even though catalytic converters are different to fuel cells, the failing 30 year old efforts (Friedman, *et al.*, 2012) to replace platinum in the former, may mean it's unlikely that platinum will be displaced from fuel cells.

17. (Kraemer, et al., 2011) show that while iPhone and iPad products, including most of their components, are manufactured in China, the primary benefits go to the US economy as Apple continues to keep most of its product design, software development, product management, marketing and other high-wage functions in the US. As demonstrated by this very case, South Africa may well decide to establish its own original equipment manufacturers in order to leverage global supply chains to her benefit. However, the experience of the Joule electric car (Cokayne, 2012) seems to suggest lack of will to pursue such an endeavour.

18. 'Appropriability' suggests the ability to appropriate or capture the benefits associated with an innovation

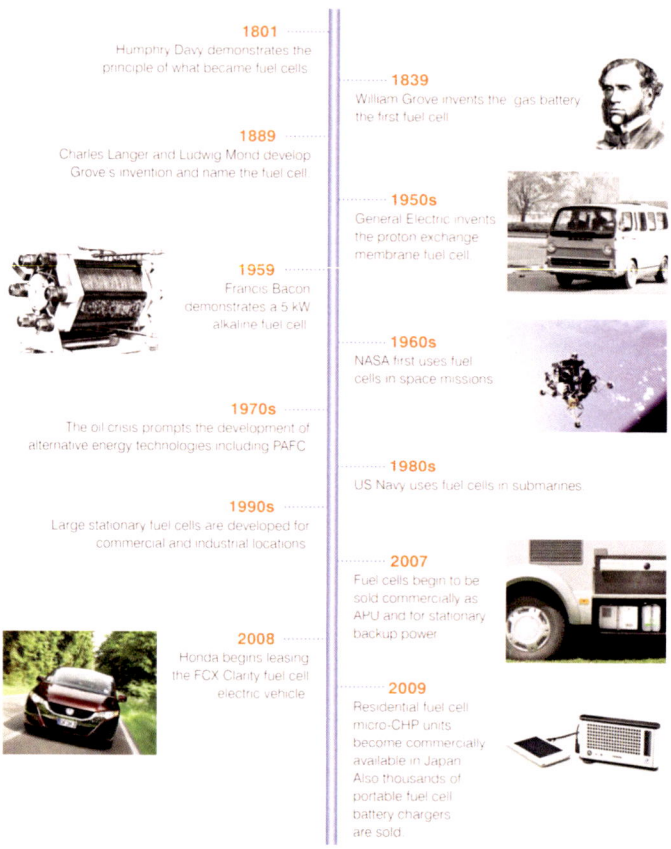

Figure 8: Some notable milestones in the development and applications of fuel cells. Source: (Fuel Cell Today, 2011). APU abbreviates Auxiliary Power Unit.

materials into a competitive edge in downstream industries are the very high (more than 95%) recyclability[19] of PGM and, especially with rapid advances in robotics and commercial space travel, PGM mining of asteroids (Elvis, 2012).

2 Global Hydrogen Fuel Cells

The focus of this report on PGM-based fuel cells necessitates emphasis on those types of fuel cells whose electrodes bear these precious minerals. To that end, the account that follows does not delve into the solid oxide fuel cell (SOFC) and the molten carbonate fuel cell (MCFC). A more comprehensive coverage – from which the following summary predominantly emanates – is

---

19. At head grades of around 2,000 grams per tonne (g/t), the recycling of catalytic converters is a much more attractive source of PGMs than the mining of virgin ores (usually less than 10g/t). The environmental impact of PGM mining also makes recycling a better option, and poses rather serious challenges to South Africa in a carbon-constrained world.

available elsewhere (Andujar & Segura, 2009).

2.1 GENESIS OF FUEL CELLS

Further detailed elsewhere (Andujar & Segura, 2009), the technical understanding of fuel cells – then referred to as 'gas batteries' – has existed since the 19th century, based on work done by Sir William Grove in 1839. More than a century would lapse before this technology would get beyond being a laboratory curiosity. Even then, the technology has remained obscure from mainstream markets[20].

In 1959, Francis Bacon presented a 5 kW alkaline fuel cell with 60% efficiency in the United Kingdom. Around the same year, a platinum-based fuel cell was invented by General Electric (GE through the work of Thomas Grubb and Leonard Niedrach) for use by the National Aeronautic Space Agency (NASA) and McDonnell Aircraft during the Gemini space program (whose mission was to fly humans into space) in the USA. Currently based at the Smithsonian Institution (Smithsonian, n.d.), a tractor using a 15 kW fuel cell was also unveiled in 1959. A chronology of some of the significant milestones in fuel cell history is represented in Figure 8.

An extensive account of the history of different types of fuel cells is articulated elsewhere (Behling, 2013). Notable in this history, and illustrative of the contingency of technology development trajectories, is the absence of the PEMFC from the horizon of historical expectations for fuel cells (Fueki, 1990). As will be seen later, this was to change as a result of breakthroughs in research and development (R&D) by Ballard and other institutions.

2.2 PEM FUEL CELL TECHNOLOGY DEVELOPMENT

William Grubb – who was then a GE employee – registered the first patent that used an ion exchange membrane-based electrolyte for the fuel cell (Grubb Jr, 1959). This design stood out from Grove's in that all of it was made of solid materials. The opportunity to displace the heavy batteries that powered the NASA's space capsules helped overcome the cost barrier for PEMFC production. NASA needed alternatives to the battery in order to be able to extend the duration of its 'manned' flights in space. Its high platinum loading and technical deficiencies – mainly of the membrane – led NASA to switch from PEMFC to the alkaline fuel cell (AFC). Nevertheless, GE

20. Godoe (Godoe, 2006) reminds us that 'telegraphy, the distant ancestor of the internet and GSM, was invented by Samuel Morse in 1838', and that 'although numerous highly successful innovations stemming from telegraphy may be observed, the development of fuel cells has been insignificant, slow, and erratic and has not yet resulted in notable positive socioeconomic effects.' However, the author uses this comparison to demonstrate the unique challenges that retard the rate of transformations in energy systems. These challenges are further elucidated elsewhere (Grubler, 2012).

Company	# of patents
United Technologies Corp./ International Fuel Cells	223
Siemens AG/Westinghouse Power Corp.	121
US National laboratories	112
Hitachi, Ltd.	52
Energy research corporation	52
Fuji Electric Co., Ltd.	46
Mitsubishi group	44
NGK Insulators, Ltd.	39
Ballard Power Systems Inc[21].	36
Exxon Research & Engineering Co.	34

Table 2: Top 10 patent holding firms in fuel cell technology from 1970–1996. Source: (Steinemann, 1999)

continued to invest in research on the PEMFC.

Despite DuPont's 1968 development of a perfluorosulfonated membrane (Nafion), which had excellent mechanical, thermal and chemical properties, NASA's space programme did not reconsider its decision to forego the PEMFC. GE then decided not to pursue commercialisation of the PEMFC.

The next milestone for the PEMFC was brought about by the Canadian Department of National Defence (DND)'s decision to purchase the technology from NASA. Established in 1979 to conduct research on lithium batteries, Canadian-based Ballard was selected (in 1983) to further develop the technology and reduce its costs for military applications. Parallel progress made by Dow Chemicals, Los Alamos National Laboratory (LANL), and Texas A&M University (TAMU) in PEMFC membranes and the lowering of platinum loadings enabled significant advances in Ballard's efforts (Srinivasan, et al., 1989).

Evident in their absence before 1990 (see Table 2) car companies began accelerating their development of fuel cell-powered prototype cars in the last decade of the twentieth century (Figure 11). This was enabled by the aforementioned breakthroughs (Mytelka, 2003), which enhanced the competitiveness of the PEMFC with alternatives. An important driver of this change was the dynamic between zero emission legislation in California and

---

21. Throughout this document, 'Ballard' stands for Ballard Power Systems Inc.

politico-economic considerations by car companies, who prefer long-term technologies (that still require significant R&D) to actual implementation of new and uncertain alternatives (van den Hoed, 2007).

Especially in Japan, interest in the use of fuel cells for stationary power generation peaked as a result of the 1970 energy crisis and subsequent concerns about environmental emissions. Based on support from energy utilities and the Ministry of International Trade and Industry (MITI), Japanese companies had built and operated about 87 phosphoric acid fuel cell (PAFC) power plants by 1994. Since natural gas was the preferred fuel, the susceptibility of PEMFCs to carbon monoxide poisoning – a problem later resolved by Ballard's process innovation that reduces carbon monoxide in reformates from carbonated fuels – ruled them out for stationary power generation (Steinemann, 1999).

2.3   HYDROGEN PRODUCTION

Hydrogen is an energy carrier and not an energy source, which means that it has to be produced. As illustrated on the supply side of Figure 1 at the outset of this chapter, there are several sources from which it can be produced. Commercially, more than 96% is generated from fossil fuels (see Table 3). Main uses are for manufacture of fertilizer (50%), petroleum refining (35%), and methanol production (Jun, *et al.*, 2013).

The minute size of the hydrogen atom poses storage problems. These include the need for special materials to stop it from leaking, and its low (volumetric and gravimetric) density (see Figure 9). Hydrogen can be stored as a compressed gas (most common), chemically in a material, or liquid in pressurised tanks.

In stationary applications, where volume may not be a constraint, it can be easily stored in its gaseous or liquid forms.

Source	Bcm*/yr	Share (%)
Natural gas	240	48
Oil	150	30
Coal	90	18
Electrolysis	20	4
Total	500	100
* Bcm: billion cubic meters		

Table 3: Annual hydrogen production share by source. From (Balat, 2008)

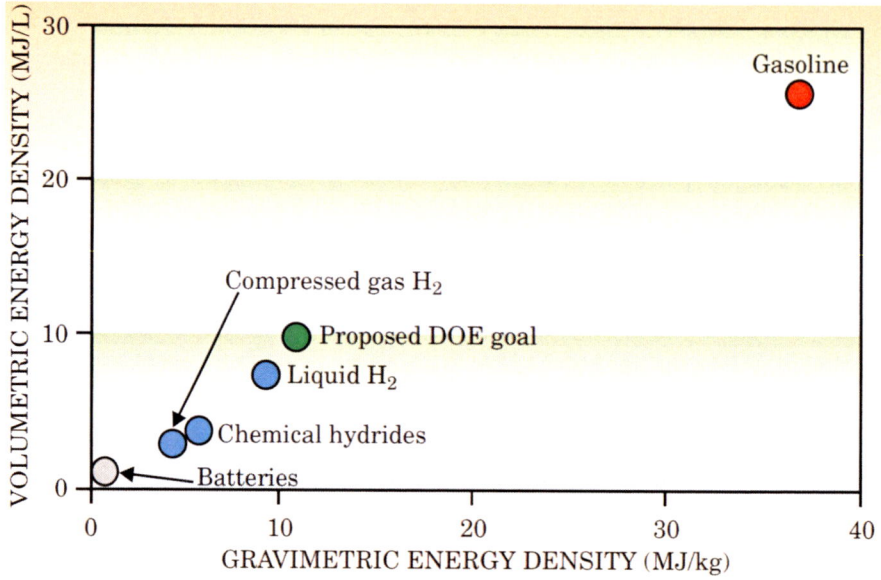

Figure 9: Gravimetric and volumetric energy densities of different fuels. Source: (Crabtree, *et al.*, 2004)

3 TECHNOLOGICAL AND ECONOMIC CHALLENGES

Spanning the whole value chain of the hydrogen system, including production[22], distribution, storage, and conversion, the specific scientific hurdles that led to the relegation of Sir William Grove's fuel cells as the internal combustion engine and oil discoveries took centre stage, may have been resolved. However, especially with progress made in alternative options and the evolving context in which they are to be applied, challenges to the hydrogen economy appear to be no less daunting[23]. A more detailed articulation of these hurdles is described elsewhere (Crabtree, *et al.*, 2004), (Bossel, 2006). Compiled elsewhere (Debe, 2012), Table 4 summarises the electro-catalyst performance, materials and process challenges to be overcome before fuel cell cars can become commercially viable.

3.1 COMPETITION

Over and above the hydrogen and fuel cell challenges just described, improvements in prevailing technologies – with which they will have to

22. Hydrogen production requires a cheap source of hydrogen as more energy is spent in producing, transporting and storing hydrogen than that derived from it (Evenson, 2004).
23. In justifying cuts to the 2010 US hydrogen budget, former Secretary of Energy and Nobel Prize winning physicist Steven Chu was recently quoted saying [regarding the four main challenges to be overcome by hydrogen]: 'If you need four miracles [in hydrogen production, distribution, storage, and fuel cells], that's unlikely: saints only need three miracles'.

Performance	• Must meet beginning-of-life performance targets at full and quarter power • Must meet end-of-life performance targets after 5000-hour or 10 years operation • Must meet performance, durability and cost targets and have less than 0.125 mg PGM per square centimeter • Corrosion resistance of both platinum and the support must withstand tens of thousands of start-up/shut-down events • Must have low sensitivity to wide changes in relative humidity • Must withstand hundreds of thousands of load cycles • Must have adequate cool start, cold start and freeze tolerance • Must enable rapid break-in and conditioning (the period needed to achieve peak performance)
Materials	• Must have high robustness, meaning tolerance of off-nominal conditions and extreme-load transient events • Must produce minimal peroxide production from incomplete oxygen-reduction reaction at the cathode • Must have high tolerance to external and internal impurities and ability to fully recover • Must have statistically significant durability, meaning individual membrane electrode assembly (MEA) lifetimes must enable over 99.9% of stacks to reach 5000-hour lifetimes • Electrodes must be designed for cost-effective platinum recycling • Environmental impact of manufacturing should be minimal at hundreds of millions of square metres per year
Process	• Environmental impact must be low over the total life-cycle of the MEAs • Manufacturing rates will need to approach several MEAs per second • MEA manufacturing quality must achieve over 99.9% failure-free stacks at beginning of life (1 faulty MEA in 30 000 for just 1% stack failures) • Proven high-volume manufacturing methods and infrastructure will be required • Catalyst-independent processes will be preferred, to enable easy insertion of new-generation materials

Table 4: Development criteria for automotive fuel cell electro-catalysts. Source: (Debe, 2012)

compete – continue apace (Ealey & Mercer, 2002). This is pertinent to the transport sector. Factors that influence the dynamics of adoption of the different technology options, and thus the outcome of the competition, include high oil prices, infrastructure costs, carbon constraints, and consumer acceptance of the different technology options. This 'spaghetti' of evolving factors and their impact on competing options is represented in the systems dynamic model illustrated in Figure 10 and various descriptions thereof are described elsewhere (Dijk, *et al.*, 2013), (Carle, *et al.*, 2007).

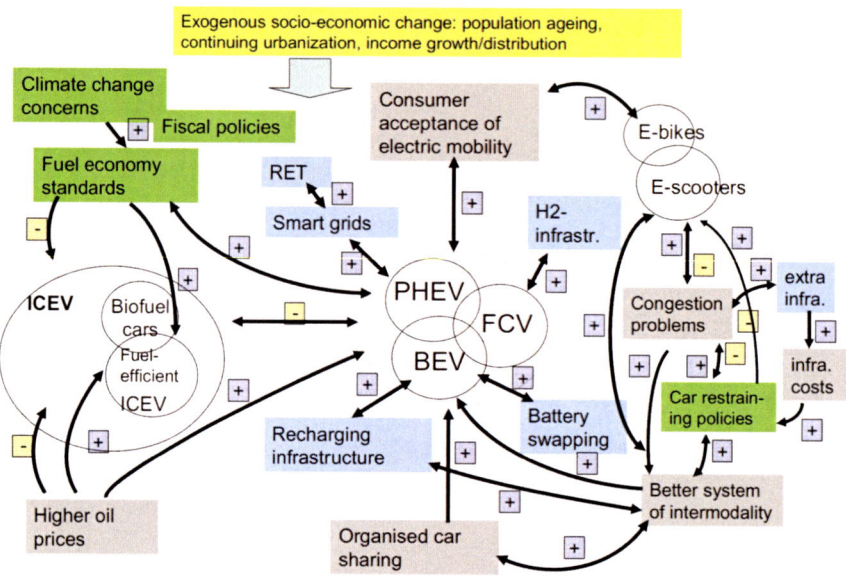

Figure 10: Some of the factors promoting (+) or detracting (-) the uptake of different power-trains. Borrowed from (Dijk, et al., 2013). RET=Renewable Energy Technology; PHEV=Plug in Hybrid Electric Vehicle; BEV=Battery Electric Vehicle; FCV=Fuel Cell Vehicle; ICEV=Internal Combustion Engine Vehicle; H2=hydrogen; E=electric.

4 PROTOTYPE AND DEMONSTRATION PROJECTS

Over and above their use as R&D tools in a trial-and-error learning curve, prototypes have been posited as instruments for generating expectations about technology (Bakker, 2010). Figure 11 plots the number of hydrogen and fuel cell technologies prototyped since the 1960s.

PEMFC has dominated total fuel cell shipments in megawatt (MW) since 2008, but the MCFC seems to be gaining ground on it at a rapid rate (Figure 7). Further described elsewhere (Fuel Cell Today, 2012), most shipments (by MW) have been for prototypes in the automotive sector, with Daimler operating a fleet of 200 fuel cell car prototypes around the world (35 of which are being leased out in California) (Fuel Cell Today, 2013). Twenty-five demonstration fuel cell electric busses were reported in the US in 2012 (Eudy, et al., 2012). Worldwide, 126 fuel cell bus demonstrations were registered by 2011 (Dicks, 2012). The global number of fuel cell cars is reported at less than 1,000 (Haslam, et al., 2012), with more than 180 having been demonstrated (cumulatively from 2005) in the US by 2011 (Wipke, et al., 2012). By numbers, PEMFCs shipments have outstripped the rest consistently throughout this period.

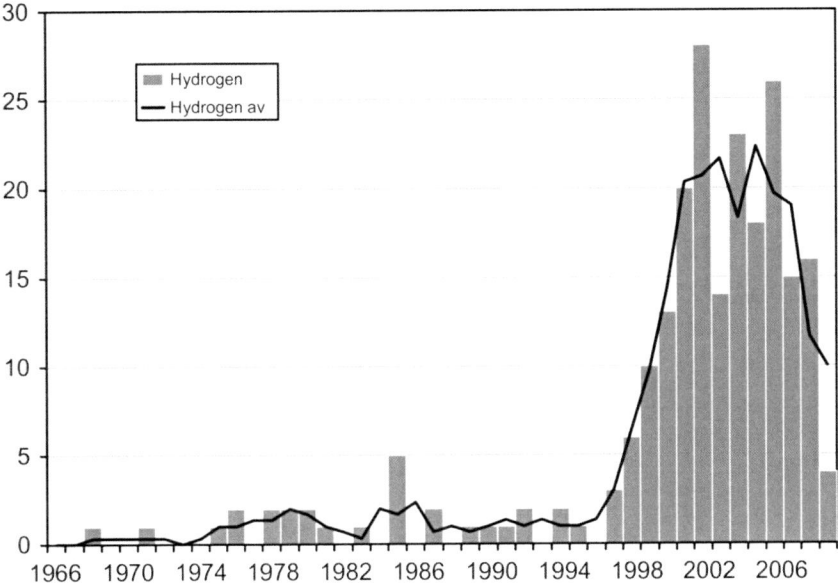

Figure 11: Number of hydrogen car prototypes per year, with 3-year-average line added. Borrowed from (Bakker, 2010)

5 Hydrogen infrastructure

Important for the successful mass marketing of fuel cell cars is the existence of refuelling stations, and codes and standards to ensure safety and efficiencies. In order to realise this, a number of governments have begun experimenting with the development of such infrastructure. The chicken and egg problem of commercialisation – with the car industry suggesting supporting infrastructure is necessary before it can industrialise fuel cell cars, and energy companies requiring confirmation of the existence of markets for fuel cells in order to invest in it – underlines the need for private public partnerships (PPP) to mitigate risks for all investors. In phases, incremental networks of roads are designated for the development of refuelling stations that will support fuel cell transport fleets.

Called 'hydrogen highways', pockets of networks of these (hydrogen refuelling stations) are being developed to further probe various aspects and implications of the roll-out of a Hydrogen Economy. By 2011, 215 such stations were known to be in operation worldwide. Most are in Europe (85) and North America (80), followed by the Asia Pacific region (47) (Fuel Cell Today, 2012). As a comparison, by 14 July 2013, there were more than 6,000 public electric car charging stations and only 10 public hydrogen fuelling stations in the US (US Department of Energy, 2013). In excess of 110 000 petrol stations were in operation by 2007 (US Department of Commerce, 2007).

6 Commercial Applications of Platinum-based Hydrogen and Fuel Cell Technologies

While fuel cell shipments (by MW) indicate their relative insignificance in the multi-terawatt (TW) global energy industry, they have begun gaining traction in stationary and material handling applications, especially for markets that prize their reliability, environmental friendliness, and relatively high energy storage density of hydrogen (compared to batteries, see Figure 9). Since no fuel cell vehicle is commercially available, and notwithstanding that fact that the transport sector[24] would be its most significant market, the focus of this section is on the sectors where platinum-bearing fuel cells seem to be gaining a foothold in the market. Such sectors include:

1. telecoms, where fuel cells are employed as back-up power systems;
2. residential combined heat and power (CHP) units, e.g. in Japan, where PEMFC units are used for prime power;
3. portable devices, where energy density requirements are raised by more power-hungry gadgets; and
4. specialised vehicles, including materials handling vehicles (MHVs).

6.1 Back-up Power

In the telecommunication industry, fuel cell technology is being used mainly to provide back-up power for base stations, with an estimated 900 of these units deployed by 2012 (Crouch, 2012). For a sense of the market yet to be penetrated, more than 800 000 3G base stations are estimated to have been deployed in China by 2011 (Dewar, 2012), and more than 300 000 by 2010 in India (Renalysis, 2011).

A number of companies in the telecoms industry have started deploying fuel cell technology in their operations, although at a limited scale. However, it is anticipated that as technology reliability improves and costs start coming down, more companies will increase their adoption of fuel cell technology for their base stations.

---

24. For the years 2010 and 2011, at least 4 of the top 10 fuel cell patentees were car manufacturers, including General Motors, Nissan, Toyota, Honda, and (in 2011) Daimler (Fuel Cell Today, 2012). Nevertheless, some in the car industry have warned against optimism for significant markets in the near to medium-term (Frenette & Forthoffer, 2009).

Recent experiences with base stations backed up by fuel cells in natural disaster-prone areas also suggest their utility as climate change adaptation technologies (Satyapal, 2012).

6.2 RESIDENTIAL CHP SYSTEMS[25]

Inspired by Canada's lead in PEMFC technology development, the Japanese Ministry of Economy, Trade and Industry (METI), designed a two-phase eight-year project (from 1992) to enhance Japanese stationary PEMFC capabilities (in 1,5-10 kW systems) so they could catch up with their foreign peers. Companies involved included Sanyo, Toshiba, and Mitsubishi Electric.

On finding it successful, a 1999 review of the project proposed a next generation project and related policy support mechanisms – including standards, certification systems and tax incentives – that would result in the commercialisation of PEMFC products. More fully described elsewhere (Behling, 2013), a series of interventions were subsequently put in place, including:

- implementation of the Millenium Project, which supported the application of PEMFC in fuel cell vehicles and residential CHP systems;
- adoption of a Science and Technology Basic Plan (2001) that prioritised fuel cell R&D in the country's science and technology policy; and,
- the establishment of an industry association – the Fuel Cell Commercialisation Conference of Japan (FCCJ) – that would enhance cooperation between government and the private sector around FC commercialisation.

In 2009, PEMFC developers – including Eneos (JX Nippon Oil & Sanyo), Toshiba and Panasonic – announced their intent to sell residential-use PEMFC units under the Ene-Farm brand, and were the first to roll out these units from automated production lines (Staffell & Green, 2013). With a declining government subsidy, more than 20 000 of these 40 000 hours (run-time) units have since been sold, with the rate of sales significantly higher following the Fukushima nuclear power plant accident (Fuel Cell Today, 2012). However, despite the relatively high Japanese electricity prices and significant subsidies to encourage their uptake, the economics of these systems continue to limit them to customers who are environmentally conscious and have an interest in advanced technology (Greene, *et al.*, 2011).

---

25. This section acknowledges the work of Behling, from which much of the information is derived (Behling, 2013).

6.3   Materials Handling Equipment

Particularly for indoor use, fuel cell-powered forklifts are one of the largest sectors for fuel cell applications. In fact, Ballard's turnaround (and enhanced prospects for becoming the first profitable fuel cell company) is attributed to their decision to forego their automotive intellectual property (IP) and focus on the forklift market (Mason, 2012). Fuel cell-powered forklifts provide significant benefits over both petroleum and battery-powered forklifts as they produce no emissions and when compared to battery-powered forklifts, have a greater range, take less time to recharge and cool before use, are not prone to voltage drops as power discharges, and do not suffer from downtime due to battery change-outs. A further advantage of fuel cells in the material handling vehicle is capability for quick refueling of the vehicle. Table 5 shows a few organisations that are adopting fuel cell forklifts into their businesses.

Company	Location	PEMFC supplier	No. of forklifts
Bridgestone Tyre	South Carolina	Plug Power	43
Central Grocers	Illinois	Plug Power	220
Fedex	Missouri	Plug Power	35
Fedex	Canada	Hydrogenics	Not available
Martin Brower	California	Oorja Protonics	15
Nestle Waters	Texas	Plug Power	32
Super Store Industries	California	Oorja Protonics	Not available
Sysco	Virginia	Plug Power	100
Sysco	Texas	Plug Power	98
Walmart	Canada	Plug Power	60–75

Table 5: US Organisations using dedicated fuel cell forklift trucks. Source: (Dicks, 2012)

Plug Power, who buy Ballard fuel cell stacks, account for 86% of the US fuel cell-powered forklift market, and is thus the dominant supplier. The company, which used to focus on residential combined heat and power (CHP) systems but later entered the forklift markets because of low demand for the former, benefited to the tune of US$9.7 million from the American Recovery and Reinvestment Act (ARRA) stimulus package for shipment of over 400 forklifts, many of which were purchased by the US Department of Defence (Behling, 2013).

For perspective, in a global market estimated to have introduced more than 900 000 forklifts in 2011 (companiesandmarkets.com, 2012), it is expected that around 3,000 fuel cell-powered forklifts would have been sold by 2012 (Fuel Cell Today, 2012).

Despite the apparent promise of the forklift market for fuel cells, financial records of Ballard and Plug Power – both of which are not profitable – reflect a puzzling trend that seems to suggest 'their products are perhaps more expensive to make than the price at which they can be sold' (Behling, 2013). Both from a cost and air pollution perspective, ventilation challenges associated with underground mining make fuel cell-powered locomotives an attractive proposition. However, this application is still at demonstration phase (Righettini & Mousset-Jones, 2004).

6.4 PORTABLE POWER

Distinct from stationary power, portable power fuel cell units are not fixed and can be used in hand-held devices. Durability, grid independence and long run-times are some of the characteristics driving fuel cell technology deployment for portable power applications. Nevertheless, markets for these are proving tricky to penetrate (Kanellos, 2013), with milliwatt-scale toys and educational kits being the market showing the best promise for commercialising portable fuel cells.

All these commercial applications notwithstanding, the list of profitable listed fuel cell companies around the world remains empty (Various, 2013).

7 HYDROGEN AND FUEL CELL TECHNOLOGIES: SOUTH AFRICA

The policy objectives of the South African government have been introduced earlier in this document. Driven by the Hydrogen South Africa (HySA) programme, the country's HFCT RDI strategy constitutes a top-down government-supported effort geared to stimulate a thriving PGM fuel cell catalyst industry that leverages and builds on existing know-how for socio-economic development.

Structured in the form of three virtual centres of competence (CoC), three universities and two science councils make up the core institutions driving the HySA programme, with the University of Cape Town and Mintek paired up to lead the catalysis CoC (HySA Catalysis); the North-West University and the Council for Scientific and Industrial Research (CSIR) constituting the (hydrogen) infrastructure CoC (HySA Infrastructure); and the University of the Western Cape leading the systems integration and technology validation CoC (HySA Systems).

Detailed elsewhere (SAASTA, n.d.), (SAASTA, n.d.), HySA is targeting a number of fuel cell markets, including those for combined heat and power, vehicles (including MHVs), and back-up power.

7.1 Private sector

Among notable recent developments related to fuel cells is the Investec US$7.7 million transaction to facilitate Ballard's acquisition of Idatech assets (Fuel Cells Bulletin, 2012). Idatech – which specialises in fuel cell-based telecoms extended run-time backup power systems, and accounts for the bulk of these in South Africa (Crouch, 2012) – was in 2006 acquired by Investec, which is listed on the Johannesburg Stock Exchange and London Stock Exchange (Fuel Cells Bulletin, 2006), and subsequently listed on the London AIM exchange. The company has since de-listed (Nolan, 2011).

7.2 Fuel cell manufacturing

Among commercial agreements announced in South Africa (in 2010) is a joint venture between the Department of Science and Technology (DST), Anglo Platinum's Platinum Growth Metals Development Fund (PGMDF), and California-based Altergy Systems. The venture – in which each of the parties holds equity – established Clean Energy Investments, 'whose principal objective is to manufacture and market Altergy fuel cell systems in the Republic of South Africa and other sub-Saharan countries'. The collaboration was reported to 'mark the launch of the South African government's Hydrogen South Africa (HYSA) strategy to develop a manufacturing based "hydrogen economy" and transform and expand uses for the country's national resources, including platinum' (Altergy Systems, 2010).

In 2012, a related announcement was made by the Minister of Science and Technology, who confirmed the sale of 18 platinum-based fuel cells to the South African mobile phone sector (Creamer Media Reporter, 2012).

7.3 FUEL CELL DEPLOYMENT

Largely driven by the need to replace cumbersome, theft-prone, energy intensive, and high maintenance batteries, telecommunication systems seem to be the only market where deployment of fuel cells has gained appreciable traction in South Africa (Crouch, 2012)[26] with the country ranking second after Indonesia. However, in view of the approximately 15 000 [27] South African base stations, the rate of penetration of fuel cell-based back-up power systems seems comparatively low.

8 CONCLUSION

This chapter has attempted to provide a snapshot view of the state of hydrogen fuel cell technologies in the world and in South Africa. The long history and further factors that influenced the evolution of these technologies were also covered briefly.

Progress in the development of fuel cells has been intermittent, with government policies and programmes – e.g. NASA's space program, ZEV legislation in developed countries, and FreedomCAR – and the energy crisis of the 1970s, playing a vital role in creating demand. On the supply side, advances in research and development were crucial for enabling progress.

At about 120 MW, the size of the global fuel cell market is still too small to draw much attention from the terawatt-scale energy market. To this extent, and in view of the numerous hurdles still to be overcome along many frontiers, niche markets have to be nurtured whereby opportunities can be explored and related know-how advanced.

Particularly within the context of an increasingly globalising world, and a locked-in, path-dependent energy system, leveraging international connections and advancing in a coordinated fashion becomes crucial if the opportunity cost of investing in this highly risky venture is to be kept in check. To this extent, and in order to ensure that its choices do not inadvertently result in its exclusion as the hydrogen economy develops, South African membership in strategic fuel cell networks should be encouraged.

---

26. While this author indicates 107 FC units were deployed by Vodacom, Ballard suggests the number is 200 units (also by Vodacom), (Ballard, 2012).
27. Personal communication with a South African telecoms expert.

Bibliography

European Commission Enterprise and Industry. (2010). *Critical raw material for the EU: Report of the Ad-hoc Working Group on defining critical raw materials*, s.l.: European Commision.

AAAS. (2011). *Beyond Technology – Strengthening energy policy through Social Science*, Cambridge: American Academy of Arts & Sciences.

Abedian, I. (2013). *The case for a PGM exchange in South Africa* [Interview] (21 May 2013).

Altergy Systems. (2010). *PR Newswire* [Online] Available at: http://www.prnewswire.com/news-releases/altergys-freedom-power-fuel-cell-systems-to-power-africa-101632243.html [Accessed 28 February 2013].

Andujar, J. & Segura, F. (2009). Fuel cells: History and updating. A walk along two centuries. *Renewable and Sustainable Energy Reviews*, 2309–2322.

Anon. (2010). *Get the Hydrogen Facts.* [Online] Available at: http://www.hydrogenfuturetoday.com/info.html [Accessed 25 november 2012].

Bakker, S. (2010). The car industry and the blowout of the hydrogen hype. *Energy Policy*, July-September, 38(11), 6540–6544.

Bakker, S., van Lente, H. & Engels, R. (2012). Competition in a technological niche: The cars of the future. *Technology analysis & Strategic management*, 24(5), 421–434.

Balat, M. (2008). Potential importance of hydrogen as a future solution to environmental and transportation problems. *International Journal of Hydrogen Energy*, pp. 4013–4029.

Ballard. (2012). *Fuel cell seminar& exposition.* [Online] Available at: http://www.fuelcellseminar.com/media/51071/b2b23-1.pdf [Accessed 27 February 2013].

Ballard. (2013). *EX-99.2 3 exhibit99-2.htm MANAGEMENT'S DISCUSSION AND ANALYSIS.* [Online] Available at: http://www.sec.gov/Archives/edgar/data/1453015/000120677413000708/exhibit99-2.htm [Accessed 27 February 2013].

Baxter, R. (2011). [Online] Available at: http://www.saimm.co.za/Conferences/JhbBranch/RogerBaxter-17Feb2011.pdf [Accessed 26 February 2013].

Behling, N. H. (2013). *Fuel Cells: Current Technology Challenges and Future Research Needs.* s.l.:Elsevier.

Bessarov, D. (2012). *Is possible to establish a Fuel Cell Industry in South Africa* [Interview] (15 June 2012).

Blewis, D. I. & Wilburn, D. R. (2010). *Platinum group metals – the world supply and demand.* US Geological survey open file report, p. 60.

Bossel, U. (2006). *Does a hydrogen economy make sense?* s.l., s.n., pp. 1826–1837.

Brandon, N. & Hart, D. (1999). *An Introduction to Fuel Cell Technology and Economics*, London: Centre for Energy Policy and Technology, Imperial College.

Breakthrough Technologies Institute. (2012). *2011 Fuel cell technologies market report*, Washington, DC: US Department of Energy.

Bullis, K. (2009). *Q&A Steven Chu.* [Online] Available at: http://www.technologyreview.com/news/413475/q-a-steven-chu/page/2/ [Accessed 3 May 2013].

Bush, V. (1945). *Science – the endless frontier*, Washington, D.C.: U.S. National Science Foundation.

Butler, J. (2011). *Platinum 2011*, s.l.: Johnson Matthey.

Butler, J. (2012). *Platinum 2012* – Interim Review, s.l.: Johnson Matthey.

Cato, J. (2009). *Back to the future: electric vehicle at the core of GM's vision*. [Online] Available at: http://m.theglobeandmail.com/incoming/back-to-the-future-electric-vehicle-at-the-core-of-gms-vision/article1091524/?service=mobile [Accessed 2 May 2013].

Chen, Y.-H., Chen, C.-Y. & Lee, S.C. (2011). Technology forecasting and patent strategy of hydrogen energy and fue cell technologies. *International Journal of Hydrogen Energy*, pp. 6957–6969.

Choi, H.-S. (1986). Science and Technology Policies for Industrialization of Developing Countries. *Technological Forecasting and Social Change*, pp. 225–239.

Christensen, C. (2006). *The Innovator's Dilemma*. New York: Collins Business Essentials.

Cokayne, R. (2012). *Why SA's electric car is not going anywhere*. [Online] Available at: http://www.iol.co.za/business/business-news/why-sa-s-electric-car-is-not-going-anywhere-1.1331580#.UZ17ZIel3rw [Accessed 23 May 2013].

Collantes, G. & Sperling, D. (2008). The origin of California's zero emission vehicle mandate. *Transportation research Part A*, Volume 42, 1302–1312.

Conceivious, H. I. (2010). The impact of customer-specific requirements on supply chain management. *Journal of Transport and Supply Chain Management*, pp. 57–68.

Crabtree, G. W., Dresselhaus, M. S. & Buchanan, M. V. (2004). The Hydrogen Economy. *Physics Today*, 39–44.

Cramer, D. (2011). *Mybroadband*. [Online] Available at: http://mybroadband.co.za/news/cellular/39627-vodacom-tests-fuel-cell-powered-base-station-at-cop17.html [Accessed 27 February 2013].

Creamer, M. (2012). *Mining weekly.com*. [Online] Available at: http://www.miningweekly.com/article/form-platinum-exchange-to-save-price-stricken-platinum-mining-iraj-abedian-2012-06-19 [Accessed 08 March 2013].

Creamer Media Reporter. (2012). *Mining Weekly.com*. [Online] Available at: http://m.miningweekly.com/article/pilot-plants-on-way-for-platinum-titanium-value-add-science-minister-2012-09-25 [Accessed 28 February 2013].

Creamer, T. (2012). *SA's beneficiation policy dialogue too narrowly framed*. [Online] Available at: http://www.miningweekly.com/article/sas-beneficiation-policy-dialogue-too-narrowly-framed-2012-02-29 [Accessed 22 May 2013].

Crouch, M. (2012). *Fuel cell systems for base stations: Deep dive study*, London: GSMA Development Fund.

Department of Energy. (2009). *Integrated Resource Plan 2010–2030*, South Africa, Pretoria: Department of Energy.

Department of Science and Technology. (2007). *National Hydrogen and Fuel Cell Technologies Research, Development and Innovation Strategy*, South Africa, Pretoria: Department of Science and Technology.

Department of Science and Technology. (2007). *National Hydrogen and Fuel Cell Technologies, Development and Innovation Strategy*, s.l.: DST.

Department of Science and Technology. (2007). *South African National Hydrogen and Fuel Cells Research, Development and Innovation Strategy*, s.l.: s.n.

Dicks, A. (2012). PEM Fuel Cells: Applications. In: *Comprehensive renewable energy*. s.l.:Elsevier, pp. 204–245.

Dijk, M., Orsato, R. J. & and Kemp, R. (2013). The emergence of an electric mobility trajectory. *Energy Policy*, pp. 135–145.

Dincer, I. & Rosen, M. (2011). Sustainability Aspects of Hydrogen and Fuel Cell Systems. *Energy for Sustainable Development Vol. 15*, p. 137.

DPSE. (2009). *State-owned Enterprises in South Africa's Developmental State: A forward-looking review*, s.l.: s.n.

Duhigg, C. & Bradsher, K. (2012). [Online] Available at: http://www.nytimes.com/2012/01/22/business/apple-america-and-a-squeezed-middle-class.html?emc=eta1 [Accessed 25 February 2013].

Duleep, K., Greene, D. & Upreti, G. (2011). *Status and Outlook for the U.S. Non-Automotive Fuel Cell Industry: Impacts of Government Policies and Assessment of Future Opportunities*, s.l.: U.S. Department of Energy.

EERE, n.d. *Types of Fuel Cells*. [Online] Available at: http://www1.eere.energy.gov/hydrogenandfuelcells/fuelcells/fc_types.html [Accessed 21 February 2013].

Elvis, M. (2012). Let's mine asteroids – for science and profit. *Nature*, 31 May, Volume 485, p. 549.

European Commission. (2003). *Hydrogen energy and fuel cells, a vision for our future*, Brussels: European Commission.

European Commission JRC. (2011). *Long term Trend in Global CO2 emissions,* s.l.: European Commission.

Evenson, W. E. (2004). The potential for a hydrogen energy economy. In: *Report on research and development of energy technologies.* s.l.:s.n., pp. 209–215.

Farley, H. R. & Mesiti, P. (2012). [Online] Available at: http://cepgi.typepad.com/files/cepgi-3d-quarter-2012-1.pdf [Accessed 22 February 2013].

Freed, J., Hodas, S., Collins, S. & Praus, S., 2010. *Creating a clean energy century – Recapturing the lead in clean tech innovation*, s.l.: third way – fresh thinking.

FreedomCAR&Fuel Partnership. (2009). *Hydrogen production – Overview of technology options*, s.l.: FreedomCAR&Fuel Partnership.

Friedman, D., Masciangioli, T. & Olson, S. (2012). *The Role of the Chemical Sciences in Finding Alternatives to Critical Resources – A Workshop Summary*, s.l.: National Academies Press.

Fuel Cell Today. (2011). *The Fuel Cell Today Industry Review*, s.l.: Fuel Cell Today.

Fuel Cell Today. (2011). *The Fuel Cell Today Industry Review 2011*, s.l.: Fuel Cell Today.

Fuel Cell Today. (2011). *The Fuel Cell Today Industry Review 2011*, s.l.: Fuel Cell Today.

Fuel Cell Today. (2011). *The Fuel Cell Today Industry Review 2011*, s.l.: Fuel Cell Today.

Fuel Cell Today. (2012). *Fuel Cell RCS Review*, s.l.: Fuel Cell Today.

Fuel Cell Today. (2012). *The Fuel Cell Industry Review 2012*, s.l.: s.n.

Fuel Cell Today. (2012). *The Fuel Cell Today Industry Review 2012*, s.l.: Fuel Cell Today.

Fuel Cell Today. (2013). *Fuel Cell Electric Vehicles: The Road Ahead,* Royston: Johnson Matthey PLC.

Fuel Cells Bulletin. (2006). UK banking group finances acquisition of US fuel cell developer. *News*, September, p. 5.

Fuel Cells Bulletin. (2012). Ballard gains IdaTech backup power range, IP, and customer base. *News*, August, p. 10.

FuelCellWorks, 2013. [Online] Available at: http://fuelcellsworks.com/news/2013/02/19/ oorja-protonics-and-hysacatalysis-forge-strategic-partnership-for-commercialization-of-methanol-fuel-cells-in-south-africa [Accessed 21 February 2013].

Gasteiger, H. A. & Markovic, N. M. (2009). Just a dream – or future reality? *Science*, 48–49.

Godoe, H. (2006). The Role of Innovation Regimes and Policy for Creating Radical Innovations: Comparing Some Aspects of Fuel Cells and Hydrogen Technology Development with the Development of Internet and GSM. *Bulletin of Science Technology Society*, 26(4), 328–338.

Greene, D. L., Duleep, K. & Upredi, G. (2011). *Status and Outlook for the U.S. Non-Automotive Fuel Cell Industry: Impacts of Government Policies and Assessment of Future Opportunities*, Oak Ridge: UT-Battelle.

Greve, N. 26. *Engineering News*. [Online] Available at: http://www.engineeringnews.co.za/ article/dea-launches-electric-vehicle-pilot-programme-2013-02-26?utm_source= Creamer+Media+FDE+service&utm_medium=email&utm_campaign=Engineering News%3A+SA+launches+electric+vehicle+pilot+programme&utm_term=http%3A %2F%2Fwww. [Accessed 28 February 2013].

Grubler, A. (2012). Grand Designs: Historical Patterns and Future Scenarios of Energy Technological Change. In: *The Global Energy Assessment*. Cambridge, UK: Cambridge University Press.

Grubler, A., et al. (2012). Policies for the energy technology innovation system (ETIS). In: *Global Energy Assessment*. s.l.:Cambridge University Press, 1551–1743.

Halper, M. (2013). [Online] Available at: http://www.smartplanet.com/blog/bulletin/want-a-cheap-hydrogen-fuel-cell-wash-out-the-platinum/13250 [Accessed 21 February 2013].

Hamilton, T. (2012). *Is the fuel-cell industry really near a tipping point?*. [Online] Available at: http://www.thestar.com/business/2012/06/08/is_the_fuelcell_industry_really_ near_a_tipping_point.html [Accessed 27 February 2013].

Harborne, P., Hendry, C. & Brown, J. (2009). Fuel cells as disruptive innovation. The power to change markets. In: *Innovation, Markets and Sustainable Energy – The Challenge of Hydrogen and Fuel Cells*. Northampton, Massachusetts: Edward Elgar Publishing, Inc., pp. 34–51.

Heistein, P. (2012). *Business Report*. [Online] Available at: http://www.iol.co.za/business/ business-news/sa-will-miss-benefits-by-pulling-plug-on-joule-car-1.1334575#.US-sxjDLPPo [Accessed 28 February 2013].

HySA Systems. (2012). *Hydrogen and Fuel Cell Technologies*. [Online] Available at: www.hysasystems.org [Accessed 5 February 2013].

IEA. (2012). *Energy Technology Perspectives 2012 – Pathways to a Clean Energy System*, Paris: OECD/IEA.

Infrastructure, H. (2012). *Solar*. [Online] Available at: www.hysainfrastructure.org.za [Accessed 6 February 2013].

IPHE. (2010). *2010 Hydrogen and Fuel Cell Global Commercialisation and Development Update*, s.l.: IPHE.

Johnson Mathey Plc. (2013). *publications: Platinum Today The world's leading authority on platinum group metals*. [Online] Available at: http://www.platinum.matthey.com/ publications [Accessed 25 February 2013].

Jones, A., Allen, I. & Silver, N. (2013). *Resource constraints: sharing a finite world – Implications of limits to growth for the Actuarial profession*, s.l.: The Actuarial Profession.

Joseck, F. (2009). *United States: Policies promoting hyrogen & fuel cells*. [Online] Available at: http://www.jari.or.jp/jhfc/data/seminor/fy2008/pdf/h20_09.pdf [Accessed 23 May 2013].

Jun, Z., Lucheng, J. & Bo, J. (2013). Hydrogen: Technologies, policies and challenges. *Applied Mechanics and Materials*, 260–261(28), pp. 28–33.

Kanellos, M. (2013). *Why Are Portable Fuel Cells Such A Flop?* [Online] Available at: http://www.forbes.com/sites/michaelkanellos/2013/01/31/why-are-portable-fuel-cells-such-a-flop/ [Accessed 24 May 2013].

Keane, A. G. (2012). *Bloomberg*. [Online] Available at: http://www.bloomberg.com/news/2012-10-16/obama-s-5-billion-slow-to-charge-electric-car-purchases.html [Accessed 21 February 2013].

Kim. (2012). *Can SA have Fuel Cell Industry* [Interview] (30 June 2012).

Kim, L. (1980). Stages of development of industrial technology in a developing country: a model. *Research Policy*, pp. 254–277.

Ko, V. (2012). [Online] Available at: http://edition.cnn.com/2012/11/25/business/eco-hydrogen-fuel-cell-cars [Accessed 21 February 2013].

Kosich, D. (2012). *Mineweb*. [Online] Available at: http://www.mineweb.co.za/mineweb/content/en/mineweb-exploration?oid=150034&sn=Detail [Accessed 20 March 2013].

Kwang, S. R., K, K. M. & Nam-Gyu Park, S. H. C. (2002). Symmetric redox supercapacitors with conducting polyaniline electrodes. *Journal of Power Sources*, 305–309.

Letourneau, A. (2013). Asteroid mining becoming more of a reality. *Forbes*, 25 January.

Lewis, D. (2001). *Does Technology Incubation Work? A Critical Review*, s.l.: s.n.

Mange, S. (2010). *Towards The Hydrogen Economy: A South African Perspective*. [Online] Available at: http://www.iphe.net/docs/Meetings/Germany_5-10/HySA_IPHE_Presentation_Rev1.pdf [Accessed 23 May 2013].

Marchetti, C. (1973). *Hydrogen and Energy. Chemical Economy & Engineering Review*, 5(1), 7–15.

Marchetti, C. (1973). *Hydrogen and Energy. Chemical Economy & Engineering Review*, 5(1), 7–15.

Mason, G. (2012). *Ballard steers away from cars toward forklifts in hunt for profit*. [Online] Available at: http://www.theglobeandmail.com/news/british-columbia/ballard-steers-away-from-cars-toward-forklifts-in-hunt-for-profit/article4249434/ [Accessed 24 May 2013].

Ministerial Review Committee. (2012). *Ministerial Review Report on the Science, Technology and Innovation Landscape in South Africa*, Pretoria: Department of Science and Technology.

Mock, P. & Schmid, S. A. (2009). Fuel cells for automotive powertrains – a techno economic assessment. *Journal of Power Sources*, Volume 190, 133–140.

Morgan, T. (2006). *The Hydrogen Economy – A non-technical review*, s.l.: United Nations Environment Programme.

Mudd, G. (2011). Key trends in the resource sustainability of platinum group elements. *Ore Geology Reviews*, pp. 106–117.

Mytelka, L. (2003). *New Wave Technologies: Their Emergence, Diffusion and Impact. The Case of Hydrogen Fuel Cell Technology and the Developing World*, Maastricht: United Nations University.

Nelson, G. & Soda, F. (2013). VW CEO *Winterkorn pans hydrogen fuel cells*. [Online] Available at: http://www.autonews.com/apps/pbcs.dll/article?AID=/20130314/OEM05/ 130319949#axzz2SD0MeDer [Accessed 3 May 2013].

Nolan, M. (2011). *IdaTech hobbles off AIM*. [Online] Available at: http://www.growthcompany.co.uk/news/1665648/idatech-hobbles-off-aim.thtml [Accessed 28 May 2013].

OECD. (2013). Economic growth in South Africa: Getting to the right shade of green. In: *OECD Economic Surveys: South Africa* 2013. s.l.:OECD Publishing, pp. 91–120.

Ohadi, M. M. & Jianwei, Q. (2008). Alternative Energy Technologies: Price Effects. In: *Encyclopedia of Energy Engineering and Technology*. s.l.:s.n., p. 36.

Perry, M. L. (2013). Durability of Polymer Electrolyte fuel cells. In: *PEM Fuel Cells*. s.l.:Elsevier, pp. 435–467.

Petersen, H. L. a. L. S. (2011). *Risø Energy Report 10, Energy for smart cities in an urbanised world*, Denmark: Scanprint.

PMG. (2012). [Online] Available at: http://www.pmg.org.za/report/20120912-hydrogen-fuel-cell-technology-research-development-programme-briefing [Accessed 21 February 2013].

Pollet, B. S. S. J. L. (2012). Current status of hybrid, battery and fuel cell electric vehicles: From electrochemistry to market prospects. *Electrochemistry Acta, manuscript submitted for publication*.

Prater, K. (1990). The renaissance of the Solid Polymer Fuel Cell. *Journal of Power Sources*, 239–250.

PricewaterhouseCoopers. (2002). Fuel Cells: *The opportunity for Canada*, s.l.: PricewaterhouseCoopers.

Rifkin, J. (2011). *The Third Industrial Revolution – How Lateral Power is Transforming Energy, the Economy, and the World*. New York: Palgrave Macmillan.

Rifkin, J. (n.d.). *The Hydrogen Economy*. s.l.:s.n.

Righettini, G. & Mousset-Jones, P. (2004). Ventilation savings with fuel cell vehicles : A cost-benefit analysis for selected US metal/non-metal mines. *SME Annual Meeting*, 23–25 February, 1–5.

Satyapal, S. (2012). *Energy.gov*. [Online] Available at: http://energy.gov/articles/calling-all-fuel-cells [Accessed 27 February 2013].

Saulny, S. (2008). *Thieves Leave Cars, But Take Catalytic Converters*. [Online] Available at: http://www.nytimes.com/2008/03/29/us/29converters.html?_r=0 [Accessed 23 May 2013].

Saurat, M. & Bringezu, S. (2008). Platinum Group Metal Flows of Europe, Part I. *Journal of Industrial Ecology*, 12(5/6), pp. 754–767.

ScotiaMocatta. (2012). *Scotiabank*. [Online] Available at: http://www.scotiamocatta.com/ scpt/scotiamocatta/prec/PGMForecast2013.pdf [Accessed 27 February 2013].

Seccombe, A. (2013). *Business Day BDlive*. [Online] Available at: http://www.bdlive.co.za/ business/energy/2013/02/28/nersa-grants-eskom-8-annual-increases-over-next-five-years [Accessed 28 February 2013].

Shinnar, R. (2004). Demystifying the hydrogen myth. *Chemical Engineering Progress*, November, 5–6.

Smil, V. (2012). Far from Electrifying: Electric car hopes never die – but electric realities keep intervening. *The American*, 26 November.

Smithsonian. (n.d.). *Allis-Chalmers Fuel Cell Tractor*. [Online] Available at: http://americanhistory.si.edu/collections/search/object/nmah_687671 [Accessed 9 May 2013].

Spath, P. L. & Mann, M. K. (2001). *Life cycle assessment of hydrogen production via natural gas steam reforming*, s.l.: NREL.

Srinivasan, S. & Ogden, J. (2006). Fuels: processing, storage, transmission, distribution and safety. In: *Fuel cells – from fundamentals to applications*. New York: Springer, 375–438.

Staffell, I. & Green, R. (2013). The cost of domestic fuel cell micro-CHP systems. *International Journal of Hydrogen Energy*, 1088–1102.

Steinemann, P. (1999). *R&D Strategies for new automotive technologies: insights from fuel cells*, s.l.: Massachussetts Institute of Technology.

Stillwater Mining Company. (2012). *Palladium Fundamentals*. [Online] Available at: http://www.corporatereport.com/stillwater/Palladiam_Fundamentals_9–26–2012.pdf [Accessed 27 February 2013].

Stroeben, J. & Sterman, J. (2008). Transition challenges for alternative fuel vehicle and transportation systems. *Environment and Planning B: Planning and Design*, 35(6), 1070–1097.

Telli, A. & Turkay, B. (2011). Economic Analysis of Standalone and Grid Connected Hybrid Energy. *Renewable Energy Vol. 36*, p. 1937.

The Economist, 2008. *Hydrogen cars – The car of the perpetual future*. [Online] Available at: http://www.economist.com/node/11999229 [Accessed 2 May 2013].

T-Raissi, A. & Block, D. L. (2004). Hydrogen: Automotive fuel of the future. *IEEE power & energy magazine*, November/December, 40–45.

Tuttle, B. (2012). *Time*. [Online] Available at: http://business.time.com/2012/09/07/is-it-time-to-declare-the-nissan-leaf-a-flop/ [Accessed 28 February 2013].

US Department of Energy. (2003). Platinum Availability and Economics for PEMFC Commercialisation. *TIAX*, p. 14.

US Department of Energy. (2003). *Basic Research Needs for the Hydrogen Economy*, USA: Office of Science.

US Department of Energy. (2005). *Roadmap on Manufacturing R&D for the Hydrogen Economy*. s.l., s.n., p. 3.

Utterback, J. (1994). *Mastering the Dynamics of Innovation*. s.l.:Harvard Business School Press.

van den Hoed, R. (2005). Commitment to fuel cell technology? How to interpret carmakers' efforts in this radical technology. *Journal of Power Sources, Volume 141*, 265–271.

van den Hoed, R. (2007). Sources of radical technological innovation: The emergence of fuel cell technology in the automotive industry. *Journal of Cleaner Production*, 1014–1021.

van Ravenswaay, J., *et al.*, (2009). *South Africa's nuclear hydrogen production development programme*. Oakbrook, Illinois, Organisation for Economic Cooperation and Development, 205–212.

Various. (2013). *Fuel cell archives*. [Online] Available at: http://www.altenergystocks.com/ archives/fuel_cell/ [Accessed 24 May 2013].

Verne, J. (1874). *The Mysterious Island*. s.l.:s.n.

Volfkovich, Y. M., et al. (n.d.). *Studies of super capacitor carbon eletrodes with high capacitance*. A. N. Frunkin Institute of Physical and Electrochemistry, Russian Accademy of Sciences, Moscow, 159–163.

Walsh, B. & Moores, P. (n.d.). *Auto Companies on Fuel Cells*. [Online] Available at: http://www.engr.uconn.edu/~jmfent/AutoCompaniesonFuelCells.pdf [Accessed 2 May 2013].

Warshay, B. (2012). *Like watching grass grow? The slow evolution of the hydrogen economy's $3 billion fuel cell market*. [Online] Available at: http://www.luxresearchinc.com/blog/ 2012/12/like-watching-grass-grow-the-slow-evolution-of-the-hydrogen-economys-3- billion-fuel-cell-market/ [Accessed 02 May 2013].

Wesoff, E. (2012). *Greentechmedia*. [Online] Available at: http://www.greentechmedia.com/ articles/read/bloom-energy-fuel-cell-financials-revealed [Accessed 26 February 2013].

Wild, S. (2013). SA set to roll out prototype hydrogen golf car this year. *Business Day*, 18 February, p. 4.

Xing, Y. & Detert, N. (2010). How the iPhone widens the United States trade deficit with the People's Republic of China. *ADBI Working Paper Series*, Issue 257.

Yang, C. (2009). An impending crisis and its implications for the future of the automobiles. *Energy Policy*, pp. 1802–1808.

CHAPTER 3

EMERGENCE OF NEW INDUSTRIES: INDUSTRY AND KNOWLEDGE FACTORS

Vítor Ferreira and Radhika Perrot

1 INTRODUCTION

Understanding how new industries emerge and how technologies evolve are central points to policy makers. The possibility of fostering a new industry simply by decree is not always feasible, since several interconnected factors play a part in the development of a new system of innovation.

In this chapter, using the theoretical lens of an innovation systems perspective, focusing on Technological Innovation Systems (TIS), we seek to study the possibility of the creation of a hydrogen fuel cell industry (HFC) in South Africa by evaluating factors that determine and shape new industries across the globe, and within the same technological field. We compare different knowledge bases within the sub-fields of hydrogen fuel cells and storage, across different countries, using bibliometric and patent data, to give us an indication of the direction and extent of search processes in each country, and how that compares to South Africa.

Several prior studies have focused on why and how new industries emerge, but so far no consensual view exists on how the process of emergence of new industries starts and develops. Moreover, there are few studies that have focused on understanding the emergence of a hydrogen fuel cells industry and the process of innovation therein, and certainly not within the context of a developing country. This is partly because the technology is considered 'new' with fragmented markets, and partly because the industry has not yet 'taken-off' despite several technological advances, leaving insufficient empirical evidence for researchers. Costly and uncertain hydrogen energy programmes, particularly in terms of their impact and success, are often not part of a country's national technological strategy, and definitely not on the agenda of many developing countries.

Using a broad theoretical approach, rooted in the systems of innovation theory, the importance of different actors, such as government, universities and private firms (a triple helix perspective) is discussed, and further built on, using network concepts of interaction (such as the TIS perspective) to characterise diverse country case studies. Such an approach validates the evolution of the knowledge bases across countries, and is useful in drawing industry and R&D policy implications for South Africa.

In this chapter we learn that industry creation is a network-based activity where knowledge creation and dissemination, national and international linkages, demand, supply (small and big companies) and government policy all play an important role. Hydrogen energy-related technologies seem to be rather knowledge intensive, thus the identification of the main nodes of knowledge and the evolution of different countries knowledge networks as a focus of this study helps in understanding the possibilities for industry evolution.

This chapter has four main sections. The first section reviews the literature that looks at the factors that explain the emergence of new industries, and the processes leading to industry formation, namely the evolution of research networks, international ties that link one country to another, and the critical role of the government in these new industries. In this section we also revisit the parameters of a technology innovation system (TIS), analysing specific countries based on these parameters. In the second section, the various data sources [and an outline of the methodological approach] are presented. The third section presents the main findings, deriving conclusions from the analysis of descriptive statistics on patent and bibliometric data, and network relations between countries. The fourth section presents the main findings and highlights key policy implications of this study. The appendix lays out the methodological approach in detail.

The interrogation in this chapter is limited to analysing global networks vis-à-vis South Africa's position using bibliometric data as indicators. South Africa's research network is small, with limited contribution to the current global knowledge pool in hydrogen fuel cells. And moreover, owing to the confidentiality of much of the detailed work currently undertaken by Hydrogen South Africa (HySA), we were unable to map and assess the network's existing knowledge pool, knowledge flows between the actors of

HySA, and the direction of its research and development (R&D) efforts.

2 THEORETICAL ANALYSIS

2.1 HYDROGEN AND FUEL CELL TECHNOLOGIES

After an earlier period of hype, hydrogen fuel cells came to be regarded as a promising technology by some in the mid-90s. But for others, and 17 years on, it was a 'hydrogen hype' that continued and that is over, or so it seems, as an industry. Governments and the public have turned their attention to the electric car in the hope of finding the clean car of the future (Bakker, 2010). Yet in fact, recent patent data published by the Clean Energy Patent Growth Index (CEPGI), and more recent announcements of fuel cell commercialisation plans by a number of automakers, suggests that interest in hydrogen fuel cells has not died off, and outstrips innovative activities in the hybrid/electric space. Figure 1 indicates that hydrogen fuel cells have had the highest patenting activities (in terms of patents granted) since 2002.

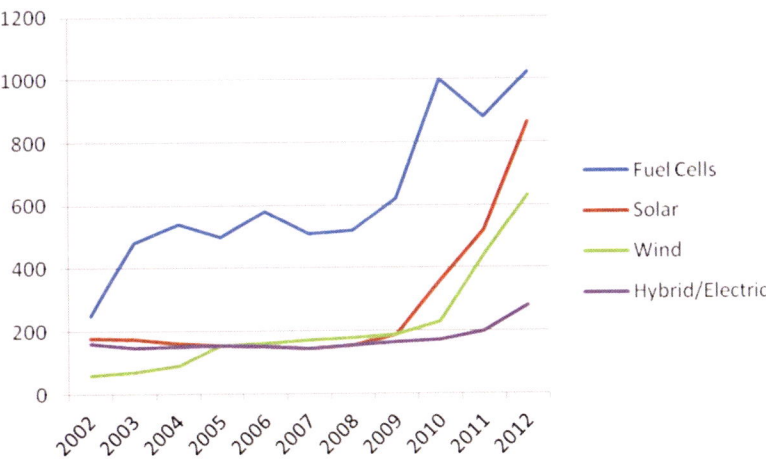

Figure 1: Patent activity in hydrogen fuel cells compared to hybrid/electric technologies

There is a need to analyse the process of technical change taking place along hydrogen technology trajectories and further evaluate the broad factors that are speeding and directing technical change. It took the fuel cell almost 200 years from concept (whose principles were discovered by William Grove in 1839) to emergence. This is because fuel cell technology has still not commercially taken off; it is still early to conclude whether it will ever emerge as the future dominant design in the transport sector (Mytelka, 2008).

Reasons explaining the long commercialisation process of hydrogen-based transportation include the chicken-and-egg[1] problems of high technology development costs and hydrogen infrastructure deployment costs. Alternative energy technologies such as hydrogen fuel cells remain expensive and seemingly non-viable. This is because past and dominant technologies and institutions have locked in economies to fossil fuel energy systems, thus making them expensive and technologically non-viable as an energy option (Unruh, 2000).

But at the core of shaping and setting the pace of technological development, including the establishment of a dominant design, lowering costs and removing technological glitches, lies the interests and preferences of both consumers and producers (Mytelka, 2004; 2008). At the core also lies the institution, and the interplay between government and business, that are shaping the process of hydrogen fuel cells (HFC) innovation, and determining the technological trajectory of hydrogen-based technologies (Mytelka, 2008).

Hydrogen technologies can be regarded as 'disruptive technologies' (Christensen, 2003; Mytelka, 2008) as they can potentially transform both the power and the transportation markets. Disruptive innovations can cause large disarrays at the systemic level, whether in the market structure, product or consumer behaviour. Moreover, the technology is evolving along a trajectory which is clearly distinct, in terms of a technological curve, from the old and conventional fossil fuel trajectory of coal, oil, natural gas and nuclear power.

Hydrogen technologies, and hydrogen fuel cells in particular, are 'new wave technologies' that are combinatorial in nature, 'anchored in the sciences and their knowledge base has developed less as a result of incremental change along a single trajectory, than through a combination of several distinct trajectories with significantly different scientific roots' (Mytelka, 2004, pg. 7). During the 1980s and the early 1990s, the convergence of IT and telecommunications, and that of the pharmaceuticals and biotechnology, happened because of the ability to combine these technologies at the basic science and engineering level.

So a transition to a hydrogen economy involves a discontinuous shift to a new trajectory (Geels and Kemp, 2007; Mytelka, 2003). Additionally,

---

1. Car companies would not invest in industrialising fuel cell cars without the hydrogen supply infrastructure in place, but nobody will invest in the latter without evidence of fuel cell cars to be refuelled.

hydrogen storage and fuel cell technologies are interrelated, such that innovations in any one of these application areas have implications on the other. The co-existence of both hydrogen fuel cells with a fledgling technology (hydrogen storage) is not only increasing the industrial and knowledge complexity of the technology, but it does in fact explain why a hydrogen fuel cell industry has not taken off yet.

2.2. EMERGENCE OF NEW INDUSTRIES

2.2.1. INTRODUCTION

BIRTH AND GROWTH OF INDUSTRIES

The development of new industries has a positive effect on general economic growth (World Bank, 1993; Stiglitz, 1996). Hung and Chu (2006) affirm that quickening the pace of the process from when a new technology has emerged to the introduction of new industries, in specific economies, is vital for economic growth, the competitive environment, and employment opportunities. Nevertheless, the birth and growth (and sometimes failure) of different industries and technologies have not always been fully understood as various factors intervene in the process.

MARKET FAILURE AND GOVERNMENT INTERVENTION

The emergence of a new industry is a complex process with numerous interacting components. Therefore, a systemic view is currently seen as an appropriate approach to a complex phenomenon that includes the role of privates firms, entrepreneurs, research institutions, demand and supply conditions, and government policies. 'Emerging technologies, or potentially radical innovations, such as hydrogen fuel cells and related technology may be subject to market failure, i.e. the market is not able to develop the technology through the co-ordination of resources between public and private actors. This implies that sometimes there are reasons to complement the market and capitalist enterprise through government intervention (Edquist, *et al.*, 2004, pg. 429).'

TECHNOLOGICAL AND SOCIO-ECONOMIC ISSUES

A new sector thus creates several challenges and generates social, economic and technologic uncertainty (Choi, *et al.*, 2011). Firms have to integrate and create new knowledge, changing their capabilities (Teece, *et al.*, 1997) while facing a much undefined environment. At the same time governments and research organisations try to mitigate the risks by supporting suitable R&D activities and stimulating collaboration between different actors, and universities play an important role by supplying human capital and building the knowledge base for the system (Freeman, 1991; Mowery and Sampat, 2005; Choi, *et al.*, 2011).

The formation of a new industry is an evolutionary process that includes a diversity of relations ranging from competitive to non-market forms, and sometimes even formal to informal forms, stemming from several heterogeneous actors in terms of their knowledge components, their firm-level capabilities (Nelson, 1994; Dosi, *et al.*., 1997; 2000; Malerba, 2002, 2004, 2006; Choi, *et al.*, 2011). And according to Dosi, *et al.* (1992) and Nelson (1992), the evolution of an industry is also influenced by the nature of the technology embedded within it.

2.2.2. From industry formation to the emergence of networks and systems of innovation

Understanding emergence of industries

Industry emergence

Industries go through different industry stages over time (Utterback, 1994; Klepper, 1997; Klepper and Graddy, 1997). Industry life cycle theory highlights the design and features of sectors in which a value chain has been established, and the components for its emergence have to a certain extent been identified. Though the value chain analysis may provide some indication of the main building blocks in emergent sectors, it is often not known when a technology has reached a certain stage of development, or a dominant design extensively commercialised (Choi, *et al.*, 2011). The emphasis here is on understanding the emergence of industries, and the preconditions that may lead to the success of new industry, in a systemic way.

The basic model of the evolution of firms, technology and industrial structures is captured in the industry life cycle (ILC) of industrial dynamics, which hypothesises that, as technology and industries evolve over the life cycle, some typical evolutionary patterns emerge (Dosi and Nelson, 1994). Industry life cycle models are considered not only to define major stylised facts and regularities in the evolution of industries, but also to build a consistent analysis of why and how industries evolve and contribute to technical change.

INDUSTRY EVOLUTION AND VALUE CHAIN

Technology has implications for the vertical structure of industries (Malerba-Nelson-Orsenigo-Winter, 2005) as it often changes the boundaries of industries and firms in terms of re-defining distribution networks or creating new ones for accessing new or existing markets. As an industry evolves, firms begin to tighten control over the value chain through integrated relationships with suppliers and distributors (Utterback and Suarez, 1993). Often as the technology matures, the focus of firm shifts from a technology-focus to a more market-orientated process (Rice and Galvin, 2006; Haupt, *et al.*, 2007) and large and established firms are generally more favoured (Schumpeter, 1943; Abernathy and Utterback, 1978; Tushman and Anderson, 1986) because of their scaling-up capabilities and extensive and dense networks of relationships between customers and vendors (Abernathy and Clark, 1985).

INDUSTRY EMERGENCE AND SYSTEMS OF INNOVATION

Malerba (2002, 2004) had introduced the idea of sectoral systems of innovation (SSI), emphasising the multidimensional, cohesive, and dynamic view of the innovation process. This concept is part of a wider research field of innovation systems that includes national systems of innovation, which has the nation as the primary unit of analysis, i.e. it is a macro-level approach to the study of innovation systems (Lundvall, 1992; Nelson, 1993). The other is the regional system of innovation (RIS), where a region – or other sub-national entities – is analysed in terms of innovations (Braczyk, *et al.*, 1998) and a technological innovation system (TIS) (Carlsson and Stankiewicz, 1991), where technology is the unit of analysis, and not solely the sector.

Malerba (2004, pg. 16) defines SSI as a 'system of innovation and production…composed of a set of new and established products for specific uses, and a set of agents carrying out activities and market and non-market interactions for the creation, production and sale of those products'.

Sectors are not created from a void but rather derived from 'numerous existing sectors over which new clusters spanned, giving examples of sectors such as Internet–software–telecoms and biotechnology–pharmaceuticals to depict the integration of knowledge and technology, new interrelations between actors, and the expansion of boundaries' (Choi, *et al.*, 2011, pg. 754). In fact, Malerba (2002, 2004) has identified three main components of an SSI: knowledge and technologies, actors and networks, and institutions which emphasised the fact that these components are essential for all emerging sectors.

Choi, *et al.*, 2011 (p. 4) cite several studies (Russo, 2003; Giarratana, 2004; Mezias and Kuperman, 2000; Murtha, *et al.*, 2001; Sapsed, *et al.*, 2007) that highlight different but relevant aspects in the emergence of new sectors, such as:

- institutional environment;
- sound technological base;
- niche-based products;
- patent systems;
- communities and organisations that bridge actors; and
- strategic alliances among firms.

These aspects are very important in the formative stages of a new technological sector, and aspects of these will be analysed below in greater detail.

NETWORKS AND INDUSTRY CREATION

Choi, *et al*, (2011) identify knowledge networks as an essential tool in the study of emerging sectors. The empirical study by Barley, *et al.* (1992), as cited in Choi, *et al.* (2011), shows for the biotech sector that firms tend to set up numerous relationships to improve competitiveness.

Choi, *et al.* (2011, p. 756) state: 'A network is a result of not only supply–demand relationships, but also the knowledge and institutional factors that are embedded in the system. The means of gaining benefits from linked actors differs between the mass-production sector and the science-based sector'. This knowledge network is rather important since it is created before the value chain, allowing the creation of knowledge and the design of the institutional setting. In emerging sectors, actors focus on forming linkages with heterogeneous actors so as to diversify their knowledge base and not only to transact their products (Choi, *et al.*, 2011).

Knowledge networks are one of the analytical lenses which can be used to study the emergence of industries. Additionally, government policies can influence networks that will impact the creation of networks, and the evolution of industry (Breschi, *et al.*, 2009) as cited in Choi, *et al.*, 2011). For example, government R&D funding can create a contractual research network, which may contribute to the development of science networks (Wagner and Mohrman, 2009 as cited in Choi, *et al.*, 2011).

Choi, *et al.* (2011, p. 755) emphasise that firms' performance relies on network structures and the modality of inter-firm ties. These linkages are common deliberate means of acquiring resources and increasing competitiveness. Actors with limited resources are able to use the network to mitigate their risks and costs by forming linkages with others in the network.

Actors widen their knowledge base and internalise external capabilities under uncertain and swiftly changing environments, using their links with other actors in the network (Dosi, *et al.*, 1997). Those actors that have technological bases and numerous capabilities different from the rest can create the necessary solutions to address specific issues and problems around technology development, overcoming several challenges of emerging sectors (Choi, *et al.*, 2011).

LARGE AND SMALL FIRMS

Large firms have been known to play an important role, even in the early stages of emerging sectors, as they are able to support higher risks and investments, and their R&D activities allow them to develop and expand their existing knowledge base (Choi, *et al.*, 2011). Larger firms play a more significant role in an industry stage that requires diverse skills, knowledge bases and wide supplier networks. As innovation demands growing complexity of knowledge bases these days, it implies that larger firms are becoming increasingly dependent on skills and knowledge bases that are external to the firm (Granstrand, Patel and Pavitt, 1997; Mytelka, 2008; Powell and Grodal, 2010; Narula and Zanfei, 2010).

However, in different national innovation systems different actors may take on this role. The early stages of an industry life cycle, or the so called 'entrepreneurial technological regime' (Winter, 1984, p. 297), are favourable to the entry of entrepreneurial start-ups, as they are often sources of new and proprietary knowledge in technology-intensive sectors (Hagedoorn, 1996b; Agarwal and Audretsch, 2001). So small firms are often firms with new ideas, innovation and technologies, that are basic, science-based and radical. Often these small firms have the potential to create or disrupt markets and technologies (Rothwell, 1984). This capacity for disruption is well described by the so called 'innovator's dilemma' (Christensen, 1997).

Small and new biotech firms played a critical role in the biotechnology industry, and later as harbingers of new knowledge and processes. They changed the structure of the entire pharmaceutical industry, including the way large pharmaceutical firms re-strategised their innovation activities to remain competitive in the industry. Nevertheless, in other cases such as the computer industry, universities took the lead role in undertaking publicly-funded basic

R&D that made immense contributions, particularly in the early stages of an industry (Malerba and Orsenigo, 1996 as cited in Choi, *et al.* 2011).

Finally, the role of public research organisations (PROs) and their contribution to networks should be emphasised. These organisations conduct basic or intermediate R&D, and they usually bridge disconnected or distanced actors, create new economic opportunities, and link the knowledge flows between the various actors of the innovation system. All these are achieved under the direct influence of government policies, as governments have an impact on the evolution of the network(s), ultimately assisting in the emergence of a new industrial sector by manoeuvering these organisations within the network (Choi, *et al.*, 2011).

TECHNOLOGICAL INNOVATION SYSTEMS

To understand a new technological field and its allied industries, we can frame the above concept of networks and actors using a technological innovation systems (TIS) approach. A TIS is 'a network of agents interacting in a specific economic/industrial area under a particular institutional infrastructure or set of infrastructures and involved in the generation, diffusion, and utilization of technology' (Carlsson and Stankiewicz, 1991, pg. 111). This view highlights the important and critical role of actors that collaborate within networks and exchange knowledge, including an understanding of the impact of institutions on innovation processes within networks (Musiolik, 2012).

The TIS perspective rests on the notion that innovations are passed, in an iterative process, through interactive networks of actor linkages and relationships, rather than through a single actor or event (Coenen and López, 2010). Therefore, innovation is considered a fundamentally social, and an interactive learning process. Furthermore, the systems approach to innovation states that organisational behaviour and strategy are fundamentally influenced (though not solely determined) by institutions. And institutions constitute the rules, laws, norms and routines, among others.

Network actors include universities and public research institutes that have diverse sets of knowledge, resources, competencies and even strategies, firms, and governmental and non-governmental agencies. As mentioned above, institutions constitute 'norms, laws, regulations, guidelines, contracts, values, culture, cognitive frames, collective expectations, etc.' (Markard, Musiolik and Worch, 2011) (Musiolik and Markard, 2011). This definition not only recognises that a TIS can be at the international, national or local levels, but that the nature of technology can be applied across various industrial sectors. For example, biotechnological methods and applications are not only used in the pharmaceutical industry, but also in the food, textile, agriculture and mining industries (Coenen and López, 2010).

Since technology is the unit of analysis in a TIS, a conceptual framework is set up, focused on the study of how actors, networks and institutions are integrated and how they help in the birth of sectors and industries (Coenen and López, 2010).

The TIS concept seems more suited to deal with the emergence of new technologies and sectors (Nygaard, 2008; Coenen and López, 2010) than the concept of a sectoral innovation system, since an SSI sets its limits on the basis of existing products (e.g. in the case of fuel cells – the technology has applications in the energy and transportation sectors). There is substantial technological and market ambiguity on the formation of a new sector, and the ex ante limit setting of the system may leave out important factors and actors that are critical in driving innovation (Coenen and López, 2010). Therefore, in recent years the TIS concept has been used as an investigative framework in the study of emerging technologies (e.g. Carlsson, *et al.*, 2002; Edquist, 2005; Jacobsson and Johnson, 2000; Markard and Truffer, 2008b; Musiolik and Markard, 2011). In the field of energy technologies, many cases from different countries have been analysed using the TIS approach ('Bergek and Jacobsson, 2003; Jacobsson and Johnson, 2000; Jacobsson, 2008; Markard and Truffer, 2008a; Suurs and Hekkert, 2009', as cited by Musiolik and Markard, 2011, pg. 1910).

The TIS concept may be used to understand the interaction of a specific population of firms and the environment where they act (Suurs, 2009). It is critical to consider that if actors are aware of being part of an emerging field or innovation community (Lynn, *et al.*, 1996) this will guide their own understanding of the TIS they are part of and lead them to deliberately engage in system building (Musiolik, 2012).

The SSI and the TIS are both firm-centric views as firms are considered the main unit that drives the process towards innovation (Coenen and López, 2010). In both these perspectives firms play a critical role in the generation, adoption and use of new technologies – SSI through sectors (Malerba, 2004), and TIS entreprenurial activities (Bergek and Jacobsson, 2003; Musiolik, 2012). And further, (Penrose, 1959), as cited by Coenen and López (2010), states that these perspectives draw on the resource-based view of the firm, in which firms are theorised as packages of knowledge bases and competences.

Musiolik and Markard (2011), and as cited in Bergek, *et al.* (2008a) and Hekkert, *et al.* (2007), identify eight main functions or dimensions of TIS: (1) entrepreneurial activities; (2) knowledge development; (3) knowledge diffusion through networks; (4) guidance of the search; (5) market formation; (6) resource mobilisation; (7) creation of legitimacy; and (8) development of positive externalities.

With this set of dimensions and framework in mind the analysis will look at various country cases that are involved in the development of hydrogen and fuel cell technologies, searching for learning outcomes that may be useful for policy design in South Africa. Through these cases it will be possible to understand and identify different development paths which an emerging industry might take, seeing that there is no single pre-determined path.

2.2.3. MARKET, INSTITUTIONS AND INDUSTRY – ROLE OF THE GOVERNMENT IN EMERGING INDUSTRIES (HYDROGEN ENERGY INDUSTRIES)

THE ROLE OF THE GOVERNMENT

A systemic transition from one socio-technical system to another, involves changes not only in artefacts, cultural habits and organisational behaviour, but also involves changes in infrastructure and institutions (Geels, 2004). For example, the creation of gasoline fuel infrastructure and a road network were critical, although not sufficient, in establishing the internal combustion engine as a dominant technology. Regulation and policies were involved as the streets became chaotic with all kinds of vehicles (bicycle, trams, cars and horse-drawn carriages) (Geels, 2004).

A critical aspect of the analysis will be to explore the government's role in allowing or restricting opportunities for technology-based businesses and other actors of the system. In a systemic approach to innovation, a main principle is that innovation is generated by the interaction between actors within the 'innovation system', i.e. in a country, a region, a sector, or within a technological field. For instance, Madsen and Andersen (2010) assert that in Europe, government plays an increasing role in hydrogen and fuel cell development. Nevertheless, they state that regions that already have well-functioning regional innovation systems (RIS) and/are already home to innovative clusters are more suited to develop TIS in this field. There are ties or links between existing hydrogen-production capacities and infrastructure, and the early adoption of this technology, thus indicating possibilities for latecomers with no infrastructure.

In an innovation systems approach, the focus is on firms and other types of actors, along with institutions and policies that influence the innovation behaviour and performance, including the emergence of new industries, through the development of new products, processes and new forms of organisation (Mytelka, 2010). And at the core of an innovation system, such as the hydrogen economy innovation system under discussion, are knowledge and information flows that link various actors, which eventually provide stimulus and support to the process of innovation (Mytelka, 2010; Lundvall, 1998; and Edquist, 1997). These flows of knowledge and information involve knowledge in manufacturing process, knowledge of product and systems integration and markets, and knowledge of the technology – which are situated within firms, and other organisations and institutions of an innovation system.

Following this systemic view, several factors have to be put in place in terms of policy such as the promotion of the linkages between actors (especially research infrastructure and companies); design of institutions; promotion of entrepreneurship; skills development and retention; support for the improvement of the innovative capacity of firms; and infrastructure to promote the speedy diffusion of technologies such as growth support for SMEs through interaction and networks (Wolfe and Gertler, 2004).

Governments play a critical role: allocating resources and thus avoiding project duplication through research portfolio variation, which allows researchers to choose dissimilar strategies (Choi, et al., 2011). Funding agencies also play an important role since they are able to manage different programmes avoiding inefficiencies (Dasgupta and David, 1994) as cited in Choi, et al. (2011).

Government intervention reduces risks and facilitates the flow of knowledge. This role, which involves policies that 'facilitate collaborative networking among firms, universities, and PROs that are involved in the formative stages' (Choi, et al., 2011, p. 757), is important in the 'emerging period during which available resources are limited and trajectories towards a dominant design remain invisible' (Choi, et al., 2011, p. 757).

Technological programmes may contribute to the creation of knowledge in a network by managing the technological trajectory using a flexible portfolio of programmes, thus avoiding some of the possible inefficiencies traditionally associated with R&D (Callon, et al., 1992; Callon, 1995) as cited in Choi, *et al*.(2011). A network of public and private actors is also an effective way to supply information about the progress trends of the different players in the field (Tijssen and Korevaar, 1997) as cited in Choi, *et al*. (2011). Government R&D projects and programmes help actors enter the knowledge network where they are able to share and exchange knowledge and information with each other (Choi, *et al*., 2011). These programmes are useful tools for firms and governments, allowing them to distribute resources not only efficiently, but to choose even an appropriate place and time for entry, particularly during the early stages of an industry (Choi, *et al*., 2011). Firms that are involved in government research programmes with the goal of promoting partnerships often have specific obligations. In this case, the trade-off for receiving support is to share some specific knowledge with their partners (Choi, *et al*., 2011).

As Choi, *et al*. (2011) state, these programmes seem to be more effective in emerging stages of technological development, when markets are seldom formed and competition is non-existent. Firms within a network have more opportunities to diminish their risks and uncertainties when they share them with the government. For most firms, and despite certain obligatory requirements that are imposed upon them, and given the positive outcomes mentioned, participating in government programmes is greatly welcomed (Tijssen and Korevaar, 1997; Polt, 2001; Audretsch, *et al*., 2002) as cited in Choi, *et al*. (2011).

Hung and Shu (2006) identify three key mechanisms in pacing up the process of industry formation that involve policy intervention and are linked with previously mentioned types of government intervention. The first mechanism includes the promotion of partnership in the commercialisation process. Governments may create 'incentives and regulations to support innovation and cooperation in R&D, partial public funding of privately performed R&D, and the setup of effective public/private partnership mechanisms' (Hung and Shu, 2006, p. 109). (This is recurring evidence in several studies.) The second mechanism is related to the promotion of entrepreneurship in the innovation system. This can be achieved by the development of the network that evinces 'dynamics of entrepreneurship, generating efficient support mechanisms for private innovation, giving incentives for entrepreneurship in transition, and for the international corporate entrepreneurship' (Hung and Shu, 2006, p. 109). The third mechanism is related to the ability to sustain a commercialisation process and the birth of new firms. This process is somewhat complex since it involves creating public support mechanisms that may solve innovation market failures, using public support to enhance the performance of new firms in the market, and designing an effective industry stimulus to technological diffusion and development (Hung and Shu, 2006).

Obviously, different types of government will have different impacts on the programmes and on the strategies for a specific field. Vasudeva (2009) shows that that knowledge-building strategies of firms can be in part explained by the differences in their respective national institutional contexts. In their study of fuel cell innovation across the US, France, Japan and Norway, they show how these countries' socio-political institutions (here considered by their levels of 'statism' and 'corporatism') contribute to differences in technological policies related to investment, collaboration, internationalisation, and diversity. These policies are bases of advantages (and disadvantages) for firms, with consequences for their knowledge-building strategies. Although the importance of co-operation involving universities, government, and business in stimulating innovation is well-known across various countries, the role of government has been found to be particularly important and direct in countries such as Japan and France, when compared to the US and Norway, where in these countries the government plays a minimal and indirect role (Vasudeva, 2009).

However, the reasons for collaborations and their outcomes varied among the countries studied. For instance, the collaborations among Japanese and Norwegian firms were stronger and more enduring, leading to denser networks, whereas in the US and France these alliances or collaborations were more cautious and often short-lived, and have more fragmented networks (Vasudeva, 2009). Vasudeva (2009) further goes to show that technological diversity differed across the various countries. Firms in Japan and Norway pursued different technologies, but these technologies embedded thenselves comfortably with each national innovation system. In the US and France technological diversity originated from competitiveness, as the possession of different technological capabilities was instrumental in separating the winners from the losers.

GOVERNMENT AND MARKET

In the case of sustainable energy, Lund (2009) shows that energy policies can contribute to the expansion of domestic industrial activities. This may be achieved through measures that enhance markets (even when the industrial base is weak). However, Lund (2009) also notes that irrespective of the market measures, investment in R&D support to industries in related technology areas may be a strong way to help firm diversification and generate export possibilities. To a certain extent it will be important to identify where in the value-chain policy, intervention can assist in creating opportunities.

Demand and supply factors are extremely important in understanding the innovation process. The rhythm of innovation in the private sector profoundly influences the growth of national output and productivity. Alongside which, the supply side (stock of knowledge and R&D laboratories) is important in the production of knowledge. In reality, and as the theory of innovation systems proposes, a variety of factors link up to promote innovation.

In terms of demand, firms and regions should have absorptive capacity to allow them 'to recognise the value of new information, assimilate it, and apply it to commercial ends' (Cohen and Levinthal, 1990, pg. 1); in other words, allowing firms and regions to accumulate knowledge in this particular field. Chiaroni and Chiesa (2006) demonstrate the complexity of a technology cluster creation (for biotech) and identify two cases. The cases were the result of a spontaneous co-existence of key technology and economic factors, and policy-driven clusters that are generated by a strong commitment of governments.

Raimund, et al.(2008) have observed that the overall functioning of an innovation system is important in the formative phases of a hydrogen fuel cell innovation system, even though the overall effect of European Union (EU) public policies seems to be too weak to enable the deployment of a hydrogen and fuel cell strategy (HFC). Further, current EU policies have not been effective in signalling long-term commitments, and lack incentives to encourage high investment in the deployment of HFC technologies.

GOVERNMENT AND CLUSTER POLICY

Clusters in their first phase may grow spontaneously or as a result of public intervention (Carlson, 2005). But path dependency and knowledge spillovers play a strong role while new technological opportunities create possibilities for spillovers.

Trippl and Tödtling (2007) show that clusters (in the case of biotech) can be created and linked, not only linked to local knowledge but also to distant knowledge sources. In the latter case, and in latecomer regions, the strong role of policy actors is seen as important. The authors show that in some cases knowledge flows through international conferences and fairs ('global buzz') and other international links (market links, formal networks, knowledge spillovers, etc.). Chiaroni and Chisesa (2006) seem to endorse this view since they identify two types of clusters – spontaneous and policy driven – thus demonstrating the importance of strong policy actors.

Mans, *et al.*, (2008) conclude that cluster policy is a useful tool that may complement existing energy R&D policies. Nevertheless, analysing whether self-declared clusters actually work as clusters (sometimes clusters are simply initiatives or policy names) and what input they provide the overall system, is focal in earning the benefits of cluster policy.

Carlsson (2005) identifies several roles for public policy in cluster creation, such as ascertaining the existence of a sufficient knowledge base, creating conditions for transfer or creation of more related knowledge. Public organisations also help by creating incentives to reinforce positive forces or to overcome negative forces (through contracts or other incentives schemes).

As mentioned above, governments also promote entrepreneurial experiments (when private initiative is lacking) (Hung and Shu, 2006; Lane, 2009; Choi, *et al.*, 2011). Governments may further help promote positive externalities by creating incentives for common labour markets, knowledge sharing, etc.

Institutional and policy reforms may help improve market conditions. Governments can further help by creating resources, such as financial or human capital. Lack of technical, financial and human resources may stop the creation of new industry clusters (Carlsson, 2005).

Carlsson (2005) identifies a set of questions that may be important to analyse this specific case, which may lead to a more appropriate response from public authorities when analysing the resource requirements to support a cluster or new industry.

With respect to the existing knowledge base, some examples are:

- *What type of knowledge exists in the system currently (scientific knowledge and know-how relating to systems, materials, components, production, design, etc.)?*
- *Are there critical gaps?*
- *Who are the carriers of knowledge, how many, how diverse, and how well connected?*
- *Is the whole value chain well represented?*

- *How does this system compare to another system elsewhere?*
- *Is the knowledge base expected to change in the next decade?*
- *With respect to other resources, how much financial capital is available (seed, angel, and venture capital, as well as other risk capital)?*
- *How many and how diverse are the sources of finance?*

THE ROLE OF EXPECTATIONS IN CLUSTER FORMATION

Konrad, *et al.* (2012) state that expectations play a role in the formation of some clusters since some organisations increase their activities in response to the perception of future importance. These expectations have importance in motivation and co-ordination efforts. The authors also state that when expectations are not met, it becomes only detrimental to the more sensitive actors. Therefore, we may conclude that 'collective expectations' of an innovation system are important in the outcome of the innovation process. The less sensitive actors seem to form the pillars for an emerging technology. This was the case of the system manufacturers in the fuel cell industry studied by Konrad, *et al.* (2012).

Firms tend to be more volatile than public research organisations and the network itself tends to adapt to the adjustment of expectations, which itself sends a signal to the expectations of the participants in a dynamic process (of hype and disappointment). Sovacool and Brossmann (2010) document that research efforts and expenditures on hydrogen technologies tend to be lagging when compared to other technologies. The lag in spending is despite the fact that few politicians, academics and companies seem to be interested in pushing the technology. Sovacool and Brossmann (2010) go on to highlight the potential of fantasy in hydrogen economics, which may affect real expectations around the technology.

2.2.4. NATIONAL VERSUS INTERNATIONAL DIMENSIONS

Several authors demonstrate that not only are local links or ties important but also the links between the national system of innovation and international systems. When a new industry is growing the possibility of a mismatch between the needs of the different actors is very high. Firms may need knowledge (through R&D and human resources) or other technological inputs that are absent in the region. Or firms may choose to acquire the technology from abroad, or base themselves abroad and thus internalise aspects of other countries innovation systems (Narula, 2002). At the same time, research institutions of a country benefit from the cross-fertilisation of knowledge facilitated by inter-country programmes and from the mobility of researchers.

Thus, the creation of a successful industry is often linked with its capability to maintain international links, whether they are knowledge exchanges or the simple ability to sell products in a larger market or buy the technology that is absent. Analysing the wind industry, Lewis and Wiser (2007) demonstrate that policies that support market size and stability, linked with policies that specifically create incentives for the manufacturing of technology locally, are most likely to result in the formation of an internationally competitive industry.

New industries of the future may result from the encounter between technological innovation and market opportunity that will assume an increasingly global character over time (Murtha, *et al.*, 2001). Entrepreneurial communities, even those that are geographically located within clusters of countries, have been observed to have international ties of a global nature with other firms, both within and outside their regional clusters (Murtha, *et al.*, 2001). Firms need to reach across national borders to access new and complementary knowledge, and tie up with partners, suppliers, and customers that are critical in the creation of new businesses or local industries (Doz, *et al.*, 2001, as cited in Murtha, *et al.*, 2001). At the same time, speed in gaining competitive advantage in a technological field, given the pace at which industries evolve these days (Nelson, 1992, as cited in Murtha, *et al.*, 2001), can fuel the benefits of clustering within a geographical region, particularly early in the industry emergence process.

The level and capacity of each country to enter international markets and improve its respective trade is closely linked to domestic capacity to take advantage of new and innovative technologies. The adoption, absorption, mastering, adaptation and application of these technologies depend on the strength and efficiency of the national system of innovation.

The benefit of engaging with foreign actors is recognised across all countries studied by Vasudeva (2009), but the method for attaining this goal varies considerably. As Vasudeva (2009, p. 1255) states, 'in the US, a large number of foreign firms participate in the national innovation system. In Japan and France, fewer foreign firms participate in the national innovation system, but firms from these countries offset this limited exposure to foreign firms by participating in other national innovation systems. These differences in approach have important consequences for firms' knowledge-building strategies and associated performance outcomes'.

Pietrobelli and Rabellotti (2011) remind us that the innovation literature inclines not to emphasise the critical impact of international or global knowledge exchange and innovation collaboration through intra-firm and inter-firm networks, and through its integration to global value chains. As the authors state, in developing countries this factor is critical, as such integration is playing an increasing and important role in enhancing learning and innovation, and in retrieving knowledge. The relation between global value chains and innovation systems is, nevertheless, nonlinear and endogenous.

Learning for the actors in a global chain can be the result of pressure to meet international standards, or the result of the engagement of value-chain leaders in periods when suppliers' competencies are found to be low (Pietrobelli and Rabellotti, 2011). Along a value chain, when actor competencies are complementary, learning among the actors becomes mutual, and interactions become intense (Pietrobelli and Rabellotti, 2011). Pietrobelli and Rabellotti (2011) confirm that an efficient innovation system helps reduces complexity and enables easy transactions between the actors of a global value chain. Nonetheless, the structure of a global chain is dynamic with the need for continuous adjustments, and the nature of the innovation systems often affects its emergence, and its eventual evolution.

The country analysis below identifies starting points of hydrogen fuel cells by several countries in the 1980s, although some among them, such as Japan and Sweden, had started basic R&D efforts in the 1970s, while efforts were erratic. Iceland, France, India and China seem to be the exception. Most countries started their programmes for environmental reasons. Nevertheless, Japan, Germany, China, France, UK, India and the US seem to have been driven also by the need to stay competitive, particularly in the automotive sector, and for energy security reasons. Most fuel cell industries have been state-driven, even though clusters of large firms and SMEs seem to have a role in several countries. In fact, in some countries where the automotive players are dominant, the industry incumbents appear to play a bigger role in research (or their suppliers – see the case of Italy); also big oil and gas firms look engaged in the development of several country industries. There seems to be very limited market competition, with either government or big firms acting as main influencers, and venture capital playing a very limited role. Notably a lot of international integration exists, especially in respect to research more or less linked to the automotive industry. Several countries are interconnected, and the actions of one or other look like they echo the

TIS Indicators → Countries	Relevant Technological Capabilities	Knowledge Networks and Ties	International Links	Entrepreneurial Activities	Market Competition	Automobile Industry	Technological Diversity	State and Institutional Support	ST&I Policy
Iceland	Weak	Medium	Very Strong	Medium	Weak	Weak	Weak	Very Strong	Weak
Germany	Very Strong	Very Strong	Strong	Strong	Strong	Very Strong	Very Strong	Strong	Strong
Norway	Weak	Medium	Medium	Medium	Medium	Weak	Weak	Very Strong	Strong
Netherlands	Medium	Medium	Medium	Strong	Strong	Weak	Medium	Medium	Weak
Canada	Strong	Strong	Very Strong	Very Strong	Strong	Strong	Medium	Strong	Very Strong
USA	Very Strong	Strong	Very Strong	Strong	Medium	Strong	Strong	Strong	Very Strong
France	Strong	Strong	Medium	Strong	Strong	Strong	Strong	Strong	Strong
Japan	Very Strong	Strong	Medium	Very Strong	Strong	Strong	Strong	Strong	Strong
Italy	Strong	Medium	Medium	Strong	Medium	Strong	Medium	Strong	Strong
Sweden	Strong	Medium	Medium	Medium	Medium	Strong	Strong	Strong	Strong
S. Korea	Strong	Strong	Medium	Strong	Very Strong	Strong	Strong	Strong	Strong
UK	Strong	Strong	Medium	Medium	Medium	Strong	Medium	Medium	Strong
China	Medium	Very Strong	Medium	Medium	Weak	Medium	Medium	Very Strong	Very Strong
India	Medium	Weak	Medium	Medium	Weak	Medium	Medium	Medium	Weak

Table: 1

Source: Compiled from various sources and review of the literature on innovation systems. Abbreviations used in the table: PROs – public research organisations; JV– joint venture; INE – Icelandic New Energy; CSIR – Council for Science and Industrial Research; IPHE – International Partnership for Hydrogen and Fuel Cells in the Economy; EU – European Union; CHP – Combined heat and power

Legend:
● Very Strong
● Strong
● Medium
● Weak

systemic relations.

In terms of funding and S&T policies, again, although some countries exhibit a certain degree of 'privatisation' of the development of the industry, most mechanisms and activities are state-driven, thus indicating the prevalent role of the State across countries.

2.3. TECHNOLOGICAL INNOVATION SYSTEMS (TIS) OF SPECIFIC COUNTRIES

Following the previous approach several case studies developed in various articles were mapped. (See table 1 on previous page).

3 DATA AND METHOD

Patents and publications analysis is a useful tool to evaluate basic and intermediary R&D and innovative activities of firms and of countries. Hydrogen and fuel cell patents and publications are publicly accessible databases; and for the purposes of this paper they were extracted for a diverse number of countries. This information evaluates a country's innovative potential and allows for conclusions to be drawn about the innovative capabilities and performances of the hydrogen and fuel cell industry, and gain a general sense of the direction of the technological trajectory.

In order to carry out the bibliometric analysis, the data was divided into two groups: hydrogen storage and fuel cells. This division allowed for a better understanding of the research focus in both areas and the evolution of both fields around the world. The analysis' main focus was to understand the evolution of research and industries in different countries over the past several years. The starting point for the bibliometric analysis was a dataset of academic papers for the period 1988 to present. The dataset was compiled from the Web of Science and included various databases: Science Citation Index includes 5,900 major journals across 150 scientific disciplines, Social Sciences Citation Index (multidisciplinary index to the journal literature of the social sciences, which indexes more than 1,725 journals across 50 social sciences disciplines), and Arts and Humanities Citation Index (all from 1988 to December 2012). The index is multidisciplinary covering the journal literature of the arts and humanities, and fully covers 1,144 of the world's leading arts and humanities journals.

A full description of data and method can be found in the appendix at the end of the chapter.

4 FINDINGS

This section is subdivided into three parts. First, we introduce basic descriptive data on hydrogen storage and fuel cells research around the world. The data is then analysed through Triple Helix lenses, wherein we analyse the importance of government, industry, and academia, and finally we explore country linkages through patent data.

4.1. BASIC DATA

As we can deduce from Figure 2, the combined number of publications and patents (in log scale) has been steadily rising since 1988. In fact, the annualised growth rate for publications for the period is 23 %, thus indicating a pronounced growth of research in this field, while the number of publications on fuel cells exceeds those of hydrogen storage (in log scale). As shown in Figure 2, and observing the number of patent applications per year, we verify that hydrogen storage, production and fuel cells have all registered peaks in 2007–2008, with the number of applications decreasing subsequently (following the international trend on the decrease of spending in R&D due to the international economic crisis). The decrease is rather striking in 2012, with all three fields registering less than half the applications of their peak years.

If we divide the number of publications into countries, we identify four main players in total (Figure 5 in the appendix). The US is the main research centre in both areas combined, with 36% of all publications, with China

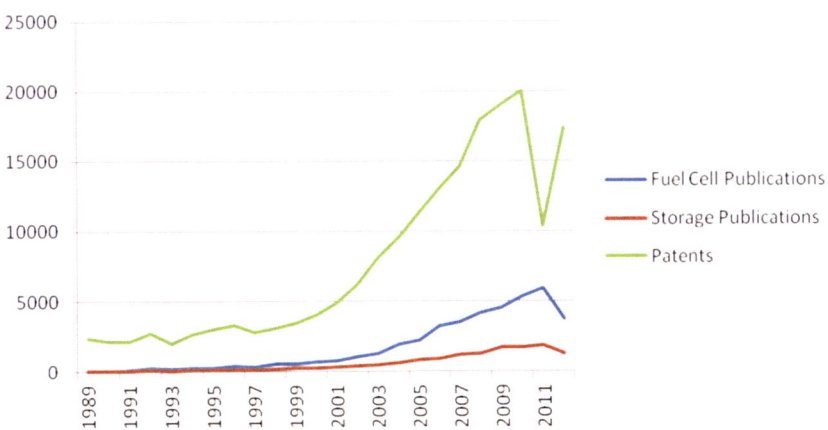

Figure 2: Combined number of publications and patents per year (1988–2011)

trailing closely, accounting for 32% of the total number of publications; Japan follows with 19% of total publications and South Korea, 13% of all publications. South Korea is the last country above 5,000 publications in the database analysed. These findings are somewhat expected, but the fast catching up of China in science and innovation, as predicted by Godinho and Ferreira (2012), puts the country in the vanguard of hydrogen technologies research.

As seen in Figure 6 in the appendix, Germany closely follows South Korea (with 4,760 publications), and then France with 3,825 publications. Finally, in Figure 6 we observe that countries like India and Brazil are catching up. South Africa, with 147 total publications as shown in Figure 7 is still in an embryonic research stage (we identified 39 publications on hydrogen storage and 108 on hydrogen fuel cells).

As for the main fields of research, using Web of Science (WoS) categories, we can see that the application of electrochemistry, energy fuels, chemistry, physical and material sciences are the most important categories for fuels cells. For hydrogen storage the most important science bases are chemistry, physical and materials science, metallurgy and electrochemistry. Hydrogen storage research seems to have a stronger bond with nanotechnology and materials science (including metallurgy).

Chinese organisations seem to be leading part of the effort in this area as they lead all the tables (Figure 10 and Figure 11) with the most prolific organisations in Fuel Cells and Storage research, but also in terms of funding and results obtained from that funding (papers per organisation).

Using a VOSviewer[2] and bibliometric data extracted from WoS we conducted several exploratory analyses, clustering and mapping several characteristics of our dataset. We did this analysis with our datasets combined (storage and fuel cells), but due to problems of computational feasibility we preserved only papers that had more than 20 citations.

The results laid out in the appendix section are presented as co-authoring with organisational affiliations (above a threshold of 30 publications), main research subjects (provided by text analysis of titles and abstracts, above a threshold of 40 counts), and authors (for storage and fuel cells). Regarding organisations, the data reaffirmed the importance of American and Chinese universities, specially the case of the Chinese Academy of Science. As we can see, and as previously mentioned in cross-country investigation, these organisations seem to co-operate (as they aggregate in groups).

---

2. VOSviewer is supported by the Centre for Science and Technology Studies, Leiden University, The Netherlands.

The text analysis on our combined data set identified several clusters of research (see Figure 15; Figure 16 and Figure 17). As for fuel cells, the main areas registering around 400 citations are centred around solid state research, environmental research, electrochemistry, power sources, and membrane science (as we can infer from Figure 18; Figure 19; Figure 20 and Figure 21 on co-citation of sources). In the case of word analysis provided by the abstracts and weighted by importance of papers, we identify clearly five different areas of research: production, catalysts, membranes, electrolytes/solid state, and transport) (see Figure 22 and Figure 23).

For hydrogen storage authors with above 50 citations, the main papers in the analysis are above or close to one thousand citations. As with text analysis, we conducted word analysis from abstracts and the results show five main areas of research within hydrogen storage.

Finally, we emulated the methodology used by Leydesdorff and Persson (2010) to create a map of distribution patterns and networks of relations among cities and countries for the top 500 cited papers in hydrogen technology. As can be seen Figure 26, the main findings in this field cross the northern hemisphere with very few interactions in the South (with the exception of Australia, Brazil and New Zealand).

4.2 TRIPLE HELIX

Academia, government, and industry make up the three main helices that contribute to an innovation process. Institutions of higher learning (mostly universities) represent academia, though other forms of educational institutions are not excluded from contributing to, and participating in, the triple helix innovation processes. Government may be represented by any level. According to the US Office of Science and Technology Policy (January 2010) the industry is represented in a triple helix innovation process through the participation of firms and can constitute private companies, joint venture and other forms of partnerships and company formations.

In the 'triple helix' concept, the institutional domains of government, university, and industry, in addition to executing their customary objectives, assume the roles of the other actors (Leydesdorff and Etzkowitz, 1998). The institutionally-defined triple helix is thus based on distinct industrial, academic, and governmental domains and the 'knowledge flows' among them. Transfer is no longer viewed as a linear process from an origin to an application. Historical designs of interaction can be rebuilt (Leydesdorff and Etzkowitz, 1998).

In order to find out how the different triple helix elements were contributing to the development of the industry an algorithm was created to classify every paper from the database in specific groups (university, government, and industry). Through this method we accounted for approximately 70% of all data points in our database.

From the triple helix division of the total publications in our database we observe that research efforts are primarily led by universities. Nevertheless, in Japan and the US, the industry seems to play a significant role in the development of new knowledge and technologies. In contrast, China, which has made tremendous efforts in catching up, has had an extremely good performance in terms of university-driven research (as one would expect from the analysis of the data on main organisations where most were Chinese governmental or university entities). But in terms of industry research, China is still lagging behind.

The same can be said of South Africa, as the number of university publications (147) is considerably higher than its industry (2) and government (1) publications. The US (10 965) and China (10 432) are on par in research efforts made by universities. These two countries are trailed by Japan (4,488), South Korea (3,280), Taiwan (2,254) and Spain (1,282) in university-driven research.

The government and the industrial sector are equally contributing in Germany in contrast to other countries where the role of industries has been negligible compared to government or university efforts. The German government has contributed to 1,809 publications in total, whereas the industry has contributed 528. Germany's industry efforts trail behind the US (1,352), Japan (1,652), and are closely followed by South Korea (438).

4.3   PATENT DATA

The first part of the analysis was descriptive, outlining existing trends and patterns in countries that are active in hydrogen fuel cell research and development. Through the WIPO database we drew some trend lines in hydrogen-related patenting activity. Patent databases are a prolific source of information (Griliches, 1990). Although information on patent counts themselves is very useful to gauge invention trends and possible innovation trends, one can do more with this data. Each patent has a classification that represents its technological field. Based on an understanding of the technical knowledge in the patent, it is possible to define which technological domain or paradigm (Dosi, 1982) it belongs to and on what prior knowledge it draws.

Looking at the main countries, and although PCT and EPO patents are aggregated, we discern that South Africa has registered significant activity in hydrogen production patents (with more than 11 000 patent applications) and hydrogen storage (with more than 490 patent applications). For the case of hydrogen production, it is striking to note that South Africa surges ahead of Japan.

In order to map the flows of knowledge in this area, we used a data set of patents and created a network based on backward patent citations. Through Ucinet and Netdraw we analysed and created the following descriptions of this network. Social Network Analysis allows the construction of a visual presentation for an innovation System (De Nooy, *et al.*, 2005).

Analysis was carried out at the four-digit level patent data. 'Circles' or 'nodes', as shown in Figure 3 and Figure 4, are countries. Size of the 'nodes' could be drawn such that they reflect the number of patents which have received such a citation from another country, and degree of 'between-ness' and/or 'Eigenvector'. 'Between-ness' and 'Eigenvector' are explained below.

To characterise the different countries, we used different measures of centrality (Borgatti, *et al.*, 2008): a measure of how network structure and position contributes to a node's importance; and how a value is associated with every node. This mathematical representation of a network is useful to quantify information that can be used to identify 'key players' in a network. These measures include 'between-ness', 'degree centrality' and 'eigenvector centrality'.

'Between-ness' is the number of shortest paths an actor is on (Conway, 2009) (brokerage, gatekeeping, control of info) (how well connected; direct influence).

'Degree Centrality' is simply the number of connections (or edges) a vertex has to other vertices. 'Eigenvector Centrality' is a measure that reflects the fact that not all connections are equal and, in fact, connections to people that are more influential are more important (Newman, 2012) (being connected to the well-connected – a popularity and power measure).

Also, to analyse the position of South Africa we separated it from the other players using an Ego network. 'Ego' is an individual 'focal' node. A network has as many 'egos' as it has nodes. 'Egos' can be persons, groups, organisations, or in this case, countries (Hanneman and Riddle, 2005). A 'neighbourhood' is the collection of 'egos' and all nodes to which an 'ego' has a connection at some path distance. In social network analysis, the 'neighbourhood' is usually one-step; that is, it includes only 'ego' and actors that are directly adjacent. The 'neighbourhood' also comprises all of the ties among all of the actors to whom 'ego' has a direct connection. The borders of 'ego' networks are defined in terms of 'neighbourhoods' (Hanneman and Riddle, 2005).

Our first analysis is of the whole global network. We defined node size by degree of Eigenvector Centrality (thus finding that the US is the main driver of innovation). This analysis is more striking in the following Figure 3 and Figure 4, where the network is laid-out through its principal components. The logic behind principal components (factor) analysis is a method for combining correlated actors into a smaller number of underlying dimensions. The algorithm searches for the most highly correlated set of actors in the network, and this becomes the first component. It then searches for a second set of actors that is uncorrelated with the first, which then becomes the second component. Because they are uncorrelated with one another (i.e. because they are orthogonal with one another) they 'can be drawn at right-angles to one another as the axes of a two-dimensional scatter diagram' (Scott, 2000, p. 154).

Following the US, we have Japan and Germany, and then France, Canada, Great Britain, Switzerland, Korea, the Netherlands, Italy, and China. Colors in Figure 3 and Figure 4 represent commonalities between nodes (k-cores).

KNOWLEDGE NETWORKS

South Africa's patent citation network is characterised by a much smaller number of countries. However, the key players in its network are also players within the larger global network, and most of them are countries that have been analysed in this study (viz. US, Germany, Canada, the Netherlands, Japan, China, Norway, Sweden, South Korea, and Denmark). A few outliers in this network are Brazil, Saudi Arabia, Venezuela, Qatar, Hungary, Philippines, Indonesia, the Czech Republic and Bulgaria.

Figure 4, which shows South Africa's position in the patent citation 'ego' network (as shown in the appendix), allows us to surmise that although South Africa's network is small, the country is connected to all the major global actors. The country exhibits both a high degree of centrality (across different measures) and a high degree of closeness.

TECHNOLOGICAL DEVELOPMENT

As observed in this study, South Africa seems to be a minor contributor to R&D in hydrogen energy and fuel cell technologies.

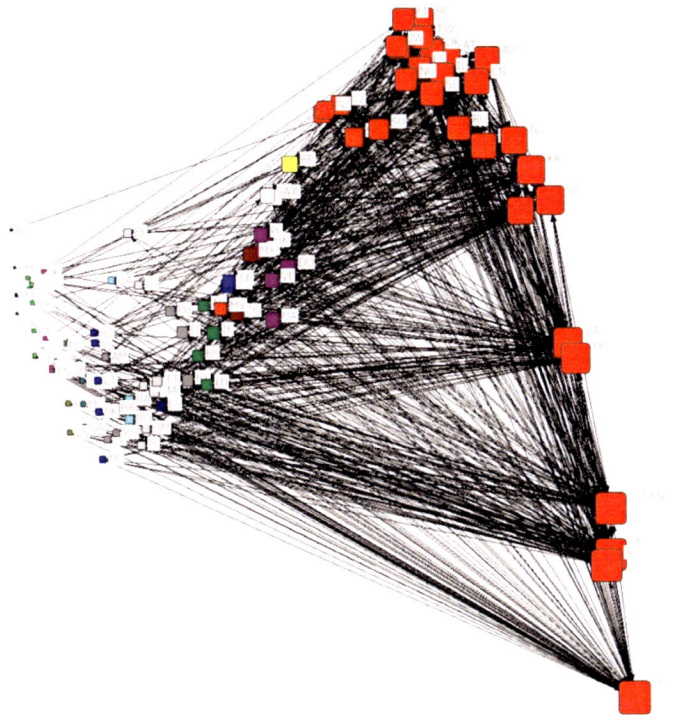

Figure 3: Backward citation patent network – PCA Layout (Source: Authors' Own)

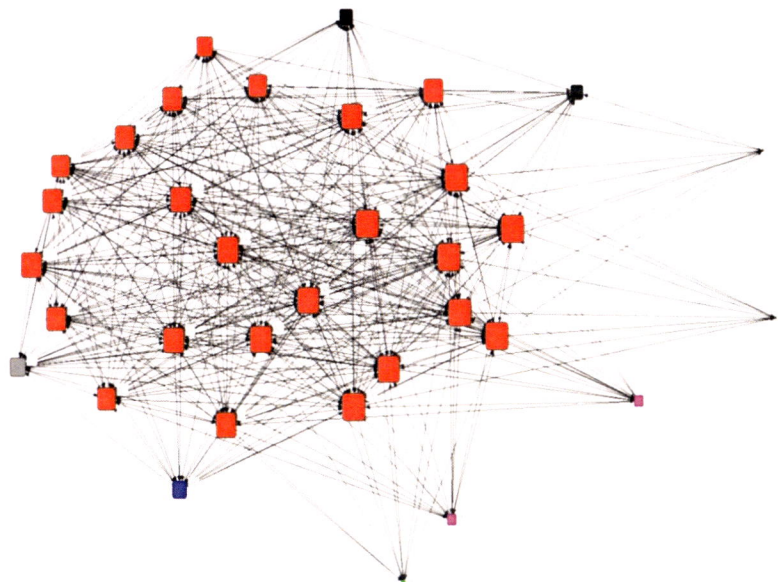

Figure 4: Backward citation patent network 'ego' South Africa network (Source: Authors' Own)

Patent analysis in this study reflects active technological development in hydrogen production in South Africa, exceeding even Japan. The main patent applicants are foreign companies. However, it can be conjectured that 39% of the patents filed have their use in the pharmaceutical industry as they have been filed by companies like Novartis (Germany), Glaxo (UK) and Astrazeneca (Switzerland).

Technological development in hydrogen fuel cells and storage has been low in the last couple of years in South Africa as is shown in Table 2 below. The main applicants of the patents that were filed in South Africa in this category are all foreign companies: Shell International (Netherlands), BASF (Germany), Motorola (USA), Johnson Matthey Plc (now UK), More Energy (Israel), Unilever (UK) and Procter and Gamble (UK), among others. South Africa as shown in the appendix shows us that most applications in South Africa stem from foreign companies and that the trend for patenting in this field has been clearly negative in the past few years.

INDUSTRY COMPETITION

In countries where there is an automotive industry, auto industry incumbents have been playing a bigger role in research and development of hydrogen fuel cells. In the absence of these, big oil and gas firms or energy

Hydrogen Production		Hydrogen Fuel Cells		Hydrogen Storage	
Main Applicant	No	Main Applicant	No	Main Applicant	No
SHELL INTERNATIONALE BV	10	BASF	3	DURACELL INC	4
BASF	10	SHELL	2	STANDARD OIL	3
TRESPAPHAN GMBH	9	MOTOROLA INC	2	JOHNSON MATTHEY PLC	1
UNILEVER PLC	6	JOHNSON MATTHEY PLC	2	FRAMATOME ANP	1
HOECHST	6	HALDOR TOPSOE A/S	2	FORETOP CORP	1
HENKEL	6	E. I. DU PONT	1	ENECO INC	1
BAYER	6	DE NORA SPA	1	CANON	1
HALDOR TOPSOE A/S	4	CHEVRON INC	1	CANADIAN HYDROGEN ENERGY	1
EXXONMOBIL	4	BLACKLIGHT POWER INC	1	ASEMBLON INC	1
BAYER AG	4	ASEMBLON INC	1	ALUMINAL GMBH	1
TOTAL	65	TOTAL	16	TOTAL	15

Table 2: Number of patents filed in South Africa between 1988–2012
Source: WIPO, 2012

companies are engaged in the development of several country industries. In all the countries analysed, there is limited market competition, with either government or big firms acting as main influencers, and venture capital playing a very limited role.

In the case of South Africa, there is no interest shown in R&D of hydrogen fuel cells and storage by its industry and energy incumbents as yet. Sasol is rumoured to be exploring the option of hydrogen production and, as an energy and chemicals company, it has been developing liquid fuels from coal for many years now. One of the by-products of such a process is hydrogen.

Anglo American Platinum, a South African-based platinum-mining company, is currently integrating Ballard fuel cells into mining equipment and a locomotive. It has invested into Ballard's R&D efforts in fuel cell development, thus linking itself to international networks. It is also experimenting with electricity generation from fuel cells as a primary source of energy for remote areas not linked to the grid.

5   Conclusion

A fundamental mechanism in an industry creation process is the creation of knowledge networks. A technological programme such as Hydrogen South Africa (HySA) is a research network that contributes to the creation of knowledge within a network. Governments typically create these programmes, playing a critical role in allocating resources. Such government-driven knowledge networks assist in avoiding project duplication through research portfolio diversification and allocation of resources and tasks among the different actors of the network. Germany, for example, has several research networks, both publicly and privately governed, which are advancing the application of hydrogen fuel cells through well-co-ordinated R&D activities among actors along an emerging value-chain. Training and integration of different suppliers are critical components that characterise the German R&D network where there is a deliberate transfer of specifications, industry standards and codes, technical knowledge, and licenses to suppliers, among others.

Similar to Norway, there exists a strong concern and willingness in South Africa to beneficiate its natural resources and build a national competence based on hydrogen fuel cells that use platinum. But such a view can often subjugate the significance of variety creation, narrowing down the research focus to platinum catalyst fuel cells research, much too early on.

But on a positive note, platinum-based catalysts in fuel cells – the PEMFC – is a dominant technology; and its current dominance is indicated by the number of global research activities, projects, and shipments of PEMFC. Such market optimism reduces technological uncertainty that often burdens decision-making. Choosing the 'right technology' is not easy and history has been stricken by many examples of failed projects, including many hydrogen fuel cell projects. Their failure was the result of a variety of factors, among which would typically be lack of funding, lowered expectations about the relevance of the technology in the future, absence of incentives and subsidies, and unco-ordinated research efforts at the national level.

The failed hydrogen fuel cell projects, in the context of Norway, have been explained in terms of 'market failure'. There was duplication of efforts: two projects (out of three) that were developing the same type of technology: SOFC based on natural gas. Instead, policy should have encouraged two different technological options, as the creation of technological variety is often critical in industry emergence, though it is often spurred by competition. So a difficult policy question is: to what extent should one allow for competition? Competition nurtures variety, yet at the same time provides a difficult ground for fledgling technologies, thus killing variety. To what extent should policy be deliberate in 'picking-the-winner'?

The second mechanism that is instrumental in speeding up the process of industry formation is related to the promotion of entrepreneurship. Entrepreneurship is promoted through effective support mechanisms that encourage private innovation and provide incentives for entrepreneurship and for international partners. For example, Canada has commercially driven policies that drive development along value chains and local clusters that provide knowledge, distribution, marketing, and supply chain channels. And leading from here is a third mechanism, which is related to the ability to sustain a commercialisation process, meaning the birth of new firms.

The sustenance of a commercialisation process is infused with a lot of issues and concerns, and only a handful of countries have arrived at the third mechanism, viz. Canada, US, Germany, South Korea and Japan. Such issues and concerns constitute fostering R&D and entrepreneurial activities and mobilising public support for post-entry performance of firms, and designing effective policies for stimulating technology uptake and development. Japan, for example, is stimulating technology diffusion by designing effective demand-side policies and regulations, and supply-side incentives to support an industry uptake.

South Africa is at a point where the creation of research networks, and the development of skills, knowledge and capabilities in hydrogen fuel cells, seems to be high-up in the national hydrogen programme. The promotion of entrepreneurship and the engagement of the industry are negligible at this stage, although attempts are being made to engage various industry incumbents, as well as some small firms, with research and development and demonstration (RD&D) work along the fuel cell value chain.

For the country, technological innovation in a new and uncertain field such as a hydrogen fuel cells industry requires garnering critical knowledge from various international sources and capabilities. As such, and as shown in the bibliometric data used in this study, South Africa compares poorly against research and development efforts in hydrogen fuel cells when compared to Japan, the US, Germany and Canada. It also compares poorly against its BASIC counterparts (Brazil, South Africa, India and China) and, among which, China is leading global efforts in hydrogen fuel cell RD&D. Therefore, international partnerships to advance R&D efforts in hydrogen fuel cells in South Africa will remain a critical factor for the growth of this industry.

In countries where the automotive players are large, industry incumbents have played a bigger role in research (or their suppliers, as in the case of Italy). Big oil and gas, and energy firms are engaged in the development of several countries' fuel cell industries. There seems to be very limited market competition, with either government or big firms acting as main influencers, and venture capital playing a very limited role. Notably, there appears to be a lot of international integration, especially with respect to research, more or less linked to the automotive industry. Several countries are interconnected, namely the EU countries, and their actions seem to echo the systemic relations that are prevalent in the industry and across the countries.

In terms of science and technology policies, although some countries exhibit a certain degree of 'privatisation' of the development of the industry, most mechanisms and activities are still state-driven, indicating the prevalent role of the government across countries.

As discussed in the literature review in this study, industry creation is a rather complex phenomenon where government policy may play a very important role. Nevertheless, as was stated, entrepreneurship, research institutions, research networks, demand (on downstream industries), and even financial support for the whole system are important factors for the development of this industry. The countries identified as main players through patent and bibliometric data, and listed as success case studies in our qualitative analysis, all share specific traits: strong research institutions, strong growth of knowledge production (private and public), strong invention output (patent count) and internationally-leveraged firms.

6 MODEL / ANALYTICAL LIMITATIONS

As OECD (2010) states, the indicators that may originate from this analysis (to a certain extent) help the mapping of knowledge, its concentration and dynamics. Nevertheless, patents are not a perfect measure of innovation (Godinho and Ferreira, 2012) and are often used as barriers or bargaining chips (hence also not a perfect measure of inventiveness if one omits their citation value). Bibliometric and patent data, although useful, should be used always in combination with qualitative approaches (case studies or expert-based methods). Even the limitations suggested in this collection of data may imply that relevant papers classified in different fields or patents from other IPC classes, that at first may seem irrelevant, may have important methods/knowledge for future inventions in this field. Policy recommendations based solely on qualitative or quantitative information should be cautiously received.

We do not provide a sophisticated causal model where the level of performance of an industry is generated by some of the factors described. Nevertheless, the statistical approach and network analysis are basic building blocks in the comprehension of the phenomenon. Future work may help identify a more specific set of causal relations.

Nevertheless, in this study we strive to complement our qualitative analysis of the birth process of this industry in several countries with a complementary quantitative analysis that helps to understand the accumulation and flow of knowledge.

7 POLICY IMPLICATIONS FOR SOUTH AFRICA

Theory and evidence in this chapter has shown that research networks are precursors to the emergence of an industry. The analysis presented represents an evaluation of knowledge and policy factors in various countries that are active in research and development, and commercialisation of hydrogen fuel cells. The evaluation can further inform the national strategic goal-setting process in South Africa, as the country's intent is to move beyond mining, and evidently towards knowledge-based activities. Such a policy shift towards value-added services of mining activities or beneficiation demands an understanding of knowledge factors that will be critical in establishing a hydrogen fuel cell industry in the country.

As South Africa aims to supply 25 per cent of global demand for PEM fuel cells by 2020, policy shifts towards a greater understanding of knowledge factors that typically drive new technologies, and subsequently new industries are required. Such an ambition necessitates the setting up of an industry. Hydrogen fuel cell development is a highly knowledge-intensive activity, and the intent to globally supply a quarter of the fuel cell catalyst would require knowledge and skills that are not only highly-competitive with leading-edge products, but would also require a precocious understanding of global fuel cell markets.

Key insights of this study point to critical knowledge and policy factors that are essential for South Africa to adopt and/or enhance if it wants to establish a local manufacturing industry based on hydrogen fuel cells.

INTERNATIONAL COLLABORATION

It has been shown in this study that networking and collaboration are the order of new technologies that are highly knowledge-intensive in nature. Research is increasingly generating networked knowledge. And this is particularly true of technologies that are 'global' in character, such as hydrogen fuel cells. And therefore strategies that are based solely on narrow local self-sufficiency can prove to be counter-productive, particularly when the policy aim is to become a major global player. Countries that are exploring hydrogen fuel cells as an energy/fuel option are those that increasingly rely on international integration, especially with respect to research, currently more or less linked to the automotive industry. The study shows that several countries are interconnected, e.g. the EU-centric countries, and the actions of each country seem to echo the systemic relations that are prevalent in this industry.

South Africa needs to foster its linkages in the research area in the main fields identified in this study, and with the identified principal knowledge producers that were mapped using patent/bibliometric data. South Africa may need to target and access the knowledge pool of countries such as the US, Japan, Germany, Canada, France, UK, China and South Korea, through partnering and technological collaborations. Complexity and the fragmentation of knowledge and intellectual property (IP) rights have indicated that no single country can effectively develop new technologies such as hydrogen fuel cells and their applications all on their own. 'Knowledge pools' may help, although sometimes such protected pools of knowledge can act as barriers to technological development and industry emergence.

ROLE OF INSTITUTIONS

It will be critical to invest in efforts that encourage both the development and access to knowledge factors by all relevant actors of a technological innovation system. The HySA network is a research network that needs to be continuously accessible to new knowledge and skills, as well as new firms, and highly qualified personnel that will work together in creating and commercialising new products based on hydrogen fuel cells, locally. The Department of Science and Technology (DST) is an anchor institution that is leveraging and assisting universities and public research labs to strengthen their research capacity around the hydrogen fuel cells value-chain activities at the pre-commercialisation stage. It is also aligning international research activities with local technological development activities.

However, DST has limited influence, as its capacity to take the technology to the South African market relies on a host of market-side factors and policies that are not within its functional jurisdiction. Although the DST has recently entered this terrain, with initiatives to establish PGM-based economic zones, there is a large deficit in the area of policies and regulations promoting the manufacture and uptake of hydrogen fuel cell products in the country.

The government should encourage the entry of private companies, as private players possess the greatest drive to commercialise technologies. The recommendation would be to create a platform that will generate 'local buzz' and awareness of the technology: one which taps into global markets and international pipelines of hydrogen fuel cell research and products and align with local technology development and commercialisation plans.

ROLE OF INCUMBENTS

As previously stated, in countries where there is an automotive industry (US, Germany, France and Japan), the incumbents have been playing a very important part in the research and development of hydrogen fuel cells. When auto industry is absent, oil and gas firms or energy companies were involved in this process. In those countries previously analysed, there was limited market competition, with government and/or big firms acting as main drivers and with venture capital playing a very limited role.

The government should encourage the involvement of large incumbents in either advancing the technology through investments in R&D, or encouraging them to be involved in the sales and marketing of imported hydrogen fuel cell technologies.

FURTHER RESEARCH

This chapter has attempted to outline where South Africa stands in the global research arena. It briefly critiqued South Africa's current approach to fuel cells development, viz. the HySA programme and DST's ambition of supplying 25 per cent of global hydrogen fuel cells. One of the fundamental questions in this regard is the extent to which South Africa should emphasise research and development (if the technology is being developed elsewhere), or whether it should consider, as part of its posture in manufacturing and marketing closer-to-market applications? The question is one about balance rather than an either/or approach. Further, a glaring weakness in the policy environment as demonstrated reveals that there is insufficient focus on creating a market for hydrogen fuel cells.

Future research should consider the policies identified in more detail, in particular the degree to which cluster-policies are based on successful beneficiation projects in South Africa

KNOWLEDGE SPILLOVERS AT HySA

It is necessary to acknowledge a gap in the analysis contained in this chapter pertaining to the details of the work being undertaken by HySA. Any knowledge-based study would have ideally analysed the knowledge and technological spillovers between actors and assessed the linkages among local networks of HySA. We would have looked at HySA's capacity to create, use, and commercialise hydrogen fuel cell technologies. However, information on recent HySA activities is confidential, and not in the public domain.

Bibliography

Abernathy, W. J. and Utterback, J. M. (1978). Patterns of Industrial Innovation, *Technology Review*, Vol. 80 (7), 40–47.

Agarwal, R. and Audretsch, D. B. (2001). 'Does start-up size matter? The impact of technology and product life cycle on firm survival', *Journal of Industrial Economics*, 49(1), 21–44.

Anderson, P. and Tushman, M. L. (1990). 'Technological discontinuities and dominant designs: a cyclical model of technological change', *Administrative Science Quarterly*, Vol. 35 (4), 604–633.

Audretsch, D. B., Carree, M. A., Van Stel, A. J. and Thurik, A. R. (2002). 'Impeded industrial restructuring: the growth penalty', *Kyklos*, 55(1), 81–98.

Bakker, S. (2010). 'The car industry and the blow-out of the hydrogen hype', *Energy Policy*, 38(11), 6540–6544.

Bleischwitz, R., Bader, N., Dannemand, P. and Nygaard, A. (2008). 'EU Policies and Cluster Development of Hydrogen Communities', Bruges European Economic Research Paper No. 14, 1–67.

Bleischwitz, R., Bader, N., Dannemand, P. and Nygaard, A. (2008). 'EU Policies and Cluster Development of Hydrogen Communities', MPRA Paper 14501, University Library of Munich, Germany.

Braczyk, H. J., Cook, P. and Heidenreich, M. (eds) (1998). *Regional Innovation Systems: The Role of Governance in a Globalized World*, London: UCL Press, 414–440.

Callon, M. (1992). 'Techno-economic networks and irreversibility', *A Sociology of Monsters: Essays on Power, Technology and Domination*, Volume 38, 132–161.

Callon, M. (1995). 'Four models for the dynamics of science', Jasanoff, S., *et al.*, 29–63.

Carlsson, B. and Stankiewicz, R. (1991). 'On the nature, function and composition of technological systems', *Journal of Evolutionary Economics*, Volume 1(2), 93–118.

Carlsson, B., Jacobsson, S., Holmen, M. and Rickne, A. (2002). 'Innovation systems: analytical and methodological issues, *Research Policy*, Volume 31(2), 233–245.

Carlsson, B. (2005). 'Entrepreneurship and Public Policy in Emerging Clusters', in Braunerhjelm, P. and Feldman, M. (eds.), *Cluster Genesis: The Emergence of Technology Clusters and the Implication for Government Policy*, New York: Oxford University Press.

Cassi, L., Corrocher, N., Malerba, F. and Vonortas, N. (2008). 'Research networks as infrastructure for knowledge diffusion in European regions', *Economics of Innovation and New Technologies*, Volume 17 (7–8), 663–676.

Chiaroni D., Chiesa V. (2006). 'Forms of creation of industrial clusters in biotechnology', *Technovation*, Volume 26(9), 1064–1076.

Choi, H., Park, S., Lee, J. (2011). 'Government-driven knowledge networks as precursors to emerging sectors: a case of the hydrogen energy sector in Korea', *Industrial and Corporate Change*, Volume 20 (3), 751–751.

Christensen, C. M. (1997) *The Innovators Dilemma: When New Technologies Cause Great Firms to Fail*, Boston, Massachusetts: Harvard Business School Press.

Coenen, L., López, D. and Fernando, J. (2010). 'Comparing systems approaches to innovation and technological change for sustainable and competitive economies: An explorative study into conceptual commonalities, differences and complementarities', *Journal of Cleaner Production*, Volume 18, 1149–1160.

Cohen, W. and Levinthal, D. (1990). 'Absorptive Capacity: A New Perspective on Learning and Innovation', *Administrative Science Quarterly*, Volume 35(1), 128–152.

Dasgupta, P. and David, P. A. (1994). 'Toward a new economics of science', *Research Policy*, 23(5), 487–521.

De Nooy, W., Andrej, Mrvar and Vladimir, B. (2011). 'Exploratory Social Network Analysis with Pajek, Cambridge: Cambridge University Press.

Dosi, G., (1982). 'Technological paradigms and technological trajectories: A suggested interpretation of the determinants and directions of technical change', *Research Policy*, Volume 11(3), 147–162.

Dosi, G. and Nelson, R. (1994). 'An Introduction to Evolutionary Theories in Economics', *Journal of Evolutionary Economics*, 153–172.

Dosi, G., Nelson, R. R. and Winter, S. G. (2000). 'Introduction: the nature and dynamics of organizational capabilities', *The Nature and Dynamics of Organizational Capabilities*, 1–22.

Doz, Y., Santos, J. and Williamson, P. (2001). *From Global to Metanational: How Companies Win in the Knowledge Economy*, Boston: Massachusetts: Harvard Business School Press.

Etzkowitz, H. and Leydesdorff, L. (1998). The triple helix of university- industry-government relations: A laboratory for knowledge based economic development', *East Review*, Vol. 14 (1), 11–19.

Geels, F. (2002). 'From sectoral systems of innovation to socio-technical systems: Insights about dynamics and change from sociology and institutional theory', *Research Policy*, Volume 33 (6–7), 897–920.

Gertler, M. S. and Wolfe, D. A. (2006). 'Spaces of knowledge flows: Clusters in a global context'. In *Clusters and Regional Development: Critical Reflections and Explorations*, London: Routledge, 218–235.

Giarratana, M. S. (2004). 'The birth of a new industry: entry by start-ups and the drivers of firm growth: the case of encryption software', *Research Policy*, Volume 33(5), 787–806.

Godinho, M. M. and Ferreira, V. (2012). 'Analyzing the evidence of an IPR take-off in China and India', *Research Policy*, 41(3), 499–511.

Gomory, R. (1992). 'The Technology-Product Relationship: Early and Late Stages'. In *Technology and the Wealth of Nations*, edited by Nathan Rosenberg, Ralph Landau and David C. Mowery. Stanford, CT: Stanford University Press, 1992, 383–394.

Grandstrand, O., Patel, P. and Pavitt, K. (1997). 'Multi-technology Corporations: Why they have distributed rather than distinctive core competence', *California Management Review*, Vol. 39(4), 8–25.

Griliches, Z. (1990). 'Patent statistics as economics', *Journal of Economic Literature*, Vol. 28, 1661–1707.

Hanneman, R. A. and Riddle, M. (2005). *Introduction to Social Network Methods* , University of California, Riverside, USA.

Haupt, R., Kloyer, M. and Lange, M. (2007). 'Patent indicators for the technology life cycle development', *Research Policy*,Volume 36 (3), 387–398.

Hekkert, M. P, Suurs, R. A. A., Negro, S. O., Kuhlmann, S. and Smits, R. E. H. M. (2007). 'Functions of innovation systems: A new approach for analysing technological change',

Technological Forecasting & Social Change, Volume 74, 413–432.

Hung, C. and Chu, Y. (2006). 'Stimulating new industries from emerging technologies: challenges for the public sector', *Technovation*, Volume 26, Issue 1, 104–110.

Jacobsson, S. and Johnson, A. (2000). 'The diffusion of renewable energy technology: an analytical framework and key issues for research', *Energy Policy* 28(9), 625–640.

Jörg, M. and Jochen, M. (2011). 'Creating and shaping innovation systems: Formal networks in the innovation system for stationary fuel cells in Germany', Energy Policy, Volume 39 (4), 1909–1922.

Klepper, S. (1997). *Industry Life Cycles, Industrial and Corporate Change*, Oxford: Oxford University Press, Vol. 6(1), 145–81.

Klepper, S. and Graddy, E. (1990). 'The Evolution of New Industries and the Determinants of Market Structure', *The Rand Journal of Economics*, Vol. 21 (1), 27–44 (1997 in text).

Klitkou, A., Nygaard, S. and Meyer, M. (2007). 'Tracking techno-science networks: A case study of fuel cells and related hydrogen technology R&D in Norway', *Scientometrics*, Volume 70(2), 491–518.

Konrad, K., Markard, J., Ruef, A. and Truffer, B. (2012). 'Strategic responses to fuel cell hype and disappointment', *Technological Forecasting and Social Change*, Volume 79(6), 1084–1098.

Lane, J. (2009). 'Science innovation: assessing the impact of science funding', *Science* 324 (5932), 1273—1275.

Lewis, J. and Wiser, R. H. (2007). 'Fostering a renewable energy technology industry: An international comparison of wind industry policy support mechanisms'. *Energy Policy*, Volume 35, 1844–1857.

Leydesdorff, L. and Olle P. (2010). 'Mapping the geography of science: Distribution patterns and networks of relations among cities and institutes', *Journal of the American Society for Information Science and Technology*, Volume 61(8), 1622–1634.

Lund, P. D. (2009). 'Effects of energy policies on industry expansion in renewable energy', *Renewable Energy* 34 (1), 53–64.

Lundvall, B. A. (1992). Introduction, in Lundvall (ed) *National Systems of Innovation: Towards a Theory of Innovation and Interactive Learning*, London: Pinter Publishers, 1–22.

Malerba, F. (2002). 'Sectoral systems of innovation and production', *Research Policy*, Volume 31(2), 247–264.

Malerba, F. (Ed.) (2004). *Sectoral Systems of Innovation: Concepts, Issues and Analyses of Six Major Sectors in Europe*, Cambridge: Cambridge University Press.

Malerba, F. (2006). 'Innovation and the evolution of industries', *Journal of Evolutionary Economics*, Volume 16 (1–2), 3–23.

Mezias, S. J. and Kuperman, J. C. (2001). 'The community dynamics of entrepreneurship: the birth of the American film industry, 1895–1929', *Journal of Business Venturing*, 16(3), 209–233.

Murtha, T. P., Lenway, S. A. and Hart, J. A. (2001). *Managing New Industry Creation: Global Knowledge Formation and Entrepreneurship in High Technology*, Stanford University Press.

Mytelka, L. (2004). 'Catching up in new wave technologies', *Oxford Development Studies*, 32(3), 389–405.

Mytelka, L. K. (2008). *Hydrogen Fuel Cells and Alternatives in the Transport Sector: A Framework for Analysis*, Tokyo, Japan: United Nations University Press, and Ottawa, Canada: International Development Research Centre (IDRC), 5–38.

Mytelka, L. (2010). 'On the Effects of Institutional Arrangements for Innovation in Clusters –A comparative case study of sugar clusters in São Paulo, the North East of Brazil and Cuba', *Lund Studies in Research Policy*, No. 2.

Narula, R. (2002). 'Innovation systems and "inertia" in R&D location: Norwegian firms and the role of systemic lock-in', *Research Policy*, 31(5), 795–816.

Narula, R. and Zanfei, A. (2010). 'Globalisation of Innovation', *Handbook of Innovation*, in J. Fagerberg, D. Mowery, R. R. Nelson (Eds.), UK: Oxford University Press.

Nelson, R. (1994). 'The co-evolution of technology, industrial structure, and supporting institutions', *Industrial and Corporate Change*, Volume 3(1), 47–63.

Newman, M. (2001)). 'The structure of scientific collaboration networks', *Proceedings of the National Academy of Sciences*, Volume 98 (2), 404–409.

Nygaard, A. M. and Andersen, P. D. (2010). 'Innovative regions and industrial clusters in hydrogen and fuel cell technology', *Energy Policy*, Volume 38(10), 5372–5381.

OECD, (2001). *Science, Technology and Industry Scoreboard: Towards a Knowledge-based Economy*, Paris: OECD Publishing.

Pavitt, K. (1984). 'Sectoral patterns of technical change: towards a taxonomy and a theory', *Research Policy*, Volume 13(6), 343–373.

Pietrobelli, C. and Roberta R. (2011). 'Global Value Chains Meet Innovation Systems: Are There Learning Opportunities for Developing Countries?', *World Development*, Volume 39(7), 1261–1269.

Polt, W. (2001). 'The role of governments in networking. Innovative Networks: Co-operation in National Innovation Systems', OECD Publishing.

Powell, W. W. and Grodal, S. (2010), 'Networks of Innovators', in Fagerberg, J., Mowery, D. and Nelson, R. R. (Eds.) *The Oxford Handbook of Innovation*, UK: Oxford University Press.

Rice, J. and Galvin, O. (2006). 'Alliance patterns during industry life cycle: The Case of Ericsson and Nokia', *Technovation*, Vol. 26, 384–395.

Rothwell, R. (1984). 'The role of small firms in the emergence of new technologies', *Omega*, Volume 12, Issue 1, 1984, 19–29.

Russo, M. V. (2003). 'The emergence of sustainable industries: building on natural capital', *Strategic Management Journal*, Volume 24(4), 317–331.

Sapsed, J., Grantham, A. and DeFillippi, R. (2007). 'A bridge over troubled waters: Bridging organisations and entrepreneurial opportunities in emerging sectors'. *Research Policy*, Volume 36(9), 1314–1334.

Schumpeter, J. A. (1934). *The Theory of Economic Development*, Harvard University Press.

Schumpeter, J. A. (1943) *Capitalism, Socialism, and Democracy*, Peter Smith Publishers.

Scott J. (2000). *Social Network Analysis: a Handbook*, 2nd ed., London: Sage.

Sovacool, B. K. and Brossmann, B. (2010). 'Symbolic convergence and the hydrogen economy', *Energy Policy,* Volume 38(4), 1999–2012.

Tijssen, R. and Korevaar, J. C. (1997). 'Unravelling the cognitive and interorganisational structure of public/private R&D networks: A case study of catalysis research in the Netherlands', *Research Policy*, Volume 25(8), 1277–1293.

Trippl, M. and Tödtling, F. (2007). 'Developing Biotechnology Clusters in Non high Technology Regions –The Case of Austria', *Industry and Innovation*, 14(1), 47–67.

Unruh, G. C. (2000). 'Understanding carbon lock-in', *Energy Policy*, 28(12), 817–830.

Utterback, J. (1994). *Mastering the Dynamics of Innovation*, Boston: Harvard Business School Press.

Teece, D., Pisano, G. and Shuen, A. (1997). 'Dynamic Capabilities and Strategic Management', *Strategic Management Journal*, Volume 18(7), 509–533.

Vasudeva, G. (2009). 'How national institutions influence technology policies and firms' knowledge-building strategies: A study of fuel cell innovation across industrialized countries', *Research Policy*, Volume 38 (8), 1248–1259.

Wolfe, D. and Gertler, M. (2004). 'Clusters from the Inside and Out: Local Dynamics and Global Linkages'. *Urban Studies*, Volume 41, 1071–1093.

CHAPTER 4

INNOVATION AND HYDROGEN FUEL CELL MANUFACTURING

Vítor Ferreira and Radhika Perrot

1 NEW SHIFT IN GLOBAL MANUFACTURING

Globally, manufacturing is entering a transformative period. The global economic slowdown has not only stirred a period of major fiscal adjustment in many countries in Europe and the US, it has also led to a slowdown in manufacturing in many of these countries, a process that had been unfolding even before the economic crisis. These countries that were once leaders in high-tech manufacturing now have to deal with market and manufacturing base shifts to emerging economies in Asia, and other low-cost offshore competitors, resulting from a need to drive down manufacturing costs.

This emerging and shifting paradigm in manufacturing is, however, complex, characterised by a 'sustained but modest growth, a renewed focus on product, materials and process innovation, and unprecedented collaboration across the value chain' (KPMG, 2012). This shift in manufacturing can only be brought about by innovation[1] – most notably process and materials innovations and business model innovations. In the 2012 KPMG survey of global manufacturers, 72 per cent of respondents believe that 'transformational innovation' (KPMG, 2012) in their manufacturing sites has begun and manufacturing businesses are 'working to extend and enhance their product lines while cutting costs via process innovation'.

---

1. Innovation is the implementation of a new or significantly improved product (good or service), process, new marketing method or new organisational method in business practices, workplace organisation or external relations (OCED Oslo Manual, 2005).

Global macroeconomic uncertainties in recent times have shown us how manufacturing had shifted every advanced country's economic base. In many advanced economies of Europe, manufacturing contributes more to innovation than to GDP growth and employment (McKinsey, 2012). In these advanced economies there is a heavy reliance on service-like manufacturing activities[2]. However, in many developing countries, rapid demand-side growth is anticipated and therefore manufacturing is expected to contribute more to GDP growth and employment. In developing countries, global manufacturers will have substantial new opportunities from the additional demand — but under a much more uncertain environment (McKinsey, 2012).

1.1 EMISSIONS REGULATIONS IN GLOBAL AUTOMOBILE INDUSTRIES

Environmental regulations and engine-fuel standards for emissions control are driving innovation in many industries, including the global automotive industry. As a result of this, suppliers of global automotive original equipment manufacturers (OEMs) are being confronted with challenges to innovate and meet numerous regulatory requirements imposed by their governments or governments of the markets which they serve (KPMG, 2010).

The global automotive industry, too, is undergoing fundamental transformation. The transformation is also driven by governments that are throughout the world requiring OEMs to comply with environmental regulations mainly for emissions control and engine-fuel economy.

Environmental regulations are driving shifts in demand and encouraging innovations in new materials, processes, information technology and operations, that are linked with improved fuel efficiency and innovations in new technologies such as fuel cell vehicles/fuel cell electric vehicles (FCV/FCEVs).

1.2 HIGH R&D INTENSIVE MANUFACTURING

Hydrogen fuel cell manufacturing is *very high* in research and development (R&D) intensity. Figure 1 shows the various industries that are intensive in R&D and 34 per cent of this group's manufacturing activities are expected to constitute value-added services or post-manufacturing activities. In the case

---

2. The traditional view is that manufacturing services is distinct from manufacturing. Manufacturing is increasingly recognised to include a range of activities in addition to production. Service-like activities such as R&D, procurement, distribution, marketing and sales, back-office support and management – have become a large part of what manufacturing companies do. According to McKinsey (2012), service type activities make up 30–55 per cent of manufacturing employment. The service and manufacturing type ratio is 55:45 for hydrogen fuel cells technologies.

of hydrogen fuel cells, manufacturing service activities will comprise product distribution, repair and maintenance, and end-use and life cycle services (elaborated in the innovation value chain of Fig. 2).

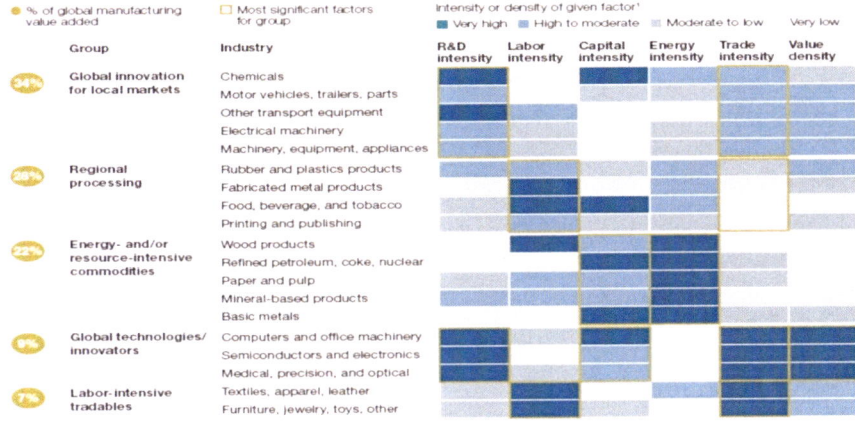

Figure 1: Global Manufacturing Activities. Source: McKinsey, 2012

2 COLLABORATIVE INNOVATION IN MANUFACTURING

Two critical factors emerged from the above 2012 KPMG study and have important implications for global hydrogen fuel cell manufacturing: *R&D and supply chain innovation collaboration* and increasing importance of *geographical proximity to customers and R&D and technological bases.* These two factors are driving manufacturers to focus on competitive cost structures, and to realign their business models with the changing market dynamics. Such characterising changes in global manufacturing point towards an increasing importance given to collaborative innovation.

The importance of interactions between partnering organisations has been recognised in innovation systems[3], and is becoming more evident today as multinationals are localising new manufacturing and R&D facilities in emerging countries in geographically dispersed areas and so with more focus on R&D collaborations with several external partners (Chesbrough, 2003; José Guimón, 2011; Perrot and Filippov, 2011).

In assessing the factors that will be critical for the success of manufacturers globally, it was emphasised in the KPMG-survey study that, 'Manufacturers

---

3. Innovation systems have been discussed variously in a few chapters of this report.

are focusing their efforts on their core competencies, both externally and internally, forming outside alliances or joint ventures with others who complement them, and driving greater efficiencies from within by analysing and transforming their supply chain and internal processes' (KPMG, 2012, p. 24).

2.1   R&D AND SUPPLY CHAIN COLLABORATION

The hydrogen fuel cell industry consists of several vertically integrated manufacturers that produce various components of hydrogen fuel cells and constitute companies such as automakers, membrane developers and catalyst manufacturers. This is an emerging space where many large oil and gas companies and large OEMs are trying their hand at capturing. Figure 3 below shows the dominant role played by large automobile manufacturers such as DaimlerChrysler and General Motors, and oil and gas companies such as Shell. This is also evidenced in a section of this report *Understanding the Emergence of New Industries* where mention has been made under country analysis, of the important and dominant role played by large oil and gas companies and automobile manufacturers in advancing fuel cell technologies in most countries.

A potential industry structure is emerging, though still fragmented, as many companies are actively engaging in developing and manufacturing hydrogen fuel cell components for the automobile, materials handling, and stationary power markets.

Figure 2 shows the different segments of manufacturing activities for transportation in which single or multiple organisations can be active. Each segment of the hydrogen fuel cell chain, namely the catalyst manufacturer or the fuel cell stack manufacturer, demands different sets of technological capabilities and competencies. The figure also shows where a single company may decide to vertically integrate various segments of the supply chain as an OEM, and manufacture a fuel cell catalyst and stacks, and supply it to the end-consumer. During the 1990s, for many companies, technological development had shifted from the fuel cell stack development to the system integration into the automobile (Steinemann, 1999).

Prior research has shown that for many automakers it was impossible to acquire competencies in a technology which is not their core, and which could not be entirely developed internally without external collaborations (Steinemann, 1999). For many years it was known that companies such as Toyota, Daimler-Benz, and General Motors were developing fuel cells

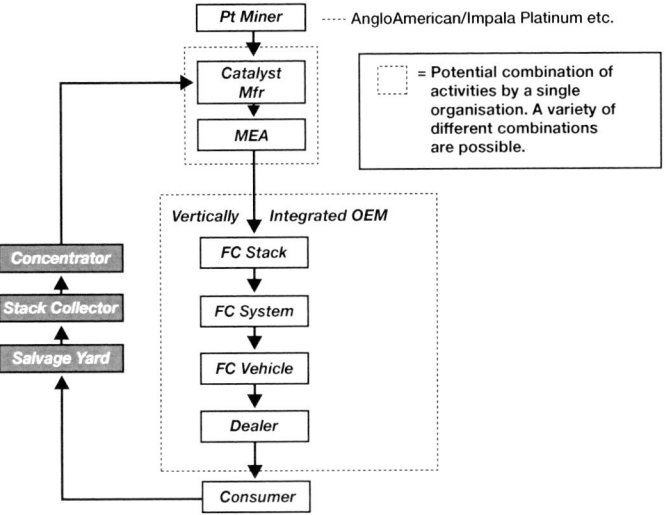

Figure 2: Segments of the fuel cell manufacturing supply chain for transportation. Abbreviations: MEA (Membrane Electrode Assemblies); FC (Fuel Cells) and Pt (Platinum). Source: Kromer, *et al.* (2009)

internally, especially during the early stages of investment, but with time, the development of fuel cells demanded more and more complex knowledge-bases, and necessitated broader fuel cell perspectives. Companies thus started collaborating with partners who contributed with skills and expertise in various fuel cell system components.

Further, according to Steinemann (1999), automobile companies with prior fuel cells knowledge in proton exchange membrane technology realised that they needed additional fuel cells know-how and expertise in the application of fuel cells in automobiles. Even further, the design of a fuel cell requires expertise in electrochemistry, electrocatalysis, and chemical fuel processing, all of which cannot be expected to be the technical know-how or core competency of a single company.

2.2 GEOGRAPHICAL PROXIMITY

Some innovation scholars claim that the rise of a knowledge-based economy is due to changes in the organisation and modularisation[4] of innovation processes that has increased the importance of geographical proximity to innovation (Sonn and Storper, 2003). In recent times, the emergence of 'modular manufacturing' has reinforced the need for proximity between

---

4. Modularisation or 'design modularity' is a manufacturing method that subdivides a system into smaller parts (modules). Each of these parts then can be separately manufactured and used in different systems to drive multiple functionalities. Modular-based manufacturing is common in the semiconductors, ICT and automobiles industries.

automakers and components-makers. There are several benefits and shortcomings of geographical proximity that have a bearing on organisational and vertical supply-chain structures, and inter-firm relationships (Frigant and Lung, 2003).

Preference for geographical proximity has become a critical consideration for manufacturers lately. Manufacturers are increasingly moving manufacturing facilities and sources of supply closer to end-markets (KPMG, 2012) and closer to technological opportunities. A recent example of locating a manufacturing facility due to availability of technological opportunities and skills is automaker Mercedes-Benz who opened a fuel cell manufacturing plant in Canada. The company ships the fuel cells to Germany for its new F-Cell cars series expected to be available by 2014 (CHFCA, 2012).

Another example of a shift in manufacturing due to customer closeness is in India where foreign manufacturing companies are increasingly focusing on local sourcing, customising their designs to meet Indian requirements, and restructuring their cost and manufacturing capacities (KPMG, 2012).

These changes are seen at a time when various supply-constraint issues are emerging in the South African catalytic convertor industry. There is a 30 per cent productivity gap between leather-seat suppliers in Eastern Europe and South Africa (Engineering News, 2012). The South African market is gradually losing out to suppliers in Central and Eastern Europe as European automobile manufacturers favour the proximity to a high-quality labour force, excellent infrastructure and technological support (Simkova, 2012). In fact, from an earlier focus on low-cost labour, there has been a strategic turn of this region's entire foreign direct investment (FDI) framework toward high value-added manufacturing[5] (Simkova, 2012).

The catalytic convertor industry in South Africa is a cluster-based industry in the Eastern Cape with a complex vertically-integrated supply chain constituting some 50 companies that are variously involved in the manufacturing of the final product: the catalytic convertor for exhaust pipes of internal combustion engines (Dewar, 2012). It is a 100 per cent export industry supplying convertors to automakers in Europe and Asia. Currently, more than 15 per cent of the global annual demand for catalytic converters comes from South Africa (Dewar, 2012; BASF, 2012). The convertor has local content in excess of 85 per cent, considered substantially higher than any

---

5. For example, Czech Republic's foreign direct investment strategy has shifted from low-cost, labour-reliant manufacturing toward high value-added sectors. An incentive scheme was introduced to promote technological and strategic service centres. Investment and tax breaks already exist for manufacturing companies.

other exported automotive component (Dewar, 2012).

In recent times, however, the competiveness of the catalytic convertor cluster is reported to be declining, including demand from foreign OEMs, for a variety of reasons:[6] a declining subsidy or government support under the Automotive Production and Development Programme and, as noted above, foreign automakers increasingly sourcing components that are in geographical proximity to their manufacturing bases such as Central and Eastern Europe.

Therefore, shifts in manufacturing are occurring because of the advantages of geographical proximity triggered by a need to be close to the customer base or to R&D and technological capabilities. R&D and technological capabilities are derived from a high-quality labour force, infrastructural base, and technological and economic support through taxes and incentives, as in the case of Central and Eastern Europe.

3 PROTON EXCHANGE MEMBRANE (PEM) FUEL CELL MANUFACTURING

A generic fuel cell industry value chain is depicted in Figure 2 showing the type of technology and competency required in each segment. The type of technology development and manufacturing activities, including services, associated with each segment of the value chain is shown in the lower section of the diagram.

The hydrogen fuel cell industry remains fragmented, but some structures that are based on components of a fuel cell system have nonetheless managed to emerge. Table 1 below shows the structures that have emerged based on the various fuel cell components. It also shows the type of organisations that are involved in manufacturing. The need for technical capabilities and competencies in each segment of fuel cell components manufacturing requires partnerships and collaboration. Mytelka (2004) argues that technologies such as hydrogen fuel cells require a pattern of precocious partnering for R&D as well as for standards setting.

A series of R&D collaborations and research alliances mark the global hydrogen fuel cell components manufacturing (Steinemann, 1999; Mytelka, 2003). Figure 3 shows the technological alliances of global fuel cell companies involved in the transfer of technologies through joint venture agreements, technological acquisitions, and licensing agreements. Small fuel

---

6. Based on an interview with the Catalytic Convertors Industry Association

cell companies that have the proprietary fuel cell technology such as a small US-based PEM fuel cell developer, Nuvera, US-based Plug Power, and Canadian company Ballard Systems are found to have active technological collaborations with large OEMs and oil and gas companies such as DaimlerChrysler, General Motors, Dow Chemicals, Shell and Total. For these large companies whose core competence is not in hydrogen fuel cells, collaborations of such nature give them access to fuel cell technology, knowledge, and other related competencies.

Figure 3 further shows that DaimlerChrysler is one of the most connected actors in the global fuel cell collaboration network, and is connected to almost every other critical player in the network, such as Nuvera, Plug Power, and Ballard that own the intellectual property (IP) and the technology. DaimlerChrysler's strategic aim was to capture and gain access to every technological development in the hydrogen fuel cell space at that time, in the hope that one of these technologies may become the dominant technology, and over which it would have a technological lead over its competitors (Mytelka, 2008).

As Figure. 3 shows, companies such as General Motors (GM) collaborate closely with Nuvera, and it is indirectly related to Ballard through its partnerships with Shell and DaimlerChrysler. In turn, Nuvera has direct strategic and technological collaborations with large companies, Shell and DaimlerChrysler.

Because of such partnering activities since the 1990s, by the end of the last decade, a number of companies from the chemical, aerospace and oil industries possessed knowledge and expertise on key components of fuel cell systems. It was apparent that technological advancement of hydrogen fuel cell technologies and the subsequent manufacturing of fuel cell components began to increasingly rely on such research (or knowledge) networks of collaborations and partnerships.

Depicted in Table 1 below, oil companies show technical expertise and competency in hydrocarbon and catalytic technologies which form the scientific basis for fuel reformers. Instead of replicating expertise and knowledge-bases, Ford and DaimlerChysryler decided to co-develop fuel reformers together with Mobil and Shell (Steinemann, 1999). Although automakers have started out by developing their own proprietary fuel cell technology in-house, especially in the period between the 1960s and 1970s, for a long time after that they had relied on Ballard to develop the fuel cells for them [7]. Ford and Daimler-Benz (parent of DaimlerChrysler) collaborated with Ballard Systems in 2003 to access its proprietary PEM fuel cell technology which put them at the technological frontier of auto fuel cells.

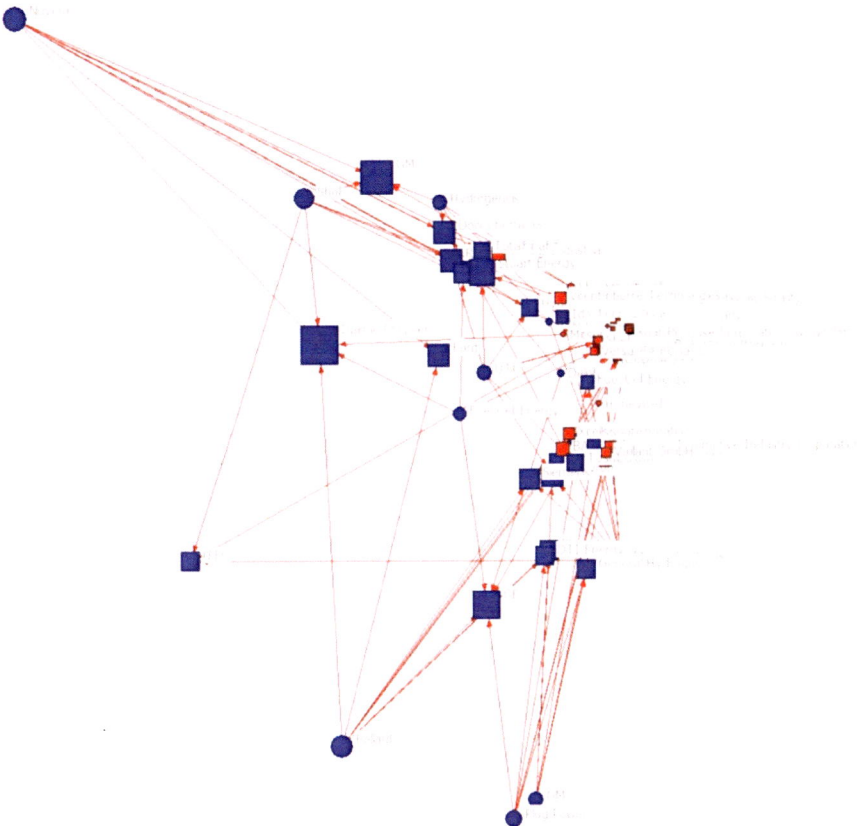

Figure 3: Technological network of global fuel cell companies (2000–2008).
Source: Authors' own

An example illustrating the increasing importance of partnerships and collaborations in manufacturing: Toshiba created a joint venture with US-based United Technologies Corporation (UTC) in 2001 to co-develop PEM fuel cell and Phosphoric Acid (PA) fuel cell technologies and commercialise them in Japan's rising market for micro-combined heat power (CHP) and fuel cell vehicles and electric fuel cell vehicles. Such ventures of joint technology development collaborations and licensing gives companies the benefits of an early mover advantage, volume production and economies of scale, and avoids high investment costs that are often associated with in-

7. In the late 1990s and early 2000s, Ballard Systems was a clear leader in auto fuel cells at the time, with a three- to four-year lead on its competitors. It boasted nearly 100 fuel cell-related patents and 100 more pending (Nauss, 1998).

Fuel Cell System Components	Type of Manufacturer	Companies	Technological capabilities and competencies
Fuel cell	Research design and manufacturer	Ballard, DeNora, H-Power, Siemens, Plug Power, Toshiba	Fuel cell technology development and design
Polymer membrane	Membrane primarily manufactured for the chlor-alkali industry	Asahi Chemical and DuPont and later Ballard, Johnson Matthey Plc, 3M, Hoechst, Dow Chemicals	Chemistry and materials technology
Fuel cell electrode/catalyst	Develop catalysts for chemical processes, chemicals and pharmaceuticals	Johnson Matthey Plc, Tanaka	Catalyst technology and electrochemical products
Fuel reformer/processor	Oil and gas companies	Shell, Chevron, Mobil, Sasol (RSA), Johnson Matthey Plc	Hydrocarbon catalysis and chemistry knowledge required for fuel processing in an on-board fuel reformer
Electric drive train	Automobile manufacturers	GM, Ford, Toyota, Honda, Renault, Mercedes, DaimlerChrysler	Integration of electric drive train and fuel cell system for hybrid vehicles
Systems integration (automobiles)	Automobile manufacturers	GM, Ford, Toyota, Honda, Renault, Mercedes, DaimlerChrysler	Need overall competence in fuel cell stacks and integration, power ancillary controls, and other components of a fuel cell system
Systems integration (backup/stationary) gensets, CHP, etc.	Oil and gas, electricity generators	Large utilities such as Osaka Gas, TOTAL, Shell, StatOil	Need overall competence in fuel cell stacks and integration, power ancillary controls, and other components of a fuel cell system

Table 1: Components of a PEM FC system and major players
Source: Adapted from Steinemann (1999)

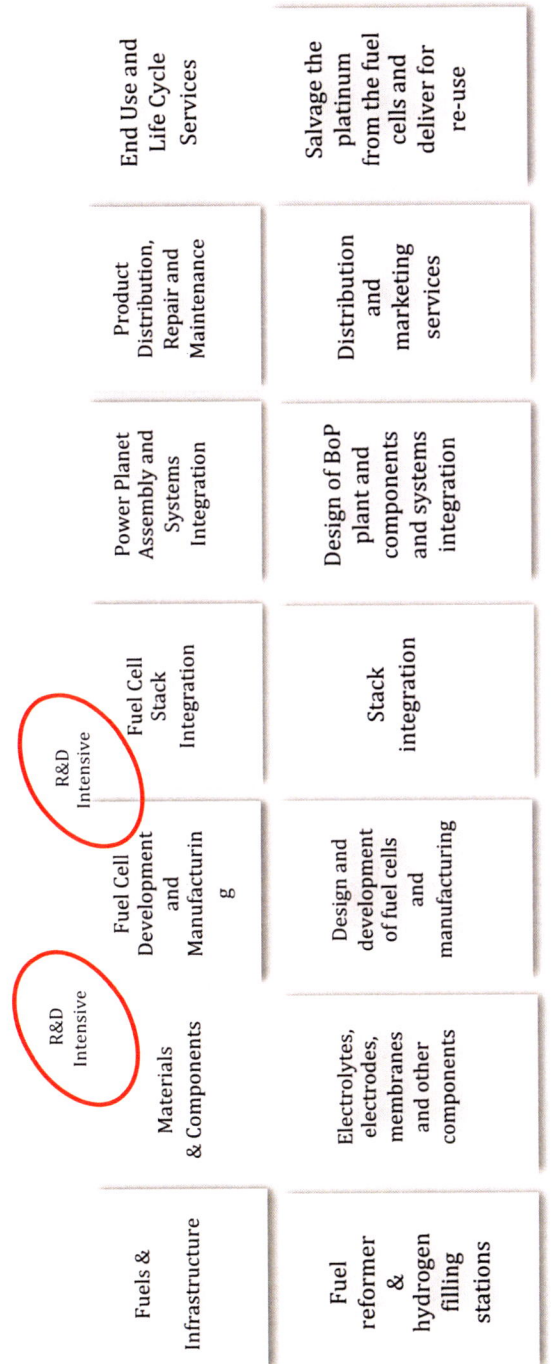

Figure 4: Value chain of R&D manufacturing of various components of fuel cells.
Source: Compiled from various sources

house development. As late as 2012, Suzuki entered a joint venture to access UK-based Intelligent Energy's air-cooled PEMFC technology through licensing.

4 FACTORS DRIVING PROTON EXCHANGE MEMBRANE FUEL CELL MANUFACTURING

4.1 PUSH-PULL FRAMEWORK

The new era of global manufacturing is driven by shifts in demand or demand-side dynamics and innovations in new materials, processes, information technology, and operations or supply-side/technology dynamics. One of the oldest schisms in innovation studies is between technology-push (Schumpeter, 1939) and market-pull (Schmookler, 1966) accounts of innovation. Starting in the 1960s, several scholars started to look at innovation from a demand rather than a supply perspective, contending that the most critical drivers of innovation are the need or demand pull forces (mainly opportunities that are derived from peoples' needs, and the market). This is in contrast to earlier-held perspectives that innovations are driven entirely by technology-push factors, i.e. technological opportunities drive innovations.

Demand for products and services are important in stimulating innovations along the innovation value-chain – the so-called demand-pull factors, where economic factors drive the rate and the direction of innovation. According to Nemet (2008), changes in market demand create technological and economic opportunities for firms, so that they invest in innovation to satisfy unmet needs. This implies that demand motivates firms to work on certain problems. The core of the science and technology-push argument is that scientific understanding and advances determine the rate and direction of innovation (Nemet, 2009). Walsh (1984) describes the polarisation into demand-pull and technology-push camps as 'crude' while Mowery and Rosenberg (1979) consider both demand-pull and technology-push as dominant drivers of innovation.

In fact, Mowery and Rosenberg (1979) pointed out that Schmookler's account of market-pull innovation seemingly applies to incremental innovations and does not address the problems of radical innovations. Radical innovations such as hydrogen fuel cells have more to do with technology-push and fundamental advances in science and technology in the

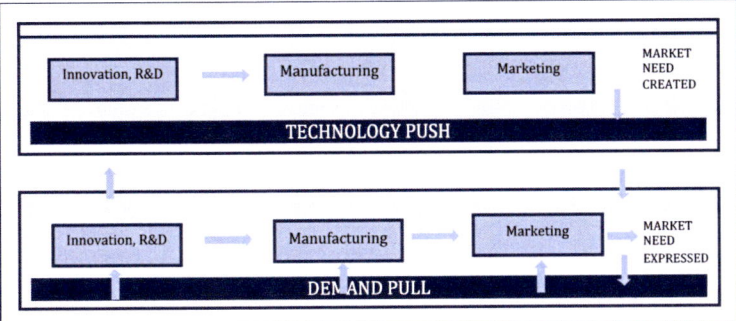

Figure 5: Technology-push and demand-pull factors as drivers of innovation.
Source: Martin, J. C. M. (1994)

initial stages of development, than market-led factors. The market knows little of the new products, their possible utility, their suppliers and manufacturers, etc. Hence, under such instances, markets need to be deliberately created.

There is, however, a need to integrate both these perspectives as both demand-pull and technology-push factors are worth considering when taking a radical innovation to the market. Figure 5 shows the multi-directional influence of these factors on innovations, driven by market needs and technological opportunities. The innovation systems perspective in fact deals with the interaction between the supply and demand of knowledge and technology.

The application of this push-pull framework to policy decisions creates a dichotomy separating government actions that affect the size of the market for a new technology from those that influence the supply of new knowledge directly (Nemet, 2009). This means that governments can encourage innovation in two (push and pull) ways: implement measures that increase the technological opportunities derived from science, technology and innovation (technology-push measures); and implement measures that increase the economic and social payoffs from these innovations (demand-pull measures).

4.2 DEMAND PULL FACTORS

Countries or regions where there is strong and proactive government support and/or even moderately generous subsidies are the ones that are fast becoming prime territories for fuel cell manufacturers. Countries such as South Korea and Japan are examples of successful fuel cell manufacturing regions despite the fact that most basic and core technological advances in fuel cells are being developed in countries of the EU and North America. Japanese and South Korean companies are rapidly acquiring external fuel cell technologies know-how through a variety of partnerships, licensing agreements, and consequently expanding their manufacturing base in fuel cells. The ability to access leading-edge foreign fuel cell technologies and build a manufacturing base is attributed to the supportive measures of the Japanese and South Korean governments. Such measures create the market demand (demand-pull) and subsequently drive activities in fuel cell manufacturing.

4.3 Government Initiatives in Supporting the Nascent Fuel Cell Market Demand

Below are a few examples of various government support measures that are boosting manufacturing activities in hydrogen fuel cells.

4.3.1 A declining subsidy scheme

As mentioned in an earlier chapter of this report, Japan launched one of the most successful schemes so far in the history of fuel cell commercialisation. In 2009, a declining subsidy scheme named Ene-Farm was introduced for residential micro-combined heat and power (micro-CHP) systems based on fuel cells: with the magnitude of the subsidy declining as more units are produced and revenue improves. The scheme resulted in the largest number of fuel cell installations in the world reaching 20 000 units in cumulative installations. For 2013, a subsidy has been allocated to fund more than 50 000 units.

Japanese gas utilities such as Tokyo Gas and Osaka Gas have been developing PEM-based CHP fuelled by natural gas since 2000, though in separate efforts. Ene-Farm became commercial only after four years of large-scale demonstrations and customer trials by these two companies. Between 2005 and 2008, around 1,300 PEMFC CHP systems were demonstrated at customer sites. In 2012, PEM CHP was dominant over the SOFC CHP installations with a ratio of 4:1.

In 2001, Osaka Gas collaborated with Ballard Generation Systems[8] to accelerate the commercialisation of residential PEM-CHP systems for continuous power generation. The collaboration involved integrating Ballard's fuel cells with the fuel processing systems of Osaka Gas. In the same period, Osaka Gas joined a consortium comprising Kyocera Corporation,

---

8. Ballard Generation Systems (BGS) was formed in 1996 to commercialise PEM fuel cell stationary power. Approx. 82 per cent of BGS is owned by Ballard Power Systems and the remaining shares are owned by ALSTOM France.

Aisin Seiki Co. Ltd., Chofu Seisakusho Co. Ltd., and Toyota Motor Corporation to jointly develop and commercialise Solid Oxide Fuel Cells (SOFC) CHP.

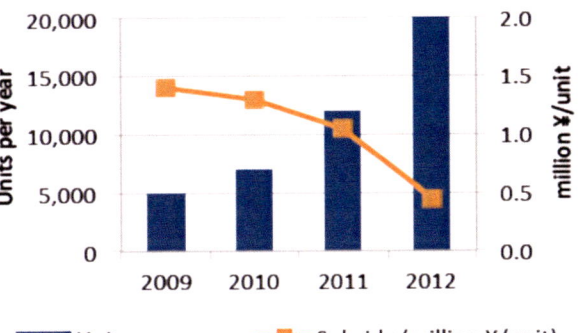

Figure 6: Declining subsidy scheme offered in Japan. Residential units were given a declining subsidy amounting to $14,987/unit in 2009 that fell to $ 4,817/unit by the year 2012.

Source: Fuel Cell Today, 2012

The Toshiba Fuel Cell Power System (TIFC)[9] supplies fuel cell systems to the Ene-Farm through the city utilities such as Osaka Gas. Toshiba and UTC have been jointly developing the phosphoric acid (PA) fuel cell and PEM fuel cell since 2001. And in 2003, Toshiba was one of four companies selected by Osaka Gas to jointly develop residential CHP systems based on PEM fuel cell technology.

There has been an increase in the demand for catalysts in residential fuel cells since the launch of the Ene-Farm scheme in Japan. Tanaka, a manufacturer of precious and industrial metals, is one of the suppliers of platinum-catalysts to fuel cell developers in Japan. Tanaka, with its extensive metal catalyst and electrochemical technology experience started developing and manufacturing catalysts for PEMFC and platinum alloy/carbon catalysts for Phosphoric Acid Fuel Cells (PAFC) some years ago.

Japanese automobile manufacturers are conducting R&D with the aim to commercialise fuel cell electric vehicles (FCEV) starting from 2015, which in turn is likely to drive the demand for platinum-catalysts.

4.3.2 FEED-IN TARIFFS AND RENEWABLE PORTFOLIO STANDARD

Feed-in Tariffs (FIT) offer a guarantee of payments to renewable energy developers for the generation of electricity based on a tariff that is guaranteed over the investment period of 15–20 years. Such tariff guarantees are expected to boost investments and new generation capacities from renewable energy sources, including fuel cells. These instruments make the

---

9. TIFC was established in March 2001 as a joint venture between Toshiba and UTC Fuel Cells.

cost of renewable energy technologies competitive or sometimes almost on par with conventional energy technologies. In 2011, 20 per cent of Germany's electricity came from renewable energy and 70 per cent of this was due to feed-in tariffs. South Africa had introduced FIT for wind, solar, biomass and hydro, but this was later scrapped before it was implemented, in favour of price setting through a competitive bidding process.

Renewable Portfolio Standard (RPS) is a regulation that obliges electricity supply companies or utilities to produce a pre-determined fraction of their electricity from renewable energy sources. In contrast to feed-in tariffs which guarantee purchase of all types of renewable energy regardless of cost, RPS is based on market competition and thereby market-selects different types of renewable energy based on competitive pricing.

4.3.3 INITIATIVES UNDERTAKEN BY VARIOUS COUNTRIES

It is unfortunate that the benefits of hydrogen fuel cell technologies are not yet reflected in government policies. This is despite the fact that the last 10 years have seen considerable technological advances in fuel cell technologies with technological performances considerably improved and costs fast declining[10]. Australia (the state of Victoria) began to provide FIT for fuel cells in 2012, while the country had long been home to a few successful hydrogen fuel cell companies, including Ceramic Fuel Cells Ltd. Ceramic Fuel Cells have been at the global forefront of SOFC-based micro-Combined Heat and Power (micro-CHP) technology since 1992. Because of attractive support policies overseas, Ceramic Fuel Cells is almost entirely in Europe, notably Germany and the UK.

UNITED KINGDOM

In the United Kingdom, FIT were enacted into law in 2008, and can be received for both the generation of electricity (a generation tariff) and for giving unused renewable energy generated electricity back to the UK national grid (called an export tariff). The UK is one of the few countries providing FIT for fuel cells. In 2012 it increased its tariff for micro-CHP fuel cells from 14.1pence (p) to 15.6p. The tariff comprises a generation tariff, which was increased from 11.0p to 12.5p for every kilowatt hour of electricity generated on-site, in addition to an export tariff of 3.1p for every kilowatt hour of electricity exported to the grid (Lamberti, 2012).

---

10. According to CBC News (2013), Ballard Systems PEM fuel cells are 60 per cent cheaper than they were three or four years ago.

SOUTH KOREA

South Korea has, for some time, been attracting fuel cell manufacturing owing to strong state support for fuel cells. Local manufacturing has become the cornerstone of the growth plans of many South Korean fuel cell manufacturers, and its geographical proximity to other fuel cell markets in Asia is an advantage. According to the scheme, fuel cells earn double renewable energy credits under RPS in South Korea, in addition to which the government offers subsidies of 80 per cent for micro-CHP fuel cell plants (Burger, 2012). The government is subsidising as much as 80 per cent of the cost to households that are purchasing fuel cells. The residential scheme is part of its 'Green Home Project' that aims to have fuel cell systems installed in one million 'green' homes by 2020 (Burger, 2012).

Prior to implementing RPS, South Korea had provided FIT guarantees between 2008 and 2011 for surplus electricity that was produced from a fuel cell unit and this surplus was purchased at premium prices. The tariff was provided for up to 50 MW of energy generated from fuel cells.

JAPAN

Japan has been serious in complying with the Kyoto Protocol and targeted a six per cent reduction of greenhouse gas emissions from the 1990 level by the year 2012, from recycling efforts to anti-idling campaigns. It began regulating automotive emissions in 1966 set by the Environment Agency in Japan. Moreover, legislation enacted in the 1990s in the US affected Japanese automakers. The California Air Resources Board (CARB) enacted a Low Emissions Vehicle Regulation which required American automobile manufacturers, and affected Japanese cars sold in the US, to make a percentage of their car sales as zero emissions vehicles (ZEV), according to Yarime, et al., (2008). Such regulations significantly tightened emission regulations surrounding carbon monoxide and nitrous oxide from vehicles.

Several technological developments were made by Japanese carmakers in response to the regulations implemented by the Environment Agency of Japan and those set by CARB for ZEVs. CARB regulations played a critical role and influenced Toyota's R&D for instance, since California was a key market for the Japanese auto industry (Yarime, et al., 2008). According to Yarime, et al. (2008), the number of patents led by Toyota, Nissan, Honda, Mazda, Mitsubishi and Fuji Heavy Industry in low-emissions vehicles increased in the early 1990s.

Since the mid-1990s Japanese carmakers filed for fuel cell vehicle patents, and the number increased sharply in the 2000s, reflecting the changes in both national and global regulations influencing the research focus in the Japanese auto industry (Yarime, et al., 2008).

To meet its Kyoto targets, Japan established further rigid fuel efficiency standards for its automakers in the 1990s and 2000s, making it the world's most stringent emissions level.

And according to a recent report by the US Environmental Protection Agency (EPA), Ford Motor Company has improved in fuel economy more than any other major US automaker. Because of its efficiency improvement programmes, Ford's fleet-wide emissions at 270 g of CO_2 per kilometre[11] in 2009 have improved more than any other US automaker (KPMG, 2010). Ford's improvements were the result of a variety of engineering innovations (of an incremental nature), some of which are the result of improved aerodynamics, weight reduction, and powering its steering with an electric motor[12] rather than a hydraulic pump.

It has become increasingly evident, therefore, that innovations in fuel cells (and other low-carbon-emission vehicles) have been primarily driven by environmental regulations and other emission-regulation programmes and enactments.

4.4   Technology Push Factors

Cluster Creation: R&D networks and manufacturing

The interest in clusters[13] emerges out of a growing understanding that the sources of firm innovation and competitiveness rest on firm capabilities and the broader industrial structure within which the firm(s) operate (Holbrook, et al., 2007). The benefits and outcomes of the creation of clusters or, most notably, research networks have been already elaborated at length in the chapter *Understanding the Emergence of New Industries*.

Canada

The Vancouver-based hydrogen fuel cell and technology 'cluster' in Canada is endowed with several favourable factors that support a strong infrastructure

---

11. Vehicular carbon dioxide (CO_2) emissions are calculated in gms of CO_2 emitted per kilometre.
12. Powering the steering with an electric motor rather than a hydraulic pump results in a 3 per cent improvement in fuel economy (KPMG, 2010).
13. According to the European Commission Report on Clusters (2008, pg. 8), 'clusters are seen as an important factor for the explanation of the empirical phenomenon of geographical concentration of economic and innovation activities. More than one definition of cluster exists, depending on its purpose and the specific context of its use. In many discussions no clear distinction is drawn between clusters as a real economic phenomenon and cluster policies and initiatives which are more of a normative function'.

for R&D and demonstration projects to its 40 lead companies. An industry structure is gradually emerging from the cluster, characterised by the co-existence and dynamic interactions of a group of firms, including public research organisations and universities. The cluster has around 2,000 highly skilled workers, and is producing more than 100 graduate students a year specialising in hydrogen fuel cells and related fields. The cluster has incubated over 18 start-up companies over a period of 8 years under its Industry Partnership Facility programme.

INDUSTRY-UNIVERSITY-GOVERNMENT LINKAGES

The Vancouver cluster started out of the National Research Council (NRC)'s Institute for Fuel Cell Innovation (IFCI) and is a public-private partnership between the NRC, Natural Resources Canada, Industry Canada, the Government of British Columbia, and the industry association, Hydrogen & Fuel Cells Canada. NRC-IFCI has partnerships with the University of British Columbia (UBC), Simon Fraser University (SFU), the University of Waterloo, the Province of British Columbia, and many more.

SKILLS AVAILABILITY

One of the critical factors for cluster and/or research networks creation has been the availability of, and access to, specialised skills and human capital. According to Holbrook, *et al.* (2010), Vancouver has a good supply of human capital for the fuel cells cluster. All the local post-secondary institutions in Vancouver have programmes that train professionals for the fuel cell and hydrogen industry (Holbrook, *et al.*, 2010). Mentioned in the same paper is a survey that shows that the majority of hydrogen fuel cell firms in Vancouver recruited between 80 per cent and 100 per cent of their employees locally.

A fundamental condition that makes it advantageous for foreign firms to engage in knowledge transfer and for domestic firms to gain access to their knowledge is local skills endowment and ability to handle knowledge (Saliola and Zanfei, 2009).

COLLABORATIONS AND INTERNATIONAL LINKS

A critical requirement for hydrogen fuel cell technology development and manufacturing is external technology collaborations with foreign companies and research organisations, and access to the global supply manufacturing

---

14. Canada has leading-edge hydrogen and fuel cell capabilities, products and services, but no domestic market. This is the result of the low cost of energy as well as the abundance of other sources of energy – fossil fuels, hydro and nuclear [Conversation with Terry Kimmel, Chairperson of the Partnership for Advancing the Transition to Hydrogen and Director of the International Association for Hydrogen Energy].

chain. The latter is particularly critical for Canada as the country exports 90 per cent of its fuel cells[14]. In fact, a Canadian and Chinese science and technology collaboration includes two universities: the British Columbia Innovation Council and the National Research Council (NRC). They are working to strengthen the global supply chain, something which the Canadian fuel cell industry needs to ensure its commercial success in China and other emerging countries. Concomitantly, to enhance its R&D networks, NRC is internationally linking Canadian companies to the rest of the world.

Government cluster policies

To develop and promote a cluster that is based on a new technology, policy makers need to adopt policy measures and subsidies, and nurture technology-based programmes. Such government intervention reduces risk and facilitates the flow of knowledge. This is of significant importance to industry in its emerging stages as resources are limited and trajectories towards a dominant design remain invisible (Choi, *et al.*, 2011)[15].

In March 2013, the Canadian government invested CAN $646,750 to develop and implement an Accelerated Trade and Investment Programme specifically designed for small- to medium-sized enterprises in the field of hydrogen fuel cells. The initiative includes training, consultation, materials development and the support of industry presence at key sector/market events.

Cluster policies may also include tax rebates and incentives that encourage the setting up of a technology-based company or a manufacturing plant often in a specific location within the country. It may also be through government funding initiatives similar to those of the US Department of Energy, which funds early-stage corporate product development in new energy technologies and acts as an 'angel investor' by bringing together the industry and national laboratories (OECD, 2009).

However, as noted by OECD (2009, p. 26), 'a frequent mistake made by policy makers and analysts is to think that clusters are synonymous with deliberate policies or deliberate co-operation in formal networks'. In a study of clusters globally, it was found that there were no specific policies to drive networks or clusters in most of these innovative clusters (Sölvell, *et al.*, 2003; Carlsson, 2005). While important, the influence of policy is often indirect, and drives infrastructure, research, education and training rather than

---

15. The role of public policies in enabling cluster creation (and its precursor – research networks) has been elaborated in the next chapter of this report, *Understanding the Emergence of New Industries*.

clusters *per se* (Uyarra and Ramlogan, 2012).

The triggering of events to a successful cluster creation and sometimes even its precursor (research networks) is random – and often constitutes an element of luck (Carlsson, 2005).

SOUTH AFRICA

As stated in other chapters, in 2006 the Department of Science and Technology (DST) developed a national hydrogen and fuel cell technology research and development programme, known as Hydrogen South Africa (HySA)[16]. As discussed specifically in *Understanding the Emergence of New Industries,* technological programmes such as HySA contribute to the creation of knowledge within a network. In such a network, firms and the governments are able to assign resources collaboratively and appropriately. They can decide the right time and country/region/place of entry during the emerging stages of an industry, when markets are seldom formed and competition is non-existent (Callon, *et al.*, 1992; Callon, 1995; Choi, *et al.*, 2011). Actors entering a knowledge network collaborate, share information and knowledge with each other, thus reducing and sharing investment risks and uncertainties (Cassi, *et al.*, 2008; Tijssen and Korevaar, 1997; Choi, *et al.*, 2011).

INDUSTRY-UNIVERSITY-GOVERNMENT LINKAGES

HySA has three Centres of Competence (COCs): HySA Infrastructure based at North-West University, Potchefstroom, in collaboration with the Council for Scientific and Industrial Research (CSIR); HySA Catalysis based at the University of Cape Town, collaborating with MINTEK; and HySA Systems based at the University of the Western Cape (DST, South Africa).

COLLABORATIONS AND INTERNATIONAL LINKS

The role of private sector partners is small but important for the HySA research network. Some of its partners are companies such as PetroSA and Impala Platinum, which provide R&D funding. A company called Hot Platinum works with HySA Systems on intellectual property contract research. A Western Cape-based manufacturing design company called TF Design is assisting with the designing of prototypes for HySA fuel cell systems.

---

16. There are five HySA R&D programmes that capture the entire value chain R&D and manufacturing: Combined Heat and Power (CHP), Portable Power Systems; Hydrogen Fuelled Vehicles; Hydrogen Filling Stations; and Renewable Hydrogen Production (through the process of electrolysis).

As discussed in the chapter, *Emergence of New Industries*, the size of South Africa's research collaboration and international links has been shown to be small but significant. It is considered significant because South Africa connects, through its international network of partners, with countries that are very active in hydrogen fuel cells such as the UK, Germany, Japan and the USA. Some achievements by HySA included the first demonstration of a fuel cell-powered tricycle, based at the University of the Western Cape, and the first metal hydride hydrogen storage unit fuel-based forklift truck in SA (PMG, 2012).

This study further highlights the fact summarised elsewhere in this report that the level and capacity of each country to enter international markets, and improve their respective trade, is closely linked to domestic capacity to take advantage of new and innovative technologies. Thereby the adoption, absorption, adaptation and application of these technologies depend on the strength and efficiency of the country's national system of innovation.

SKILLS AVAILABILITY

There are no full-term specialised courses offered in the field of hydrogen fuel cells and storage in South Africa. HySA organises a number of workshops each year, and trains participants from across the country in various aspects of hydrogen fuel cells. The participants are from the energy industry and post-graduate students from across the country.

Although Wits University in Johannesburg is not part of the HySA research network, it is active in advancing the use of these materials in fuel cells and energy storage technologies, shown in Table 2. Students working in some of these projects are awarded a post-graduate degree upon completion. Technologies such as fuel cell membranes, novel photovoltaic materials, batteries and ultra-capacitors, and heat transfer media are a few examples where progress has been made at the university, linked to material sciences such as nanotechnology.

5 STRATEGIES OF FIRMS IN SOUTH AFRICA

It will be essential to comprehend the behaviour and objectives of firms as different firms have different motivations, fuel cell development, and market strategies and interests. Below we explore the strategies of firms in South Africa in the area of hydrogen fuel cells.

RESEARCH AREA	SECTOR	RESEARCH TITLE	LEVEL	SCHOOL	PRINCIPAL RESEARCHER
Hydrogen Storage	INDUSTRY/ RESIDENTIAL	Examination of metallic hydrides and diboride nanostructures	5 lead staff	MERG: The Materials for Energy Research Group: Physics Chemistry and Engineering	Prof. A. Quandt Dr L. Chown Prof. D. Billing Dr D. Wamwangi
Hydrogen Generation	INDUSTRY/ RESIDENTIAL	Several methods explored for their ability to use solar energy to split H_2O and generate hydrogen	4 lead staff	MERG: The Materials for Energy Research Group: Physics Chemistry and Engineering	Dr S. Durbach Dr S. Mhlanga Prof. D. Biling P. Franklyn
Fuel Cells	INDUSTRY	Pilot plant for production of ion exchange membrane for fuel cell that uses natural gas process fuel	Post-graduate	Chemical & Metallurgical Engineering	Prof. Sunny Iyuke; Alain Mufula

Table 2: Research areas in the field of hydrogen storage, generation and fuel cells
Source: Author's own compiled for the Wits University Energy Forum (2012), South Africa

CLEAN ENERGY INVESTMENTS

In 2010, Anglo American Platinum's Platinum Growth Metals Development Fund, the government of South Africa through the Department of Science and Technology, and US-based Altergy Systems created a joint venture called Clean Energy Investment. The joint venture was created with the objective of manufacturing and marketing Altergy fuel cell systems in South Africa and southern Africa. A huge market potential is anticipated in the region for PEM fuel cell-based back-up power telecommunications, data centres, and other critical applications that need uninterrupted power supply.

Initially, Clean Energy Investments was expected to establish a distribution network for Altergy's fuel cell products and, after a few years, it would establish a fuel cell manufacturing and assembly plant in South Africa, the rights for which would be licensed by Altergy (Behling, 2013). Early in

---

17. The reformer extracts hydrogen from a fuel (in this case methanol and water) and feeds the hydrogen into the fuel cell system, thus eliminating the nearly constant supply of hydrogen cylinders required for off-grid locations.

2013, Clean Energy Investments partnered with a company called H9 for its on-site hydrogen generation fuel processor or reformer[17], which is then to be integrated with the fuel cell systems of Altergy.

ANGLO PLATINUM

In a separate initiative, Anglo American Platinum entered into a product development agreement with Canada's Ballard Systems to develop and prototype PEM-based electric generators. The two companies have been variously demonstrating and applying fuel cell technologies within the mining environment – such as mining headlamps, mining locomotives and fuel cell electric generators. Under the agreement, a 150 kW fuel cell system was 'demonstrated' at the COP17 conference in South Africa[18] in 2011.

Anglo Platinum has been providing funds to Ballard Systems for the development and testing of home generator systems using the company's proprietary fuel cells and a methanol-based reformer for on-site hydrogen generation. The aim is to use these systems to supply primary power for rural electrification of households. Anglo Platinum's argument is that on-grid power is too costly[19] to be supplied to remote communities by the local utility Eskom.

Further in March 2013, Anglo Platinum further invested US$4 million[20] in Ballard Systems through its PGM Development Fund for the development of a prototype of home-generation system, to be applied in remote housing communities. Ballard's partner in China, Azure, has separately invested US$2 million in Dantherm Power[21].

BALLARD POWER SYSTEMS AND INVESTEC

In 2012, Ballard Systems acquired IdaTech, a methanol-based fuel cell company that provides back-up and prime power to the telecoms market. These markets, which often require back-up power, have seen a preference for methanol fuel cell systems to replace traditional diesel generators.

As mentioned in another chapter, Ballard Systems acquired IdaTech's intellectual property licence and its manufacturing plant in Mexico. To date, IdaTech installed over 250 units[22] in South Africa for Vodacom at base

---

18. The demonstrated generator was a proof-of-concept system based on existing technology from Ballard and Dantherm Power.
19. It has not yet been estimated by how much the cost of a grid-connection by Eskom would surpass the cost of installing a fuel cell electric generator in a community.
20. The investment is said to be a five-year non-interest bearing loan repayable in Ballard common shares. The PGM development fund has the option of repayment on or before the maturity date.
21. Dantherm is Ballard's telecom back-up power subsidiary.
22. An industry analyst's feedback.

telecommunication sites and shipped its 500th unit worldwide in 2013. South African specialist bank and asset management company, Investec, is IdaTech's principal funder. Investec holds the majority of the Ballard shares issued under this transaction.

OTHER STATIONARY FUEL CELL SUPPLIERS

There are few other hydrogen fuel cell companies that are present in South Africa, targeting the prime power telecommunication market in southern Africa. The uptake and installations have been positive but slow in South Africa, and are currently limited to base telecommunication sites of telecommunication companies such as Vodacom. Two companies involved in this are ReliOn developer of methanol-based fuel cells (using Ballard Systems stacks in their fuel cells) and Helion, part of the AREVA group that designs, develops and manufactures PEM fuel cells.

Many companies, however, are struggling in the absence of a market in South Africa or a market that favours demand-side policies and other government procurement schemes. Vodacom, one of the few market players, has been installing these systems as a value-proposition strategy in the place of diesel generators that are prone to diesel theft.

6 RESEARCH NETWORKS AS A LOCALISATION STRATEGY

As mentioned in another chapter, research networks,[23] such as HySA, represent a local content or localisation strategy that is conscientiously and deliberately building local capabilities in the field of hydrogen fuel cells, storage and the hydrogen infrastructure space. HySA brings together researchers and companies in an elaborate network that works collaboratively and across the hydrogen fuel cell, storage and infrastructure value chain. Various capabilities have been pooled together, bridging capabilities and filling knowledge gaps.

The Renewable Energy Independent Producer Procurement programme of the Department of Energy stipulates local content requirements with the aim of ensuring that a portion of project spending is captured in the local economy, and supports the development of local manufacturing and local industries (Coovadia and Rennkamp, 2013). Further, the Green Economy Accord of South Africa highlights that: 'Without localisation, we will bear

---

23. Choi, et al., (2011) identify knowledge networks as an essential tool in the study of emerging industries. The analysis regards a network as not only the result of supply-demand relationships, but of knowledge and institutional factors that are embedded within a system.

much of the cost of greening our society without reaping an important benefit in the form of job creation' (Green Economy Accord, 2011, p. 7).

The use of local content requirements (LCR) or localisation as a policy mechanism is essentially to build local capabilities and to create markets for new energy technologies (Pegels, 2010; Lewis & Wiser, 2007; Coovadia and Rennkamp, 2013). According to Lewis & Wiser (2007), governments can use localisation policy mechanisms to encourage localisation of manufacturing. Various policy mechanisms that can achieve objectives of localisation of manufacturing include local content requirements, financial and tax incentives, customs duties, export credit assistance, quality certification and research and development agreements (Lewis & Wiser, 2007).

The HySA research network is a form of a local content policy mechanism that has the objectives of localisation. Only once the desired level of technological capabilities and skills have been locally developed in hydrogen fuel cells and achieved by these networks, can research networks such as HySA be complemented by financial and tax incentives and favourable custom duties, among other interventions.

7 Policy implications

The declining competiveness of South Africa and its inability to become a leading supplier of other components of the automotive industry is attributed to the gaps in productivity and supply constraints. Policies have been unable to mandate the type of components-based skills (e.g. in steel manufacturing) which foreign car manufacturers in South Africa require.

Moreover, South Africa has been unable to channel knowledge and technology spillovers into building its own indigenous automobile industry. This is despite the fact that it is a manufacturing base for several foreign automakers, namely BMW, Volkswagen, Toyota, Mercedes-Benz and Renault, among others. In contrast, India and China's indigenous automobile manufacturing industry and its associated capabilities are outstanding examples of knowledge and technology spillovers from foreign companies.

To support a manufacturing industry in hydrogen fuel cells and to enable it to compete globally would require policy makers to have a comprehensive and thorough understanding of the diverse hydrogen fuel cell industry segments, as well as an understanding of the wider global trends and factors affecting each manufacturing segment. Such an understanding will enable policy makers strategically and effectively to address issues that may arise, namely supply constraints, productivity and skills-based gaps.

Given its geographical distance to automobile manufacturers, and the absence of a local automobile manufacturer, policy makers must ensure that South Africa remains an attractive components manufacturing destination. According to KPMG (2012), policy makers should also recognise that their goals for job creation and growth are best met by providing critical stimuli to manufacturers, such as investing in modern infrastructure and providing incentives and subsidies.

Strategies to attract manufacturing plants to South Africa will need to be effective and aggressive from the start. South Africa is not a low-cost manufacturer, and manufacturing companies in South Africa face the same issues that companies in mature markets are facing. There are stronger and lower-cost manufacturing contenders with advanced technological bases to South Africa – namely China and India. And in terms of aggressive demonstration and marketing strategies and policies to access global markets and build a local manufacturing base, Canada serves as a good example.

For instance, China has strong science and technology (S&T) and industrial policies that protect R&D and manufacturing bases in many technological areas within the country and it aggressively invests in research, development, demonstration, and manufacturing of hydrogen fuel cells[24].

As discussed above, there are no cluster-specific policies per se and so policies must directly and indirectly influence the development of infrastructure, research, education and training.

A key priority should be on education and the development of skills in hydrogen fuel cells and storage technologies. In addition to supporting ongoing efforts to improve education in schools – particularly the teaching of maths and analytical skills – policy makers must ensure that skills learned in higher-educational institutions meet the needs of the industry. This is exemplified in the case of the reported Vancouver hydrogen fuel cell cluster where the majority of firms recruit between 80 per cent and 100 per cent of their employees locally.

Local businesses must be equally involved in the development of skills as often there is insufficient human capital in the specific area of application. Companies need to build their R&D capabilities as well as expertise in hydrogen fuel cells. For this they need well-qualified, specialised technical

---

24. Please refer to the chapter on Understanding the Emergence of New Industries in this report to see the levels of R&D activities and research linkage activities of various countries, in particular China's aggressive patenting activities and research outputs around hydrogen fuel cells and storage technologies.

skills working on various segments of R&D and manufacturing value chains.

It is also critical to assist local companies in forging international links with foreign companies and research organisations, as local organisations will need access to technologies, skills, and growing global emerging markets. Policies can enable co-operation and networking among firms, universities, and public research organisations that are involved in the formative stages of industry development. The science and technology department can be said to be a pioneer in the field through its forging of links at the R&D level for joint technological development and commercialisation work. However, such links have to be substantiated with concrete marketing and branding strategies to access global markets.

Supply side (technology-push) and demand side (demand-pull) factors should be aligned to promote R&D and manufacturing in South Africa. This requires a unified strategy and goal – one that will match the technological opportunities leveraged by the HySA research network with the market needs created through demand-side policies.

Moreover, the goals of all actors, including the behaviour of firms, must be aligned for a successful diffusion of hydrogen fuel cell technologies. All initiatives leading up to fuel cell manufacturing of all the relevant actors must be co-ordinated to take South Africa beyond current efforts that consist of fragmented strategies with unaligned goals.

As discussed earlier and elsewhere in this report, governments play a critical role allocating resources and ensuring that duplication of R&D and manufacturing is avoided. This can be ensured through well-designed research networks and industry clusters. Research networks, after all, are a form of localisation strategy that can strategically build local capabilities and skills in hydrogen fuel cells and storage. However, current efforts need to be further elaborated through more aggressive policies and incentives, extended support to HySA, and wider awareness among all stakeholders.

Bibliography

BASF. (2012). 'BASF Mobile Emissions Catalyst Division.' Information available at www.basf.co.za [Accessed 8 May 2013].

Behling, N. H. (2013). *Fuel Cells: Current Technology Challenges and Future Research Needs.* Amsterdam, Oxford, Massachusetts, San Diego: Elsevier B.V.

Callon, M. (1992). 'Techno-economic networks and irreversibility.' *A sociology of monsters: Essays on Power, Technology and Domination,* Volume 38, pages 132–161.

Callon, M. (1995). 'Four models for the dynamics of science.' Jasanoff, S., *et al.,* pages 29–63.

Carlsson, B. (2005). 'Entrepreneurship and Public Policy in Emerging Clusters.' In Braunerhjelm, P. and Feldman, M. (Eds.) *Cluster Genesis: The Emergence of Technology Clusters and the Implication for Government Policy.*

Chesbrough, H. (2003). 'The Era of Open Innovation Chesbrough.' *Sloan Management Review,* Vol. 44(3), pages 35–41.

CHFCA. (2012). 'Mercedes-Benz Opens World's First Automated Fuel Cell Manufacturing Plant in Vancouver.' www.chfca.ca [Accessed 7 May 2013].

Choi, H., Park, S. and Lee, J. (2011). 'Government-driven knowledge networks as precursors to emerging sectors: a case of the hydrogen energy sector in Korea.' *Industrial and Corporate Change,* Volume 20 (3), pages 751–751.

Coovadia, Y. and Rennkamp, B. (2013). 'Can local content requirements boost technological capabilities in South African industries? Evidence from wind energy under the REIPPP.' Masters Research Report, The Graduate School of Business, University of Cape Town.

Dewar, K. (2012). 'The Catalytic Convertor Industry in South Africa.' In: Fifth International Platinum Conference 'A Catalyst for Change': The Southern African Institute of Mining and Metallurgy Platinum 2012.

Freitas, I. and von Tunzelmann, N. (2012). 'Alignment of innovation policy objectives, a demand side perspective', Paper presented at DRUID 2012, Copenhagen, Denmark.

Frigant, V. and Lung, Y. (2003). 'Geographical proximity and supplying relationships in modular production.' *International Journal of Urban and Regional Research,* Vol. 26 (4), pages 742–755.

Green Economy Accord, 2011. Available at www.nedlac.org.za [Accessed 13 May 2012].

Guimón, José. (2011). 'Global trends in R&D-intensive FDI and policy implications for developing countries.' High-level policy workshop organised by the World Bank, 'Innovating Through the Crisis.' Croatia, June 2011.

Holbrook, A., Arthurs, D. and Cassidy, E. (2012). 'Understanding the Vancouver Hydrogen and Fuel Cells Cluster: A Case Study of Public Laboratories and Private Research.' *European Planning Studies,* Volume 18(2), pages 317–328.

KPMG. (2010). 'The Transformation of the Automotive Industry: The Environmental Regulation Effect.' Switzerland: KPMG.

KPMG. (2012). 'Global Manufacturing Outlook: Fostering Growth through Innovations.' Economic Intelligence Unit, *The Economist,* UK.

Kromer, M. A., Fred J., Rhodes, T., Guernsey, M. and Marcinkoski, J., 2009. 'Evaluation of a platinum leasing program for fuel cell vehicles.' *International Journal of Hydrogen Energy,* Volume 34 (19), pages 8276–8288.

Kunihiro N. and Hirai, K. (2009). 'Commercialization of a residential PEM Fuel Cell CHP "ENE FARM".' www.igu.org [Accessed 20 July 2012].

Lamberti, A. (2012). 'Ceramic Fuel Cells welcomes govt. plans to raise electricity feed-in tariff.' Article available at www.proactiveinvestors.co.uk [Accessed 29 April 2013].

Lewis, J. J. and Wiser, R. H. (2007). 'Fostering a renewable energy technology industry: An international comparison of wind industry policy support mechanisms.' *Energy Policy*, Vol. 35, pages: 1844–57.

Martin, M. J. C. (1994). *Managing Innovation and Entrepreneurship in Technology-Based Firms*, New York: Wiley.

McKinsey. (2012). 'Manufacturing the future: The next era of global growth and innovation.' McKinsey Global Institute.

Mowery, D. and Rosenberg, N. (1979). 'The influence of market demand upon innovation: a critical review of some recent empirical studies.' *Research Policy*, 8(2), pages 102–153.

Nauss, D. W. (1998). 'Ballard battling to break through with Daimler and Ford behind it, Vancouver firm sets fuel cell agenda.' Wards Auto. www.wardsauto.com [Accessed 4 April 2013].

Nemet, G. F. (2008). 'Evaluating the demand-pull hypothesis.' In: *Innovation for a Low Carbon Economy: Economic, Institutional and Management Approaches.* (Eds.) T. Foxon, J. Kohler, and C. Oughton. Pages 87–143. Cheltenham, UK: Edward Elgar.

Nemet, G. F. (2009). 'Demand pull, technology push, and government-led incentives for non-incremental technical change.' *Research Policy*, 38(5), pages 700–709.

OECD. (2009). 'Clusters, Innovation and Entrepreneurship.' Paris: OECD.

Perrot, R. and Filippov, S. (2011). 'Localisation strategies of firms in wind energy technology development,' *Journal on Innovation and Sustainability*, Volume 2(1), pages 2–12.

PMG (Parliament Monitoring Group). (2012). 'Hydrogen & fuel cell technology research & development programme: briefing by Department of Science and Technology.' Date of Meeting: 12 September 2012, South Africa.

Saliola, F. and Zanfei, A. (2009). 'Multinational firms, global value chains and the organization of knowledge transfer.' *Research Policy*, Volume 38 (2), pages 369–381.

Schmookler, J. (1966). *Invention and Economic Growth*. Cambridge, MA: Harvard University Press.

Schumpeter, J. A. (1939). *Business Cycles: A Theoretical, Historical, and Statistical Analysis of the Capitalist Process*, Philadelphia: Porcupine Press.

Simkova, O. (2012). 'Central and Eastern Europe: Moving up the value chain.' PriceWaterhouseCoopers (PwC). Article available at www.pwc.com [Accessed 8 May 2013].

Sölvell, Ö., Lindqvist, G. and Ketels, C. (2003). T*he Cluster Initiative Greenbook*, Stockholm: Bromma tryck AB Stockholm.

Sonn, J. W. and Storper, M. (2003). 'The Increasing Importance of Geographical Proximity in Technological Innovation: An Analysis of U.S. Patent Citations, 1975–1997.' Paper prepared for the Conference: What Do We Know about Innovation?

Steinemann, P. (1999). 'R&D Strategies for New Automotive Technologies: Insight from Fuel Cells.' International Motor Vehicle Program (IMVP), Massachusetts Institute of Technology.

Thurston, J. B. and Stewart, R. R. (2005). 'What drives innovation in the upstream hydrocarbon industry?' *The Leading Edge*, pages 1110 – 1116.

Uyarra, E. and Ramlogan, R. (2012). 'The Effects of Cluster Policy on Innovation.' Nesta Working Paper No. 12/05, National Endowment for Science, Technology and the Arts (NESTA), UK.

Venter, I. (2012). 'Absence of appropriate steel grades a constraint, Mercedes-Benz SA warns.' *Engineering News*, Creamer Media, South Africa.

Walsh, V. (1984). 'Invention and innovation in the chemical industry: Demand-pull or discovery push.' *Research Policy*, Vol. 13 (4), pages 211–234.

Yarime, M., Shiroyama, H. and Kuroki, Y. (2008). 'The strategies of Japanese auto industry in developing hybrid and fuel-cell vehicles.' In: Mytelka L. K. and Boyle, G., *Making Choices about Hydrogen: Transport issues for developing countries.* Tokyo: UNU Press, pages 187–212.

CHAPTER 5

EXPLORING PATHWAYS TO A HYDROGEN FUEL CELL TRANSITION IN THE SOUTH AFRICAN ROAD TRANSPORT SECTOR: A PRE-STUDY

ENERGY RESEARCH CENTRE
SYSTEMS ANALYSIS & PLANNING GROUP

Adrian Stone, Bruno Merven & Mamahloko Senatla

1 INTRODUCTION

The harnessing of fossil fuels has brought humanity many benefits. Coal, oil and natural gas have been the primary drivers of a chain of industrialisation that has brought countless useful products and amenities into being that have literally revolutionised human existence. There have, however, been other, negative, repercussions including toxic emissions, greenhouse gas emissions and massively increased human populations with a growing appetite for energy services that are currently dependent on a finite resource. Whether the depletion of fossil fuel stocks is imminent or not, does not change the fact that this resource of cheap, and above all concentrated, energy will be exhausted within a scale of decades rather than centuries at current consumption rates. That research, development, and planning around alternative energy sources and carriers is necessary, is not debatable. The timing of investments is of course another matter.

This chapter explores the possible role of an alternative energy carrier, hydrogen, in the future of the South African road transport sector. Hydrogen is referred to as an energy carrier rather than a primary energy source like coal, because, although the most abundant element in the universe, it does not readily exist in elemental form on earth and has to be extracted from hydrogen-rich compounds such as water or indeed fossil fuels themselves, by processes that consume energy and can release potentially harmful emissions.

This chapter attempts to answer two questions:

1. How will fuel, infrastructure and technology prices or other critical parameters need to develop for hydrogen fuel cells to penetrate the market significantly? What factors close out opportunities for a hydrogen economy in the future?
2. What is the greenhouse gas mitigation potential of a significant shift to hydrogen fuel cell vehicles in South Africa?

An emerging world hydrogen economy is of particular significance for South Africa, being the world's dominant supplier of platinum, because many types of fuel cells, the demand-side technology that seems most likely to drive this economy, require platinum as a catalyst. South Africa produced 75% of the world's platinum in 2011 and is estimated to have 85% (Johnson Matthey Plc, 2012) of the world's deposits of platinum group metals with Zimbabwe and Russia being the only other countries producing any significant share of global supply. It therefore makes sense that South Africa would be active in promoting the industrial use of these metals, not only through scientific research but by providing opportunities for the industrial application of platinum-rich devices in the early stages of commercialisation. The fuel cell devices that rely on Proton Exchange Membranes (PEMs) are loaded with platinum, especially at the cathode, the function being to combine oxygen molecules with electrons and hydrogen ions to form water, a sluggish reaction and a bottleneck in the operation of the fuel cell.

Environmental concerns have seen considerable research focus in the last 20 years on fuel cells for use in power production and most prominently in transport, with Mercedes-Benz producing a prototype vehicle, the NECAR-1 as early as 1994. Mercedes Benz brought out a hydrogen-fuelled commercial passenger car, the B-Class F-Cell, in 2010, commercialising it in limited areas of the United States in 2012.

Transport is a large consumer of energy in South Africa and vital for economic development. Recent statistics show that the transport sector consumes 28% of final energy, the bulk of which, 97%, is in the form of liquid fuels (DOE, 2009). As the population grows and becomes wealthier, so the demand for passenger transport and private vehicles increases; similarly, rising gross domestic product (GDP) drives the demand for freight transport. Supply interruptions are costly to the economy and careful long term planning is required to ensure that there is sufficient infrastructure to support the efficient functioning and growth of the transport sector in the future. Evaluating the potential for a hydrogen economy, its possible time of onset, trade-offs with electrification, potential contribution to decarbonisation and costs are essential factors to a long-term planning strategy for the country. This chapter will attempt to lay a modest foundation towards this effort through a data gathering and modelling pre-study focusing on the hydrogen economy with particular attention to fuel cell vehicles.

2 What is hydrogen in physical and energy terms?

Hydrogen is the most abundant element in the universe and is the fuel and dominant substance of stars, including our own sun. Elemental hydrogen does not, however, occur in significant amounts on earth and energy is thus required to extract it from abundant compounds of which it is a constituent, like water or fossil fuels including natural gas.

Hydrogen cannot therefore be said to be a primary energy source like solar energy, coal, crude oil or nuclear fuel but is rather a (so-called) secondary energy source or vector as are refined liquid fuels such as diesel. As such, it is essentially a means of storing and distributing energy. This energy therefore needs to be transformed by some process from which there will be consequences for the energy system including energy losses, costs and possibly emissions. In order to supply energy services, the hydrogen needs to be distributed, which again induces costs and, possibly, losses and emissions.

There are both practical challenges and advantages to working with hydrogen that are a consequence of its properties as summarised on the next page:

Property	Hydrogen	Natural Gas (mostly Methane)	Units
Ideal gas constant R[1]	4.124	0.5182	kJ/(kg.K)
Molar mass[1]	2.016	16.043	kg/kmol
Boiling point[1]	-252.8	-161.5	deg. C
Density at standard conditions	0.084	0.667	km/m3
Density of liquid hydrogen	70.8[2]	450[3]	g/litre
HHV	142[2]	55.3–59.7[3]	MJ/kg
LHV	120[2]	49.8–53.8[3]	MJ/kg
Octane rating	120–140[4]	120[5]	

Table 1: Comparison of hydrogen properties with natural gas
1. (Cengel & Boles, 1998)
2 (Rand & Dell, 2008)
3 (IEA, 2005)
4 (Topinka, 2003)
5 (Heywood, 1988)

Mpa	bar	atm	psi	Energy Content (MJ/litre)
20	200	197	2,901	2.53
55	550	543	7,979	6.96
70	700	691	0 155	8.86
80	800	790	1 606	10.12

Table 2: Energy density of hydrogen at different levels of compression

Notably, while hydrogen has an energy density by mass nearly three times higher than natural gas or gasoline, it is around one-eighth as dense at ambient conditions. Even liquid hydrogen is less than a tenth as dense as gasoline and has a very low boiling point (cryogenic storage is at this temperature) implying high energy inputs to liquid storage and high losses to boil off. Hydrogen therefore presents storage challenges in terms of volumetric energy density and extremes of temperature and pressure compared to the energy carriers that currently predominate. The implication of replacing the current liquid and gas fuels economy with hydrogen is therefore higher storage and distribution costs, and less space efficiency in the face of growing global demand for energy.

On the other hand, hydrogen is an incredibly versatile and clean fuel. It burns readily and can serve as a heating fuel like natural gas and, like natural gas, it has a high-octane rating making it a good fuel for spark ignition combustion engines, with the caveat that range may be limited by storage difficulties. Hydrogen is also the fuel of choice for fuel cells which can be used for stationary power production or, a far more efficient alternative to combustion engines, to power vehicles. In all these cases the only emissions are water, and it is the upstream emissions only that have the potential for environmental impact.

2.1 HYDROGEN PATHWAYS

A great many pathways for hydrogen have been identified as shown below in Figure 1, many of which find limited application in the current industrial production of hydrogen.

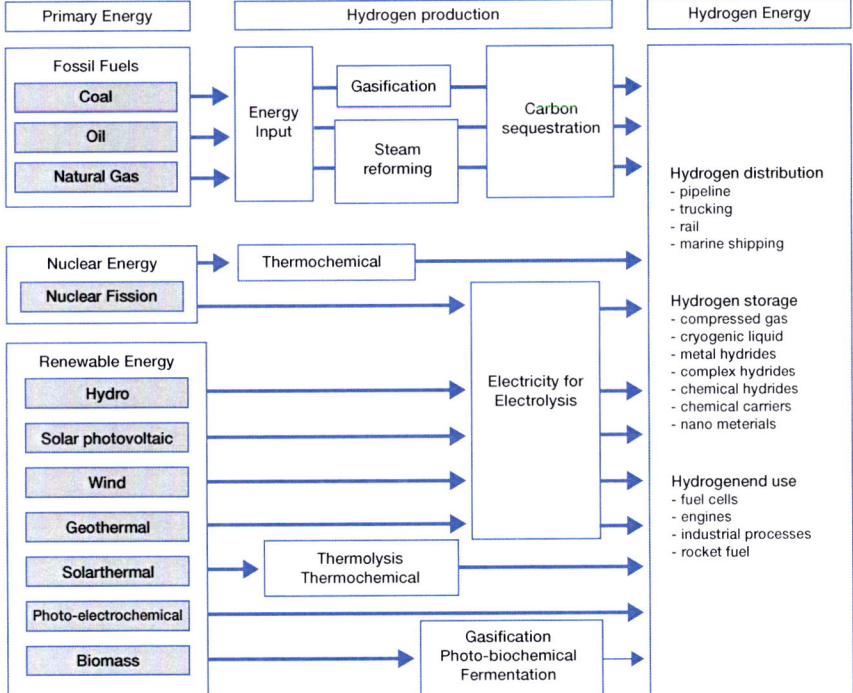

Figure 1: Potential future hydrogen pathways (Gupta, 2009)

Of the above, four processes currently dominate hydrogen production for, amongst other uses, ammonia production and liquid hydrogen fuel for the space industry:

- steam methane reforming (SMR) accounting for about 48% of global demand
- oil gasification accounting for about 30% of global demand
- coal gasification accounting for about 18% of global demand
- water electrolysis accounting for about 4% of global demand

It has been argued that it is important to think conceptually of the two secondary energy sources – hydrogen, and electricity – as complementary, with hydrogen providing a means of storing electricity through water electrolysis at relatively high energy densities and then supplying energy services at zero emissions. Certainly, this is the great vision of the hydrogen economy, but the small share of water electrolysis of world production above, is indicative of the gap between the seductive vision and the raft of challenges to implementation.

3 CHALLENGES TO A HYDROGEN ECONOMY

3.1 PRODUCTION

It is generally recognised that the most economical hydrogen mass production method is steam methane reforming (SMR) from natural gas feedstock. This is probably not the case in coal rich South Africa, however, where indications are that coal gasification would be cheaper. Both processes produce CO_2 emissions, however, and it is envisioned that a hydrogen economy would transition to water electrolysis supplied by renewable power, and hydrogen would supply vehicles with significantly greater range than battery electric vehicles with a zero-emission pathway fuel.

Current processes, as can be seen from the supply-side assumptions section below, lose quite a lot of energy in transforming the feedstock energy carrier to hydrogen energy. Investment and variable costs are also significant, all of which, when combined with the logistical and engineering challenges of transporting hydrogen to where it needs to be, place a lot of pressure on fuel price, initiating and justifying infrastructure rollout. This is why many people envision a natural gas economy as a transition to a hydrogen and renewable electricity economy. It's hard to see hydrogen supply and demand technologies competing with conventional technologies in a low or middling gas price environment.

3.2 DISTRIBUTION

Cost parity of fuel cell power plants at passenger car scale with conventional internal combustion engines is only one prerequisite for the success of fuel cell vehicles (FCV) as a technology. Hydrogen fuel also needs to approach price parity with gasoline, although the greater efficiency of FCVs gives some relief to this constraint. More challenging yet, is the problem of hydrogen storage and distribution, as hydrogen has low energy density by volume both as a pressurised gas and cryogenically stored as a liquid.

When high-carbon steel pipelines are subjected to pressure cycling in pure hydrogen, embrittlement and decarburisation of the steel and welds can occur that can accelerate crack propagation and failure (Rand & Dell, 2008). Solutions include internal coatings, special low carbon steel pipeline and more recently polyethylene sleeves and tubing systems (EIA, 2008). In the hydrogen economy such as it exists now, hydrogen pipelines have been generally kept short because it is less expensive to transport the feedstock such as natural gas through existing pipe networks. Typical costs for welded steel hydrogen pipeline have cost approximately US$1.2 million per transmission mile and US$0.3 million per distribution mile (EIA, 2008). The United States, for example, had 1,200 miles of hydrogen pipeline network by 2006 but this is dwarfed by the 302 000 miles of natural gas transmission piping (EIA, 2008).

This gives rise to the question of what type of production will be most economical, whether large centralised production with its economy-of-scale offset by high distribution costs, or smaller distributed production on the dispensing site, that only requires feedstock distribution through existing and conventional infrastructure. Comprehensive studies of the viability of the hydrogen economy have also typically considered an in-between option of medium-size production, where distribution is at least shortened and some compromise made with economies of scale.

3.3 THE EVOLUTION OF ALTERNATIVE TECHNOLOGIES IN THE VEHICLE PARC[1]

The driver of technological change in the vehicle parc is the market share that new technologies can command. While this is primarily cost and

---

1. The vehicle parc is the population of operating vehicles, usually on-road vehicles, of all types in a specified region.

specification driven for commercial vehicles, private passenger car market share is determined by other factors that can perhaps be encapsulated as the cult of the car whereby wasteful and dangerous over-specification in terms of vehicle mass, power, speed and acceleration can actually be a virtue. These specifications have, however, given way by stealth to some extent, as manufacturers have conceded to fuel economy targets in the form of the CAFE standards in the US and the ACEA targets in the EU, that became legislated fleet targets on the CO_2 emissions of passenger cars in 2009 with further amendments for 2015 (European Commission, 2012). Passenger cars remain, to a large degree substantially over-specified for their practical day-to-day tasks, but have generally become smaller and more efficient.

The battle for market share of new technologies will, however, be fought over range and roll-out of distribution and dispensing infrastructure. The issue of range is discussed in some detail in the discussion on the relative performance of fuel cell vehicles below. Briefly though, while the range of current FCVs operating on a leased basis in the US is about double that of most commercial battery electric vehicles, it is around half that of conventional technology, and less still than hybrid electric vehicles. Range limitations are a consequence of the low energy density of hydrogen by volume even when highly pressurised.

Potential future distribution and dispensing infrastructure includes distribution, storage and dispensing of natural gas or hydrogen and fast charging or battery swop for electric vehicles. The problem of sweeping technological change in the vehicle market is that there are now hundreds of millions of vehicles that operate in distributed fashion across virtually the entire surface of the globe and an immense refuelling network has developed organically to sustain operation. The challenge of ingress into this network with an alternative fuel is difficult enough, but is compounded by the problem of convincing consumers to purchase a new technology that is catered for by limited infrastructure even if the alternative fuel is cheaper. To break this deadlock requires incentives, coercion or plain necessity in the form of depleted fossil fuel reserves. The risks have been comprehensively demonstrated by the slow penetration, even failure, of natural gas vehicles in the US (Sterman, 2008) and so the financing model is of utmost importance, and it seems unlikely that an optimal transition could be led entirely by the energy and motor industries. Who will pay, and when to take the plunge, are critical questions.

4 National research programmes in South Africa and elsewhere

If a hydrogen economy is imminent, even only in the coming decades we would expect to see national research programmes in the world's large economies, because of the scale of infrastructure, planning and state/private sector collaboration across industrial sectors that is required. Indeed, the last decade has seen a number of such initiatives and two that have contributed much information into the public domain are briefly discussed below.

4.1 The NREL/DOE programmes in the US

The Bush era in the US saw a great deal of high-level interest in a hydrogen economy and substantial investment in research co-ordinated by the Federal Department of Energy (DOE) with much of the research led by the National Renewable Energy Laboratories (NREL), a national laboratory of the US Department of Energy, Office of Energy Efficiency & Renewable Energy.

This included an extensive modelling exercise as well as extensive testing and development of prototypes with Original Equipment Manufacturers (OEMs). Over a seven-year period, the Learning Demonstration fleet of fuel cell vehicles from several manufacturers accumulated 3.6 million miles (Wipke, Sprik, Kurtz, Ramsden, Ainscough, & Saur, 2012).

The approach included a lot of benchmarking and target setting of metrics like SMR and water electrolysis efficiency, fuel cell stack cost and fuel cell stack lifetime. The data generated from this empirical- and target-orientated approach is an invaluable resource for decision making.

4.2 The UK SHEC project

This is a very recently completed modelling study that explores the possible transition of the global transport sector not just to hydrogen fuel cells but other emerging technologies such as hybrids, battery electric vehicles and flex-fuel internal combustion engines as well as combinations of these in hybrid form.

This study is very useful because it also centres on an energy-environment-economy-engineering (E4) model from the MARKAL family, like ERC's SATIM model and so the data is already in a useful format. Many costs need to be moderated, though, because of relative cost premiums in the UK.

4.3 HYDROGEN AND FUEL CELL TECHNOLOGIES (HFCT) RESEARCH, DEVELOPMENT, AND INNOVATION (RDI) STRATEGY

The Hydrogen and Fuel Cell Technologies (HFCT) Research, Development, and Innovation (RDI) strategy is an initiative of the South African Department of Science and Technology (DST) launched in 2008 that has the overall vision of employment and wealth creation through the development of high-tech industries that beneficiate South Africa's mineral wealth, specifically in this case platinum as used in fuel cells in a future hydrogen economy. As discussed in earlier chapters, the programme has come to be known as 'Hydrogen South Africa' or HYSA and now includes three centres of competence that are currently undertaking research:

- HySA Systems Integration & Technology Validation Competence Centre based at the University of the Western Cape
- HySA Infrastructure Centre of Competence a partnership of the CSIR and North West University
- HySA Catalysis Centre of Competence a partnership between MINTEK and the University of Cape Town

Research activities include both supply- and demand-side technologies and a number of prototypes and technology demonstrations have been developed including a fuel cell golf cart and portable power systems.

It makes sense that as planning for a hydrogen economy takes shape, the hands-on experience of these centres would inform the critical inputs, like future costs and efficiencies of technologies, in addition to the available international data. The implementation of a hydrogen economy includes technical and engineering challenges at every level of the supply chain and this local expertise will be critical to stakeholders in a balanced decision of how and when to proceed.

This chapter does not reflect on the detail of the work being undertaken by these institutions. It seeks merely to identify some of the critical issues that may rather prove a useful starting point for a stakeholder interaction around future infrastructure planning, by highlighting the major questions and gaps in knowledge.

5 The performance of fuel cell-powered vehicles

Fuel cell-powered vehicles have features broadly common with other technologies. The wheels are driven by an electric drive in the same way as a battery electric vehicle, but the electricity is supplied by the fuel cell. Hydrogen fuel needs to be stored in a tank on board like a gasoline or gas vehicle although, as is implied by the discussion on properties above, there are unique challenges in this regard. Fuel cells remain competitive as a future drivetrain because they offer distinct advantages. The main drawback is cost which is discussed in a separate section below. This section will critically compare FCVs with other technologies.

The main attractions of fuel cells for vehicles are as follows:

- They have much better part-load efficiency than combustion engines. Combustion engines are highly inefficient at low load because frictional losses remain essentially the same as the high load case but output is much lower (Heywood, 1988). Fuel cell-powered vehicles are therefore substantially more efficient than conventional vehicles in urban driving conditions.
- This is also the case with battery electric vehicles but the range penalty for fuel cell-powered vehicles is much lower.
- Fuel cells are potentially more scalable for use in commercial vehicles and buses.
- Charging time for electric vehicles is still in the region of four to eight hours but FCVs can potentially be refuelled in minutes.

Vehicle Model	Engine / Type	Range	Price	Comb. Fuel Con.	Urban Fuel Con.	Highway Fuel Con.	Urban Driving Penalty	Relative to MB C250
		(km)	(US$)	litres/100km[1]	litres/100km[1]	litres/100km[1]		
2012 Ford Focus BEV	BEV/ compact	122	39 200	2.0	1.9	2.2	-13%	-77%
Honda FCX Clarity	H₂ FCV /mid-size sedan	386	leased	3.4	3.4	3.4	0%	-60%
Mercedes Benz B-Class F-CELL	H₂ FCV/ Hatch	306	?	3.9	3.9	3.8	2%	-54%
2012 Honda Civic Hybrid	Gasoline Hybrid EV/mid-size sedan	841	24 050	4.8	4.8	4.8	0%	-44%
2012 SMART ForTwo 1.0 l	Gasoline ICE/ Compact	454	12 490–14 690	5.9	6.2	5.6	11%	-31%
2012 Honda Civic	Gasoline ICE/mid-size sedan	612	15 755–24 055	6.7	7.6	5.4	40%	-21%
2012 Honda Civic Natural Gas	NG ICE/mid-size sedan	354	26 155–27 655	6.9	7.9	5.6	40%	-19%
2012 VW Golf 2.0 Diesel	Diesel ICE/mid-size sedan	714	17 995–29 440	6.9	7.8	5.6	39%	-19%
2012 Mercedes Benz C250 1.8l	Gasoline ICE/mid-size sedan	630	34 800–35 220	8.5	10.1	6.9	47%	0%
2012 BMW X5 xDrive 35d 3.0L	Diesel ICE / SUV	717	56 700	10.7	12.4	9	38%	26%
2012 BMW X5 xDrive 35i 3.0L	Gasoline ICE / SUV	619	47 500–57 700	11.2	13.3	9.2	44%	32%
2012 Cadillac CTS Wagon 6.2 L	Gasoline ICE / SUV	365	39 015–49 750	15.2	17.7	11.9	50%	79%

Table 3: Selected vehicles with fuel economy, range and price data from the US EPA fuel economy database (EPA, 2012)

COMB: Combination

CON: Consumption

H₂ FCV: Hydrogen Fuel Cell Vehicle

NG: Natural Gas

ICE: Internal Combustion Engine

BEV: Battery Electric Vehicle

SUV: Sport Utility Vehicle (usually 4x4)

1: All converted to litres/100 km diesel equivalent as measured on the EPA Federal Test Procedure.

Comparison Vehicle	Fuel Economy	Range
vs Compact car	-42%	-15%
vs Diesel hatch	-51%	-46%
vs Mid-range gasoline sedan	-49%	-37%
vs High-end mid-range gasoline sedan	-60%	-39%
vs Hybrid	-29%	-54%
vs Electric	70%	216%

Table 4: Fuel economy and range comparisons between Honda Clarity FCV prototype and other vehicle types

A number of insights can be gained from this selection of vehicles that are commercially available in the US (see Table 3 and 4). CO_2 emissions have not been listed for the first three technologies as these depend on the mix of technology supplying the electricity grid, but for the purposes of this discussion, CO_2 emissions can be assumed roughly proportional to fuel consumption:

- Large internal combustion engine-(ICE) powered vehicles can have very poor fuel economy relative to smaller vehicles and diesel engine technology has a fairly small impact on mitigating this, relative to the span of fuel economies in the passenger car market.
- Conventional technology has a very high efficiency penalty for stop-start urban driving except for ultra-compact vehicles like the SMART. Fuel Cell vehicles and hybrids have little or no urban driving penalty and battery electric vehicles are actually less efficient for highway driving, presumably because the average power demand is higher causing more heat loss from the motor.
- The price premium on hybrid IC/battery electric vehicles is no longer high.
- FCVs offer a potential jump up in efficiency from hybrids, but are still significantly less energy efficient than battery electric vehicles.
- FCVs have far better range performance than battery electric vehicles,

however, although not as good as conventional vehicles in the same vehicle class.
- A basic electric vehicle has costs still in the range of larger upmarket conventional cars.

A scenario for a low carbon future in passenger cars immediately suggests itself from the above analysis:

> It seems probable that, in the absence of some new disruptive technology and given increasing oil prices and carbon taxes, the range advantage offered by hybrid vehicles will result in ICE/EV and/or FCV/EV plug-in hybrid vehicles coming to dominate the upper and even mid-range price segments of the passenger car market. Compact and ultra-compact conventional vehicles and eventually BEV vehicles may come to dominate the entry level segment of the market and these consumers will then have to live with the range and space limitations.

The following factors may drive variations in such a scenario:
- Given a global environment of increasingly stringent legislation limiting CO_2 emissions from vehicles, compact conventional technology may persist for some time.
- The transition period has the potential to include gasoline, diesel, natural gas, hydrogen and electricity which introduces real practical challenges in energy distribution. Energy distributors and retailers are likely to move to simplify such a situation, and in the case of the large energy corporations, use their capital to shape the outcome.
- In a situation where the electric vehicle battery packs themselves are leased apart from the vehicle and designed for rapid exchange, consumers can charge up by simply swopping batteries at a retailer. The barriers to entry for this type of energy retailer on new sites are quite low because the environmental and safety standards of retail sites will be far less stringent than for liquid and gas energy carriers.
- The technical complexity of hydrogen storage and distribution is likely to retard the penetration of FCVs as part of hybrid vehicles or otherwise, for some time to come.
- The scale of a hydrogen economy means state involvement and a nationally mandated decision to embark on infrastructure investment, the costs of which will ultimately be borne by the consumer. Sustained

high prices may significantly alter the demand for private transport which has hitherto been shaped by cheap and abundant oil.

Current battery electric vehicle technology is completely impractical for road freight vehicles, particularly long-haul applications because of range limitations. Large diesel trucks can, however, be substituted by electric rail in most parts of the world, and increasing diesel prices would naturally drive this. Fuel cell trucks have been mooted, both for primary and auxiliary power supply, but commercially viable models would face the twin challenges of the high energy efficiency of electric rail freight and the range limitations consequent on the difficulties of storing hydrogen. Current work is exploring the potential of metal hydrides and high pressure storage, but it's not clear if the current range of long haul vehicles can be matched. Medium and heavy commercial vehicles with reasonable range and rapid refuelling times (to assist operators to pay back their asset) will in any event be needed to distribute goods from railheads in urban areas and throughout rural districts.

6 THE FUTURE COSTS OF FUEL CELL VEHICLES

6.1 LEARNING RATES

Many energy technologies have shown consistently reduced costs from introduction to the market over a period until they mature as technologies. This is often called 'learning by doing' and is quantified by modellers using the learning rate (LR) which is the percentage drop in cost of a technology for every doubling of cumulative production as defined by Equations 1 and 2 below (Schootsa, Feriolia, Kramer, & van der Zwaan, 2008) (McDowall, 2012).

$$C_t = C_0 \left(\frac{P_t}{P_0} \right)^{-b} \qquad \qquad \text{...Equation 2}$$

Where C_t = Cost at time t

        C_0 = Cost of initial production batch at time 0

        P_t = Cumulative number of units produced at time t

        P_0 = Number of Units in initial batch at time 0

LR = 1 – 2-b   ...Equation 2

Where LR = The Learning Rate

Schwoon lists some fuel cell learning rates from the literature below in Table 5.

Reference	Learning rates used for simulation (in %)	Base fuel cell costs in US$/kW	Base cumulative production
Rogner	10; 20; 30; 40	2,500; 4,500; 10 000	2 MW
Lipman and Sperling	15; 20; 25	1,800; 2,000; 2,200	5 MW
Gritsevskyi and Nakicenovic	20	n.a.	10 MW
Lovins	20–30	100–300 by year 2010	
Schlecht	20; 30; 40	129–516	10 000 units
Sørensen, et al.	10; 20	392 (€/kW)	50 000 units
Tsuchiya and Kobayashi	26	167	50 000 units

Table 5: Learning rates, initial FC costs and initial number of units
Source: (Schwoon, 2008)

6.2 CURRENT AND PROJECTED FUEL CELL VEHICLE COSTS COMPARED TO OTHER NEW TECHNOLOGIES

The US DOE/NREL test programme indicates that given sufficient volumes, fuel cell vehicle technology is approaching competitiveness on a cost per kW basis (US DOE, 2013). As is discussed below in Platinum and Fuel Cell Costs, most of the anticipated cost reductions are for materials and conventional engineering operations like metal forming and cutting and so these projections of the project consultants, while possibly optimistic, probably at best reflect a broad trend. A concern is still durability with only a few of the vehicles in the NREL test programme exceeding 2,000 hours of operation, while 5,000 hours of reliable operation will be required to compete with conventional technology. Even optimistic extrapolations of the NREL work show current reliability to approach 3,800 hours if a 30% stack voltage loss is acceptable.

University College, London have reviewed this and other work in the compilation of future vehicle technology investment costs, including fuel cell vehicles, for the UK SHEC Markal model (McDowall & Dodds, 2012). The resulting table of costs has been converted into costs relative to conventional gasoline ICE technology to make the numbers more meaningful.

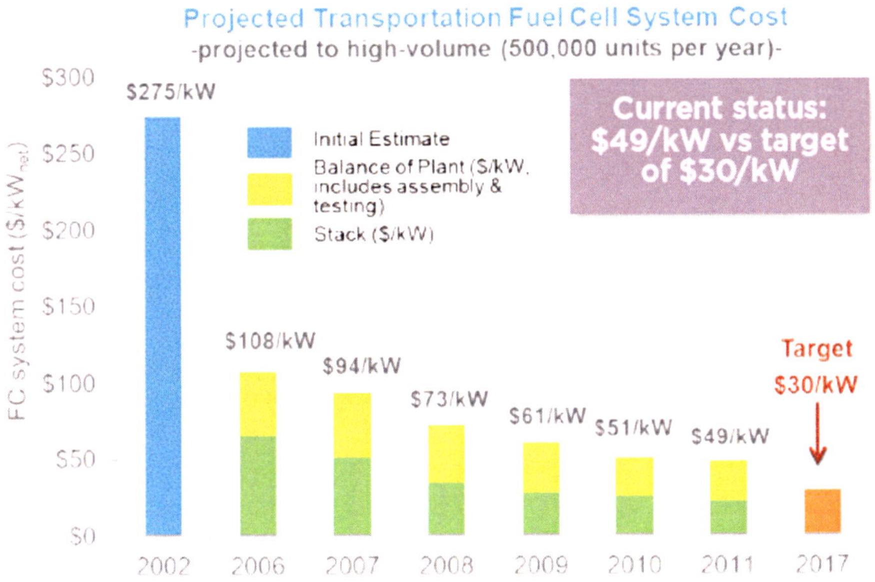

Figure 2: Projected automotive fuel cell costs assuming pass production at 500 000 units per annum (US DOE, 2013)

Year	Diesel ICE NH	Petrol HEV	Diesel HEV	Petrol PHEV	Diesel PHEV	FCV NH	FCHV	FC PHEV	BEV
2000	1.03								
2005	1.03								
2010	1.03	1.33	1.36	2.25	2.27	5.77	5.14	5.22	3.16
2015	1.06	1.10	1.13	1.51	1.54	2.13	2.02	2.20	1.82
2020	1.05	1.06	1.09	1.26	1.29	1.49	1.46	1.54	1.48
2025	1.05	1.05	1.08	1.22	1.25	1.43	1.40	1.46	1.42
2030	1.06	1.04	1.07	1.20	1.24	1.36	1.34	1.39	1.36
2035	1.06	1.03	1.07	1.18	1.22	1.31	1.29	1.34	1.32
2040	1.06	1.03	1.06	1.16	1.20	1.26	1.24	1.29	1.27
2045	1.06	1.02	1.05	1.15	1.18	1.21	1.19	1.24	1.22
2050	1.06	1.01	1.05	1.13	1.16	1.15	1.14	1.18	1.18
% Change	3%	-24%	-23%	-50%	-49%	-80%	-78%	-77%	-63%
% p.a.	0.1%	-0.5%	-0.5%	-1.4%	-1.3%	-3.2%	-3.0%	-2.9%	-2.0%

Table 6: Investment cost assumptions for vehicle technologies in UKSHEC II expressed as a ratio with conventional gasoline internal combustion engine technology (McDowall & Dodds, 2012)   /cont.

ICE: Internal Combustion Engine
NH: Non-Hybrid
HEV: Hybrid Electric Vehicle
PHEV: Plug-in Hybrid Electric Vehicle
FCHV: Fuel-cell Hybrid Vehicle
BEV: Battery Electric Vehicle

The UK SHEC II model therefore assumes that the costs of hybrid electric vehicles will drop to only a 1% premium relative to conventional technology by 2050. Hybrids, by design are, however, two technologies, ICE engines and battery electric drive, combined into one and thus more complex than either so this assumption credits economies of scale with a very large influence.

An analysis of current prices for hybrid vehicles in the US and South African markets was undertaken as a check of one of the emerging technologies listed above. The EPA vehicle database was used to compare prices between 584 models of conventional gasoline technology with 33 models of hybrid vehicles in the Small Car, Family Sedan and Hatchback categories between 2010 and 2012 (EPA, 2012). The results shown below indicate a surprisingly low premium for hybrid vehicles of only around 20% for the US market.

Year	2010	2011	2012
Hybrid price premium (US$/$)	19%	23%	19%

Table 7: US gasoline hybrid vehicle average price premium relative to conventional gasoline ICE technology
Authors' calculations using data from fueleconomy.gov (US EPA)

In contrast, a similar calculation using just Toyota South Africa's website indicates that the Toyota Prius Hybrid retails at just over double the price of the average of five comparable conventional sedans and hatchbacks.

6.3 PLATINUM AND FUEL CELL COSTS

One of the major barriers to fuel cell competitiveness appears to be cost, stemming from the amounts of platinum required to downsize a fuel cell for use in a vehicle. Much of the current research appears to be directed at reducing the amount of platinum used in fuel cells or even engineering platinum-free fuel cells with comparable efficiencies using iron or nickel-based catalysts (Lefèvre, Proietti, & Jaouen, 2009). It is possible, therefore, that a long-term future for fuel cells may not include any platinum group metals or substantially reduced quantities thereof.

For the present, however, particularly for automotive fuel cells, platinum seems to be established as a fundamental component of the fuel stack and sophisticated production techniques involving slurries and 'inks' that combine the platinum with a durable substrate that is practical to apply in mass production, have been developed. The fuel cell manufacturing technical contractor for the US DOE/NREL study published detailed cost and technical data for this process for their multi-manufacturer test fleet (James & Kalinoski, 2009) and, as shown below in Table 8, the platinum-containing catalyst ink starts to approach a third to half the fuel cell cost once production rates reach competitive levels. This is because costs for activities like metal-working operations fall by a factor of 10 or more in the transition from prototype to mass-produced good.

Technology Level Year	Annual Production Rate (Units)					Catalyst Loading Anode	Catalyst Loading Cathode
	1,000	30 000	80 000	130 000	500 000	(mg/cm²)	(mg/cm²)
2008	14%	27%	33%	36%	43%	0.07	0.18
Predicted 2010	14%	29%	35%	38%	45%	0.09	0.21
Predicted 2015	11%	21%	26%	29%	36%	0.04	0.16

Table 8: Estimated catalyst ink & application costs as a percentage of total fuel cell stack cost for different rates of production
Source: Authors' calculations from data in (James & Kalinoski, 2009)

A breakdown of prototype costs therefore obscures the impact of platinum cost on fuel cells as a competitive product. As can be seen above, NREL predict quite modest reductions for catalyst loading on the cathode by 2015, thus this picture may not change much in the medium-term.

7 RESULTS & DISCUSSION

The results of greatest interest are the sensitivity analyses, CO_2 mitigation measures and interaction with the power sector. The main scenarios are, however, first profiled below in terms of technology and fuel market penetration to convey an understanding of the model and the drivers of outcomes.

7.1 COMPARISON OF THE REFERENCE AND ALTERNATIVE (HYDROGEN) SCENARIOS

A first useful comparison is between our two foundation scenarios: the Reference (REF) with no future hydrogen economy and the Alternative (ALT) scenario for which hydrogen is introduced from 2024 and fuel cells achieve a nearly one third market share of passenger car, mini/midi bus and large bus sales by 2050.

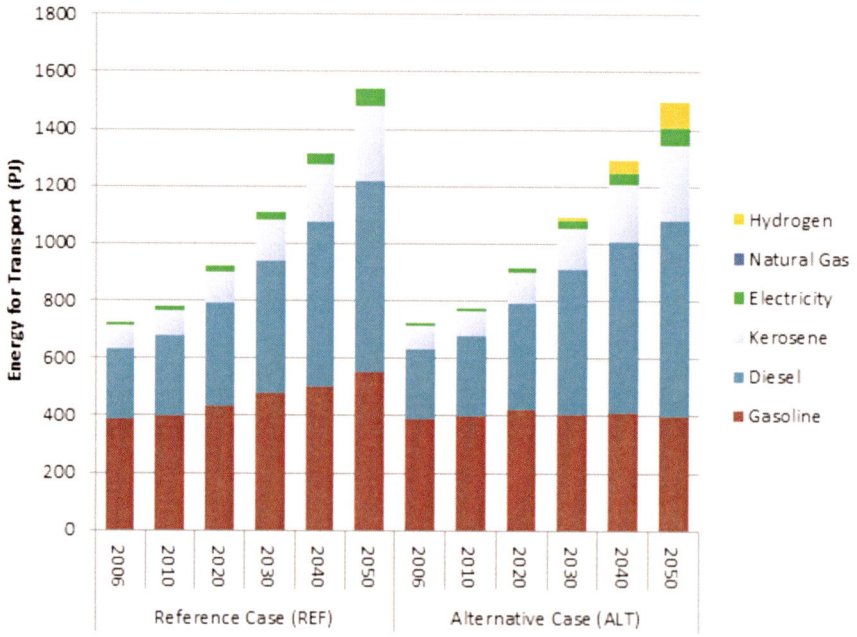

Figure 3: Final energy use in the whole transport sector by fuel for the reference and alternative scenarios

As can be seen from Figure 3 above, the emergence of hydrogen as a passenger fuel has a small but significant effect on total energy use (3% reduction) with hydrogen share of total (6%) in the model by 2050, being diluted somewhat by robust growth projected for freight and air traffic.

Fuel cell vehicles, along with gasoline hybrid electric vehicles, however become a dominant feature of the passenger mode of road transport in the model for the alternative scenario, accounting for nearly 30% of demand for vehicle-km by 2050 as shown below in Figure 4.

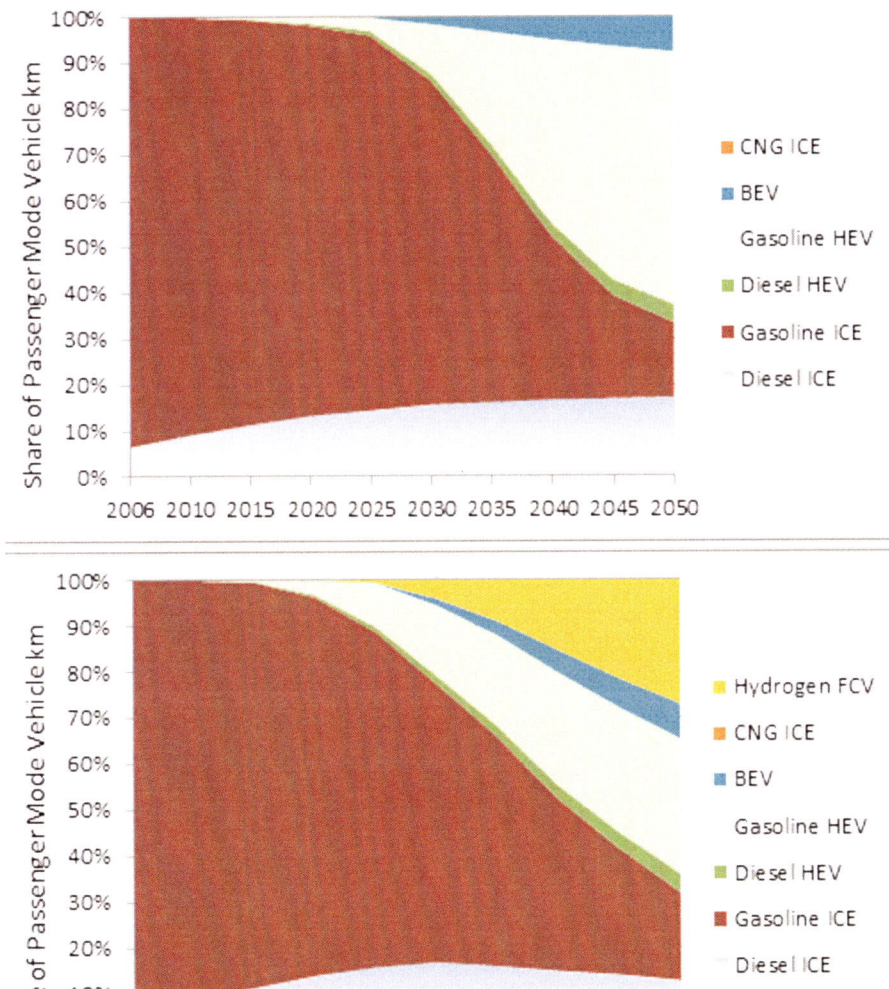

Figure 4: Share of total road vehicle passenger mode demand (vehicle km) by technology for the reference (top) and alternative (bottom) scenarios

The increasing hydrogen FCV share of demand in the model is driven mainly by passenger car market penetration with sales increasing steadily from 2024 to 30% of sales by 2050 in a similar pattern to the above as shown in Figure 17 in Appendix D. The resulting evolution of technology in the vehicle parc and the split between existing technology in the base year and new technology given scrapping assumption in the model is presented in Figure 18 in Appendix D. As shown by Figure 18, the model assumes consistent growth in the private vehicle parc from just over 5.3 million vehicles in 2006 to just over 17.5 million vehicles in 2050, a growth rate of 2.8% per annum.

As shown in Table 9 below, hydrogen FCVs are assumed to penetrate the public passenger modes more quickly because these tend to be captive urban fleets on predictable routes. This and much lower sales numbers compared to passenger cars, makes fuel switching easier because far less alternative fuel infrastructure is required per vehicle and this gains market share more quickly. For minibuses and buses the model also optimised on the upper bounds of the assumed market share envelope, while for cars and SUVs it optimised on the lower bounds. This is because the minibuses and taxis operate at far higher mileages so the energy savings achieved by FCVs contribute more to system cost and because they have longer lifetimes (18 years for buses versus 12 for cars) in the model, which means the high FCV investment costs are discounted over a longer period.

Vehicle Type	2025	2030	2035	2040	2045	2050
Passenger cars	2%	6%	12%	18%	24%	30%
SUV	2%	7%	14%	21%	28%	35%
Minibus taxi	26%	33%	44%	40%	50%	60%
Large bus	37%	50%	57%	65%	73%	80%

Table 9: Hydrogen FCV market share of new sales for passenger modes – alternative scenario

The assumed demand for all scenarios in this study assumed an ongoing increase in private car motorisation in South Africa with no mode switching interventions. This is, in other words, a pessimistic outlook of transport energy demand growth and thus the energy footprint of the public modes diminishes with time. Therefore, with increasing share of the passenger car market, hydrogen FCV passenger cars quickly come to be the dominant consumers of hydrogen in the model, as shown below in Table 10 and thus of primary interest.

Vehicle Type	2025	2030	2035	2040	2045	2050
Passenger cars	34%	48%	62%	70%	74%	78%
SUV	6%	8%	11%	13%	14%	15%
Minibus taxi	47%	36%	23%	15%	11%	7%
Large bus	13%	8%	4%	2%	1%	1%
Total Demand (PJ)	1.5	11.3	25.6	45.9	69.1	92.23

Table 10: Passenger mode share of total hydrogen consumption (PJ/PJ) – alternative scenario

7.2 UNCONSTRAINED SCENARIO – PURELY OPTIMISED TECHNOLOGY PENETRATION WITHOUT EXOGENOUS BOUNDS ON MARKET SHARE

An important aspect of this study, given all the uncertainties, is sensitivity analysis. While the exogenous envelope of technology penetration possibly presents a more realistic picture of the future it does not allow for the model to respond to sensitivity analysis of technology costs, and dispensing and distribution costs. Running the model without constraints on technology share gives us an indication of when a pure least-cost decision to switch technologies would occur. Therefore, an unconstrained scenario with no exogenous bounds on technology penetration was run for use in sensitivity analysis of costs in the energy chain. It also served as a check on how consistent the assumed exogenous bounds on technology penetration were with the cost data and efficiency data that were loaded for each technology in the Alternative Scenario. The share of passenger mode demand in vehicle km by competing technologies for the Unconstrained Scenario is shown below in Figure 5.

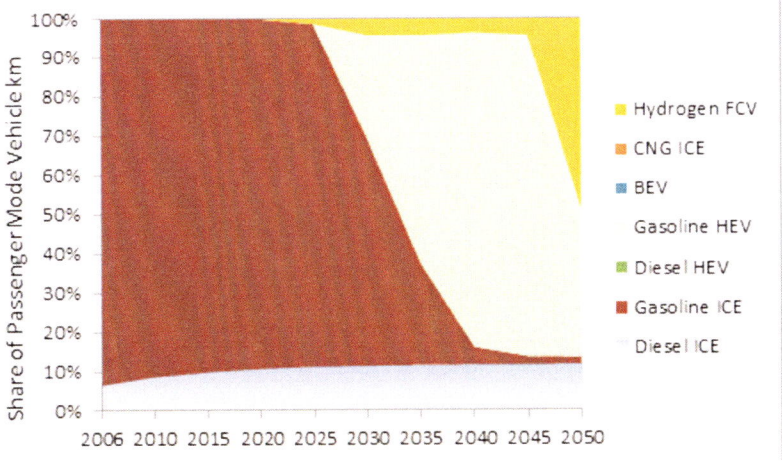

Figure 5: Share of total passenger mode demand for vehicle km – unconstrained scenario – purely optimised technology penetration without exogenous bounds on market share

As discussed above, a least-cost decision will favour buses early on because of the higher mileage and longer life, and this explains the hydrogen FCV dominance of public modes at the outset for the unconstrained scenario as shown below in Table 11. Hydrogen FCV is not selected for SUV because the high price premium of this type of vehicle diminishes the contribution of fuel savings to system cost, while hydrogen passenger cars are selected from 2045 and saturate sales by 2050.

Vehicle Type	2025	2030	2035	2040	2045	2050
Passenger cars	0%	0%	0%	0%	3%	100%
SUV	0%	0%	0%	0%	0%	0%
Minibus taxi	100%	100%	100%	100%	100%	100%
Large bus	100%	100%	100%	100%	100%	100%

Table 11: Hydrogen FCV market share of new sales for passenger modes – unconstrained scenario

The unconstrained model therefore switches to hydrogen more abruptly but meets nearly 50% of passenger demand by the end of the time horizon as compared to 30% for the Alternative Scenario and thus our Alternative Scenario is, in a manner of speaking, an approximate, but more realistic average of the unconstrained scenario. The cumulative hydrogen consumption of the two scenarios is quite similar as shown below, and thus our bounds on market shares seem approximately internally consistent with our cost and efficiency assumptions, adjusting for realistic rates of market penetration.

Scenario	Cumulative Hydrogen Consumption (PJ)
Alternative scenario	1037
Unconstrained scenario	845

Table 12: Cumulative hydrogen consumption of model 2006–2050 for the alternative and unconstrained scenarios

7.3 SENSITIVITY OF THE UNCONSTRAINED SCENARIO TO COSTS IN THE ENERGY CHAIN

Given the costs and efficiency assumptions, the model was selecting significant capacity of hydrogen FCV technology without any imposed market penetration. This 'Unconstrained Scenario' could therefore be subjected to a sensitivity analysis of costs in the energy chain to assess the impacts on the growth in the hydrogen technology and penetration of technology on a cost optimised basis. Six cases were run on the relative costs of distribution and dispensing, the investment costs of FCV vehicles, and the investment cost of hydrogen production technologies such as steam methane reforming and water electrolysis.

Case Name	Investment Costs of H_2 Production Technologies	FCV Costs in 2050[3]	H_2 Dist/Disp Costs[1]	Cumulative Hydrogen Consumption (PJ) 2024 – 2050		Cumulative CO_2 Emissions (Gt) 2024 – 2050[7]
NOCON*		–	–	845.03		13.16
NOCON1		–	Double	367.60	-56%	12.76
NOCON2		–	Triple	281.48	-67%	12.68
NOCON3		+10%[4]	–	326.84	-61%	13.30
NOCON4		+20%[5]	–	43.73	-95%	13.32
NOCON5		+40%[6]	–	43.73	-95%	13.32
NOCON6	+30%[2]	–	–	798.75	-5%	13.13

Table 13: Impact of sensitivity analysis on hydrogen energy chain costs on cumulative hydrogen consumption by the model for the unconstrained scenario
* This is the Unconstrained Scenario discussed above
1: This is the cost for H_2 relative to other motor fuels like gasoline, diesel and natural gas
2: All production technology costs were escalated by this amount
3: Baseline premium in 2050 relative to gasoline ICE is 15%
4: Premium in 2050 relative to gasoline ICE is 27%
5: Premium in 2050 relative to gasoline ICE is 38%
6: Premium in 2050 relative to gasoline ICE is 61%
7: For the Power, Upstream and Transport Sectors Combined

The contribution of the hydrogen chain to the system cost was most sensitive to hydrogen FCV costs with 10% and 20% increases on the baseline assumption for 2050 reducing the energy footprint of the hydrogen economy by 61% and 95% respectively, as determined on a least cost basis. A further increase of 40% had no further impact, this being as we shall see below, the large bus share which is much less investment cost sensitive because of high mileage and lifetime. Relative distribution and dispensing costs for hydrogen also had a large impact. In the unconstrained scenario all motor fuels are modelled with the same variable cost reflecting distribution, and dispensing costs. A more realistic doubling of relative costs for hydrogen distribution and dispensing reduced the energy footprint of the hydrogen economy by more than half as determined on a least-cost basis. A tripling of costs had a significant but smaller effect. The system was relatively insensitive to uncertainties around the investment costs of hydrogen production technology.

System CO_2 emissions decreased with the initial reductions in hydrogen demand because without constraints, the cheapest production technology was coal gasification which has very high emissions. For very large reductions in hydrogen demand, however, CO_2 increases again because ICE emissions start to offset the reduced coal gasification emissions. It is apparent that for the hydrogen economy to have an impact on CO_2 emissions, a less polluting production technology is required, but that has relative cost implications as explored further below. The impact of the costs sensitivity analysis on the market share of hydrogen demand technologies in the model are shown below. Passenger car market share was sensitive to all the costs in the chain, except for investment in production. SUV is not selected on a least cost basis for any case. Minibus taxi market share falls off sharply for increases in vehicle investment costs of 20% and above, while hydrogen FCV buses remained the least cost technology even at a 40% relative price increase on the 2050 baseline cost.

Case Name	FCV Passenger Cars			FCV SUV		
	2040	2045	2050	2040	2045	2050
NOCON	0%	3%	100%	0%	0%	0%
NOCON1	0%	0%	4%	0%	0%	0%
NOCON2	0%	0%	3%	0%	0%	0%
NOCON3	0%	0%	0%	0%	0%	0%
NOCON4	0%	0%	0%	0%	0%	0%
NOCON5	0%	0%	0%	0%	0%	0%
NOCON6	0%	2%	100%	0%	0%	0%

Table 14: New sales share of FCV cars and SUVs – energy chain cost sensitivity analysis cases

Case Name	2025	2030	2035	2040	2045	2050
NOCON	100%	100%	100%	100%	100%	100%
NOCON1	20%	100%	100%	100%	100%	100%
NOCON2	0%	100%	100%	100%	100%	100%
NOCON3	0%	100%	100%	100%	100%	100%
NOCON4	0%	0%	0%	0%	0%	0%
NOCON5	0%	0%	0%	0%	0%	0%
NOCON6	14%	100%	100%	100%	100%	100%

Table 15: New sales share of minibus taxis – energy chain cost sensitivity analysis cases

Case Name	2025	2030	2035	2040	2045	2050
NOCON	100%	100%	100%	100%	100%	100%
NOCON1	100%	100%	100%	100%	100%	100%
NOCON2	100%	100%	100%	100%	100%	100%
NOCON3	100%	100%	100%	100%	100%	100%
NOCON4	100%	100%	100%	100%	100%	100%
NOCON5	100%	100%	100%	100%	100%	100%
NOCON6	100%	100%	100%	100%	100%	100%

Table 16: New sales share of large bus – energy chain cost sensitivity analysis cases

7.4 HYDROGEN PRODUCTION TECHNOLOGIES – COST VERSUS CARBON INTENSITY

As discussed in the section on Modelling methodolgy (Appendix A), a levelised cost analysis of the three hydrogen production technologies was performed outside the model. This was used to estimate at which fuel prices and CO_2 tax levels the model would switch between the cheapest but most polluting technology, coal gasification and the less polluting steam methane reforming (SMR) or the non-polluting (at source) water electrolysis. These estimates were a means of validating the decision of the model and finding critical points for sensitivity analysis. As shown in Modelling methodology, the assumed coal price in the model is substantially lower than the gas price

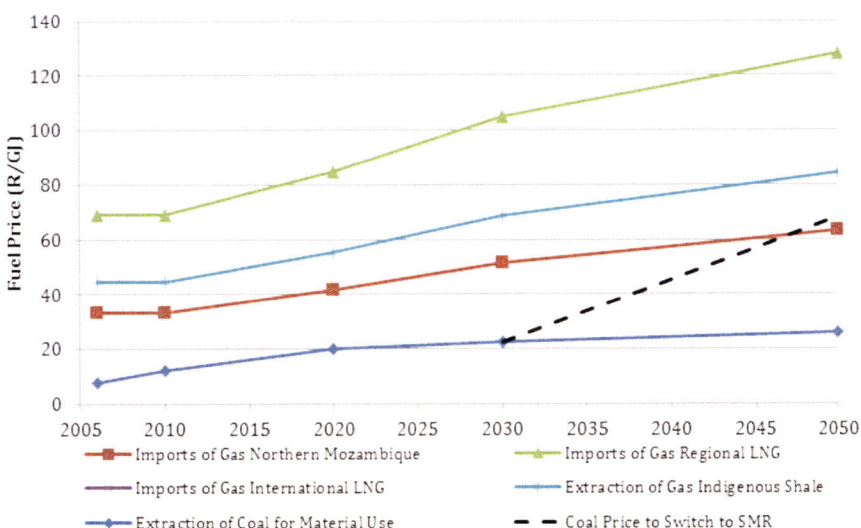

Figure 6: Model baseline gas and coal price assumptions showing coal price trajectory required for a switch from coal gasification to steam methane reforming for hydrogen production

with the gap growing over the time horizon. A sensitivity analysis was performed on the alternative scenario to see how much higher the coal price would need to be to cause a switch from coal gasification to SMR production. It was found that the 2050 estimate of coal price would have to escalate more than 2.5 times to R68/GJ, as shown in Figure 6 below, for SMR to be the least cost option by 2050 and no earlier. An earlier switch would require an even higher coal price.

Even if hydrogen were to be produced by zero emitting water electrolysis as can be seen below, electricity demand by electrolysis is substantially greater than renewable power production by 2040 and four times greater by 2050. Supplying the water electrolysis capacity with dirty power defeats the objective of a hydrogen economy. Clearly the model will select more renewable electricity technology capacity if it is cheaper and we can see below that the assumption of optimistic renewable costs gives rise to a cost optimised grid that more than covers hydrogen production demand by 2050 despite falling short in 2040. An optimistic cost evolution for renewable electricity technology is thus a probable prerequisite of an effective hydrogen transport economy.

Production & Demand	2025	2030	2035	2040	2050
Hydrogen demand (PJ)	1.5	11.3	25.6	45.9	92.2
Electricity demand of H_2O electrolysis (PJ)	2.3	17.4	39.4	70.7	141.9
RE Elec. production – conservative REIPP costs (PJ) – (alternative scenario)	48.5	55.2	47.4	37.3	36.9
RE elec. production – optimistic REIPP costs (PJ)	48.5	55.2	47.4	54.9	400.0

Table 17: Balancing electricity demand of water electrolysis production of hydrogen with renewable power production for conservative and optimistic cost cases

Three cases of the alternative scenario were then run with a 30 PJ cap first on coal gasification, and then SMR capacity; thereafter forcing electrolysis production to assess the potential costs and benefits as shown below.

Case	REIPP Costs	Hydrogen Production	Cumulative CO$_2$ Emissions (2024 – 2050) (Gt)[2]		CO$_2$ Emissions 2050 (Mt)		H$_2$ Prod. 2050 (PJ)		System Cost (bill. 2010 R)
ALT	Conservative	COG	12.18		523.1		92	3878	
ALT 1	Optimistic	COG	11.65	-4.4%	426.0	-19%	92	3874	-0.10%
ALT 2	Optimistic	COG capped at 30 PJ from 2035. Switching to ELE. SMR capped at 0 PJ [1]	11.72	-3.8%	433.2	-17%	91	3879	0.02%
ALT 3	Optimistic	SMR capped at 30 PJ from 2035. Switching to ELE. COG capped at 0 PJ [1]	11.63	-4.6%	428.4	-18%	91	3880	0.05%

Table 18: CO$_2$ and system cost impacts of forced switching to electrolysis H$_2$ production given higher renewable power share due to optimistic cost outcomes

1 Electrolysis comes online in 2040 and reaches 70% share of production by 2050
2 All sectors in the model
COG: Coal gasification
ELE: Water Electrolysis
SMR: Steam Methane Reforming

What is immediately evident is the small impacts of these interventions on cumulative CO$_2$ emissions and system costs over the time horizon. Annual CO$_2$ emissions are, however, substantially reduced by 2050 due to the large increase in the share of renewable electricity technologies which is 26% for cases ALT1 – ALT 3 compared to 3% for ALT. This means that the ongoing impacts of renewable power will be large. Somewhat surprisingly, our interventions to make hydrogen production less polluting in ALT2 and ALT3 have actually slightly increased emissions, relative to ALT1, although we can see a modest benefit of SMR over Coal Gasification in the 5 Mt (million tons) reduction from ALT2 to ALT3 for 2050. The reason for this becomes evident when we break down these emissions by source as shown below.

Source	ALT1	ALT2	ALT3
Power sector	270.9	293.1	292.9
Refineries (ex. process)	8.6	5.5	4.1
CTL refineries	16.7	16.3	15.7
Crude refineries	15.9	15.9	16.0
Hydrogen production SMR	0.0	0.0	2.4
Hydrogen production electrolysis	0.0	0.0	0.0
Hydrogen production coal gas	16.7	4.9	0.0
Transport	97.3	97.4	97.4
Total	**426.0**	**433.2**	**428.4**

Table 19: Breakdown of CO_2 emissions in 2050 (Mt) by source for forced electrolysis transition cases

As can be seen above, while annual hydrogen production CO_2 emissions have been reduced by 12–14 Mt by 2050 through a forced transition to electrolysis, power sector emissions have gone up more than 20 Mt because, although renewable share has increased, the grid as reflected in our model is still carbon intensive and the increased electricity use has pushed up emissions.

7.5 REDUCING THE CARBON INTENSITY OF HYDROGEN PRODUCTION BY POLICY INSTRUMENTS

One rationale behind a hydrogen economy is that, in theory, a zero emissions pathway can be achieved, for example, if water electrolysis is supplied by renewable power. In our model, coal gasification is significantly cheaper than the less polluting steam methane reforming (SMR) and both are significantly cheaper than the zero emission (at site) water electrolysis which is not part of an optimal solution without the imposition of constraints. Two such constraints are examined below in the form of a carbon tax or a carbon cap.

7.5.1 CARBON TAX

As discussed above in the sections on fuel prices, cheap local coal means incentives or changes to the system need to be very large to make SMR and especially water electrolysis (ELE), cost optimal. This, too, is the case with a carbon tax and just the transition to SMR from coal gasification (COG)

requires an unrealistically high tax as shown below. The different rates of fuel price increase tend to create some swapping between COG and SMR at transition levels of tax, but a transition to SMR from COG starts at around R400/ton CO_2 and is complete at R600/ton CO_2. Similarly SMR transitions to ELE by 2050 at R2,000/ton CO_2 and at 2035 at R2,300/ton CO_2. Carbon taxes were applied from 2015 on in the model.

Case*	Tech.	2025	2030	2035	2040	2045	2050	Tax (R/ton)
ALT	ELE	0%	0%	0%	0%	0%	0%	
ALT	COG	100%	100%	100%	100%	100%	100%	R 0
ALT	SMR	0%	0%	0%	0%	0%	0%	
CO_2A	ELE	0%	0%	0%	0%	0%	99%	
CO_2A	COG	0%	0%	0%	0%	0%	0%	R 2,000
CO_2A	SMR	100%	100%	100%	100%	100%	1%	
CO_2B	ELE	0%	0%	0%	0%	0%	0%	
CO_2B	COG	0%	0%	0%	0%	0%	0%	R 600
CO_2B	SMR	100%	100%	100%	100%	100%	100%	
CO_2C	ELE	0%	0%	0%	0%	0%	0%	
CO_2C	COG	74%	42%	77%	88%	92%	100%	R 400
CO_2C	SMR	26%	58%	23%	12%	8%	0%	
CO_2D	ELE	0%	0%	0%	0%	0%	0%	
CO_2D	COG	0%	0%	0%	0%	0%	26%	R 500
CO_2D	SMR	100%	100%	100%	100%	100%	74%	
CO_2E	ELE	0%	0%	36%	99%	66%	100%	
CO_2E	COG	0%	0%	0%	0%	0%	0%	R 2 300
CO_2E	SMR	100%	100%	64%	1%	34%	0%	

Table 20: CO_2 Tax cases of the alternative scenario showing effect of tax on hydrogen production technology market share

COG: coal gasification
ELE: water electrolysis
SMR: steam methane reforming
* All cases except ALT were run with high nuclear cost assumptions

The carbon tax in the model was assumed to be global across sectors and so it acts on the upstream and power sectors driving very high renewable and nuclear shares, even with higher nuclear costs. As shown below these changes to the power sector reduce CO_2 emissions massively to annual levels almost five-fold lower than the reference scenario, but at very high system cost.

Case	2010	2015	2020	2025	2030	2035	2040	2050
ALT	2%	8%	7%	8%	8%	6%	3%	3%
CO_2A	2%	17%	17%	23%	35%	35%	36%	36%
CO_2B	2%	11%	20%	31%	44%	48%	48%	41%
CO_2C	2%	9%	8%	22%	42%	48%	49%	54%
CO_2D	2%	9%	17%	25%	42%	46%	43%	35%
CO_2E	2%	18%	18%	24%	36%	35%	36%	36%

Table 21: Share of renewable electricity technology capacity for CO_2 tax cases

Case	2010	2015	2020	2025	2030	2035	2040	2050
ALT	4%	3%	3%	3%	3%	0%	0%	0%
CO_2A	4%	3%	2%	2%	4%	14%	19%	38%
CO_2B	4%	3%	2%	10%	9%	12%	23%	32%
CO_2C	4%	3%	3%	2%	2%	3%	14%	20%
CO_2D	4%	3%	2%	2%	2%	6%	20%	37%
CO_2E	4%	3%	2%	2%	4%	14%	20%	38%

Table 22: Share of nuclear capacity for CO_2 tax cases

Case	System Cost (billion 2010 Rands)	System Investment Cost (billion 2010 Rands)	System Cost Difference with REF	Investment Cost Difference with REF	Cumulative Consumption H_2 (PJ)	Cumulative Reduction in CO_2 (2024–2050)[1]	Cost CO_2 Mitigation (R/ton)	Annual CO_2 Emissions 2050 (Mt)
REF	3866	1749	0	0	0	0	0	610
ALT	3878	1773	0.3%	1.4%	1037	-9%	R 10	523
CO_2A	6171	2339	59.6%	33.7%	1040	-71%	R 243	131
CO_2B	5140	2264	33.0%	29.5%	988	-68%	R 139	142
CO_2C	4768	1979	23.3%	13.2%	988	-50%	R 133	159
CO_2D	4980	2039	28.8%	16.6%	988	-53%	R 156	146
CO_2E	6378	2383	65.0%	36.3%	1042	-71%	R 262	130

Table 23: CO_2 and system cost impacts for CO_2 tax cases
1 Relative to Reference Scenario

One of the causes of the high relative system costs other than the fact that the tax is included, is that because wind and solar renewable technologies are assumed to be intermittent, a large capacity of gas peaking plant is required to maintain a system reserve margin such that total installed capacity is double the reference case for the most extreme carbon tax example. Currently, SATIM is quite conservative in this regard, and correctly so, given current knowledge and best practice, allocating low capacity credits to wind and solar technologies. Projects are underway to review these assumptions, and the model may be more forgiving in future.

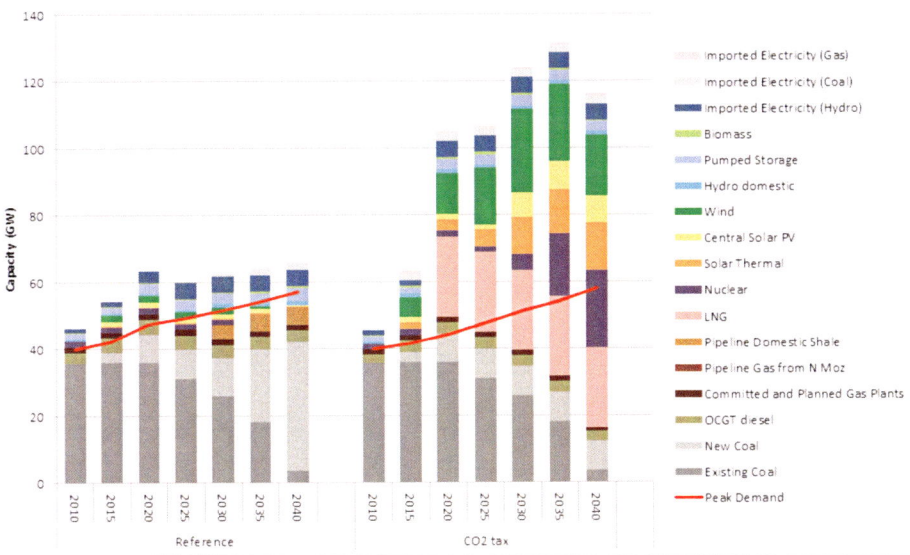

Figure 7: Installed electricity production capacity for case CO_2E compared to the reference scenario

7.5.2 CARBON CAP

From a modelling point of view, a carbon cap is a more flexible instrument as the outcome can be set and the model has more freedom to optimise a solution to achieve that. In this study, the cap was not used to try and decarbonise the hypothetical hydrogen energy chain specifically, but rather used to evaluate how an optimal solution for the cap might interact with the hydrogen chain. The results were far more realistic in cost and rate-of-change terms than the carbon tax cases. The imposed caps are compared to the reference case emissions below in Figure 8.

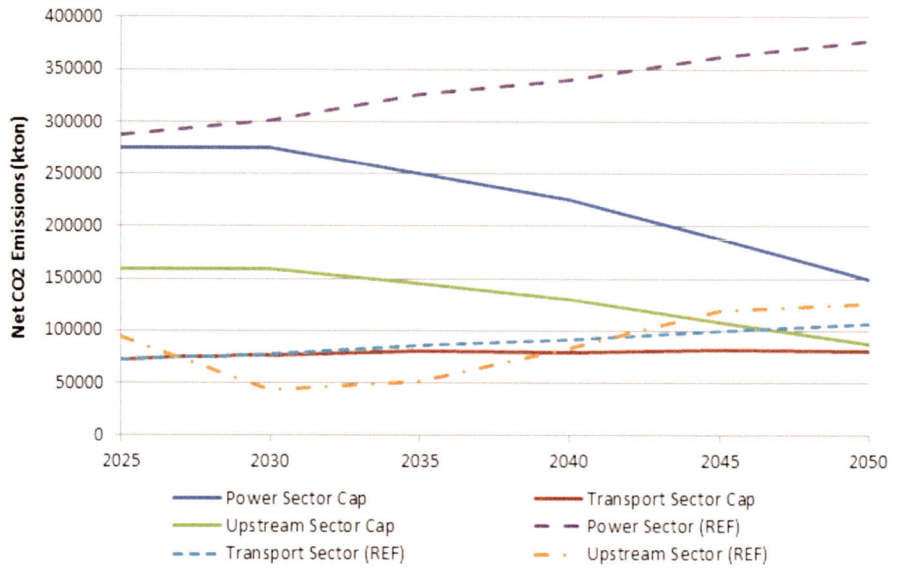

Figure 8: Sector carbon caps compared to reference (REF) scenario emissions

The cap is most profound for the power sector, the most carbon-intensive sector of the economy. The upstream sector cap is a scaled version of the power sector cap, but because CTL refinery capacity is being selected on and off by the optimiser causing a swing in CO_2 emissions for the reference case, it appears mismatched. The transport cap appears modest, but in fact it was derived iteratively, and is at the limit of what the model can solve.

The carbon cap had to be applied to the Unconstrained Scenario without bounds on the market penetration of FCVs or the model would not solve for a constant number of vehicles. Given market share constraints and a carbon cap, the model effectively runs a scrapping programme by lowering the activity levels on the existing fleet to negligible levels and investing in large amounts of new low carbon capacity. The scale of this was completely

Case	Description	FCV Investment Costs	H₂ Dist/Disp Costs
NOCON	Base Model with no bounds on vehicle sales		
CAP1	NOCON + carbon cap		
CAP2	Sensitivity on carbon cap	+20%	
CAP3	Sensitivity on carbon cap		triple other fuels

Table 24: Carbon cap cases

unrealistic, and thus the Unconstrained Scenario was used. Two sensitivity cases were applied as follows:

The CO_2 emission and cost outcomes for these cases are presented in Table 25 below.

The carbon cap case (CAP1) found a cost-effective solution that met the cap, reducing annual CO_2 emissions in 2050 to almost half the Reference

Case	System Cost (billion 2010 Rands)	System Invest- ment Cost (billion 2010 Rands)	System Cost Difference with REF	Invest- ment Cost Differ- ence with REF	Cumu- lative Consump- tion H$_2$ (PJ)	Cumu- lative Reduction in CO_2 (2024– 2050)[1]	Cost CO_2 Miti- gation (R/ton)	Annual CO_2 Emissions 2050 (Mt)
REF	3866	1749	0	0	0	0	0	610
NOCON 1	4046	1720	4.7%	-1.6%	845	-5%	R 266	570
CAP1	3830	1797	-0.9%	2.8%	2726	-27%	-R 10	324
CAP2	3859	1830	-0.2%	4.6%	524	-31%	-R 1	268
CAP3	4089	1798	5.8%	2.8%	2424	-28%	R 60	321

Table 25: CO_2 and system cost impacts for CO_2 cap cases
1 This case is effectively the baseline for the sensitivity cases.

Case, making extensive use of hydrogen technology. This was, however, very sensitive to FCV investment cost with a 20% price increase (CAP2) reducing hydrogen uptake in the system to well below the baseline. The uptake of hydrogen in meeting the cap was, however, much less sensitive to distribution and dispensing costs of hydrogen at levels triple that of competing fuels (CAP3) with a modest reduction in hydrogen uptake but a significant increase in system cost. For CAP1 and CAP3 all the extra hydrogen was produced by Coal Gasification with the system still remaining under the cap.

As shown in Figure 9 (next page) and Figure 10, in the case of CAP1, hydrogen FCVs meet 90% of passenger vehicle km demand by 2050 with share rising rapidly from 2035. For 20% higher FCV costs, however (CAP2), FCVs remain a public transport phenomenon only, and battery electric vehicles become dominant.

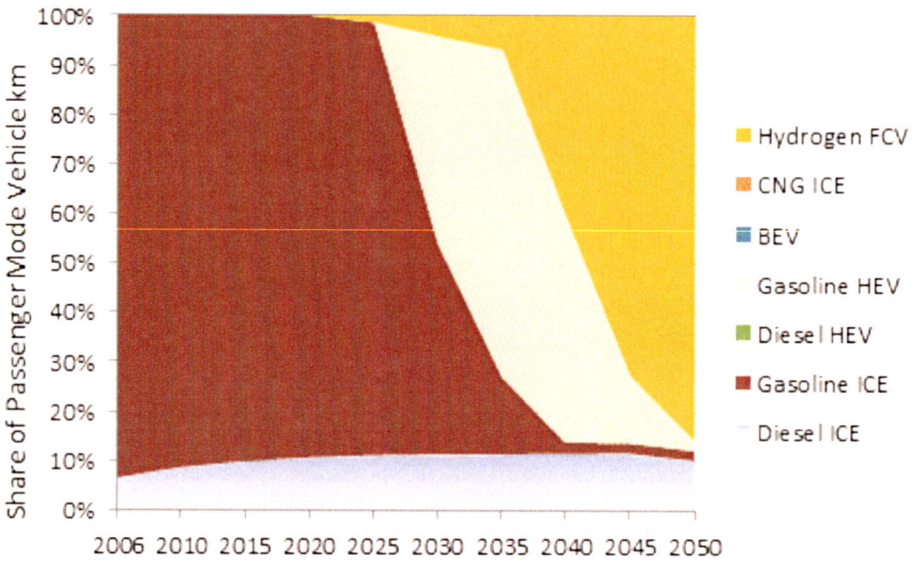

Figure 9: Share of total road vehicle passenger mode demand (vehicle km) by technology for CO_2 cap case (CAP1)

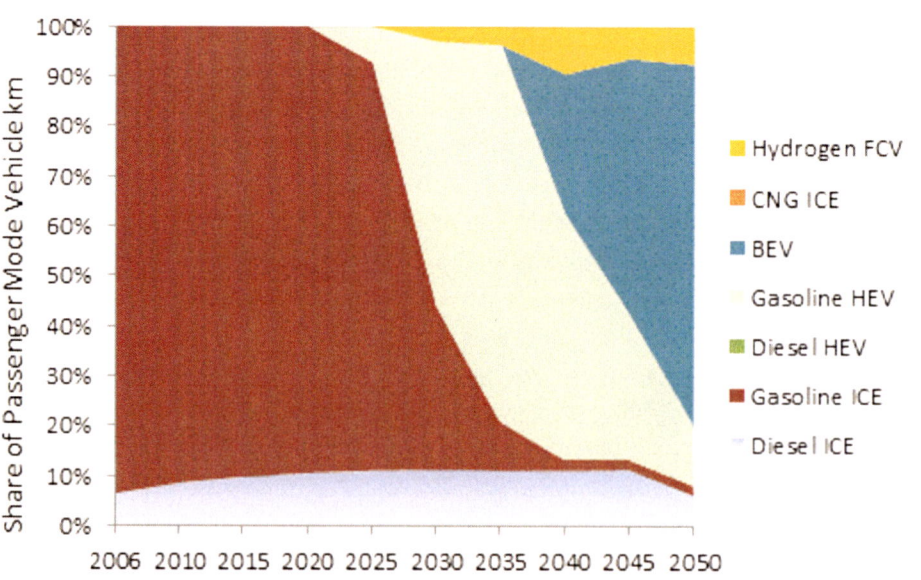

Figure 10: Share of total road vehicle passenger mode demand (vehicle km) by technology for CO_2 cap case with 20% higher FCV costs (CAP2)

The high share of demand for hydrogen FCV in CAP1 is driven by a dominant share of passenger car sales in the model from 2040. In CAP1 hydrogen FCV SUVs gains a 30% new market share from 2050 while 100% for public passenger vehicles are hydrogen FCV from 2025.

Case	2025	2030	2035	2040	2045	2050
NOCON	0%	0%	0%	0%	3%	100%
CAP1	0%	0%	6%	76%	100%	100%
CAP2	0%	0%	0%	12%	0%	0%
CAP3	0%	0%	0%	79%	71%	100%

Table 26: Hydrogen FCV market share of new passenger cars for carbon cap cases

The cap in the power sector was met with significant increases in the share of renewable and nuclear power, but more modest than that driven by a carbon tax. All the carbon cap cases assumed conservative renewable costs and high nuclear costs relative to IRP2010.

Case	2010	2015	2020	2025	2030	2035	2040	2050
REF	2%	8%	7%	8%	8%	6%	3%	3%
NOCON	2%	8%	7%	7%	8%	6%	3%	3%
CAP1	2%	8%	7%	8%	15%	21%	22%	25%
CAP2	2%	8%	7%	8%	15%	21%	22%	24%
CAP3	2%	8%	7%	8%	15%	21%	22%	25%

Table 27: Share of renewable electricity technology capacity for CO_2 cap cases

Case	2010	2015	2020	2025	2030	2035	2040	2050
REF	4%	3%	3%	3%	3%	0%	0%	0%
NOCON	4%	3%	3%	3%	3%	0%	0%	0%
CAP1	4%	3%	3%	3%	3%	3%	8%	22%
CAP2	4%	3%	3%	3%	3%	3%	9%	25%
CAP3	4%	3%	3%	3%	3%	3%	8%	22%

Table 28: Share of nuclear capacity for CO_2 cap cases

The disparity in installed capacity between the carbon cap case and the reference scenario is also far smaller as shown below.

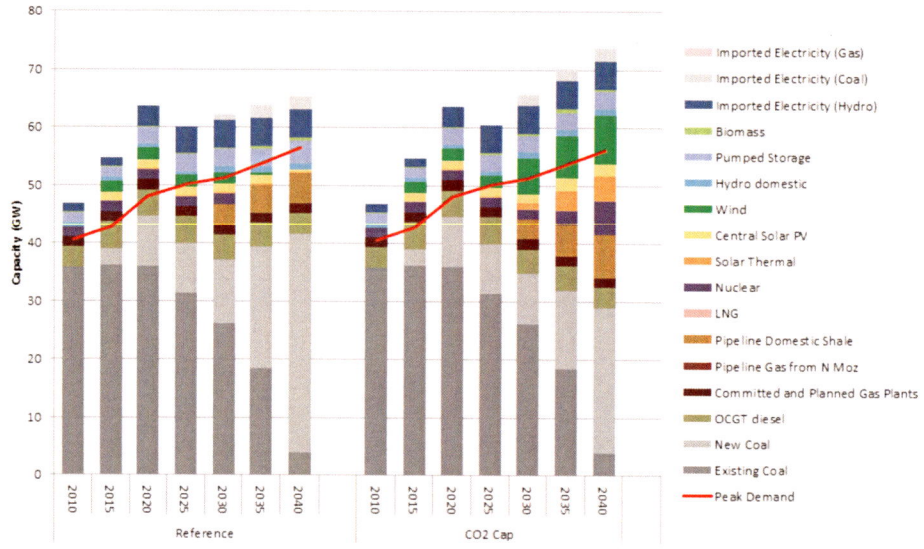

Figure 11: Installed electricity production capacity for case CAP1 compared to the reference scenario

8 FINDINGS AND CONCLUSIONS

Some key findings of the modelling exercise with some conclusions drawn by the authors are presented below:

- The outcome of the model was quite insensitive to uncertainties around hydrogen production technology investment cost with an increase in 30% having little impact.
- A levelised cost analysis validated the model outcome that coal gasification would be substantially cheaper than steam methane reforming (SMR) and water electrolysis given South Africa's cheap coal. This is in spite of the availability of natural gas in our model from Northern Mozambique, Ibhubezi and Shale source from 2020 at much lower prices than imported LNG.
- Steam methane reforming was, however, far cheaper in our model than water electrolysis, which of course offers the advantage of zero emissions at site. The emissions of the latter depend of course on the carbon intensity of the South African power technology mix.

- Very large shifts in the model's assumed feedstock prices for coal and natural gas or unrealistic carbon taxes (R600–R2,300 per ton) would be required to make SMR or water electrolysis cheaper than coal gasification.
- It appears, therefore, that even given quite large uncertainties in our assumptions, that coal gasification COG will likely be the cheapest option for hydrogen production in South Africa for some time to come.
- The modelling exercise showed little benefit to system emissions and cost in trying to force transition to SMR or water electrolysis. Increased electricity demand from South Africa's carbon-intensive grid tended to increase total system emissions, offering no benefits for increased costs.
- Given the imposition of sector-specific carbon caps on the supply sectors and transport sector, one of the pathways the model took was to more than double hydrogen uptake by passenger vehicles which was supplied entirely by hydrogen produced by the coal gasification technology.
- Current water electrolysis technology is significantly more expensive, particularly in the light of likely electricity prices, driven by South Africa's extensive power plant build programme. One of the ultimate goals of a hydrogen economy, however, is to provide a zero emissions energy pathway with renewable power supplying water electrolysis hydrogen production. While this may be beyond 2050 in South Africa, it should remain an ultimate goal.
- Only large central hydrogen production was investigated in this model. Further work should model the dispensing and distribution system in more detail so that the advantage of cost reduction on distribution and storage afforded by small distributed electrolysis could be investigated. This may offset high capital and variable costs and low efficiencies for this technology.
- There is a large coal to liquids (CTL) and gas to liquids (GTL) industry in South Africa. Further work should investigate more realistic costs for South Africa given this large base in petrochemicals and especially coal gasification. It is possible that the quite modest hydrogen demands in the first 10 to 20 years of a growing hydrogen economy could be supplied from existing retrofitted plants at lower costs than our assumptions.

A sensitivity analysis of dispensing & distribution costs was undertaken by modelling one technology for both dispensing and distribution with a single cost per unit energy supplied. The imposition of dispensing and distribution costs for H₂ supply that were two and three times higher than the current liquid fuels infrastructure, cut H₂ uptake by 56% and 67% respectively unless a carbon cap was imposed.

- The cost optimal selection of hydrogen FCV passenger car vehicles in the model was very sensitive to investment costs, especially in the light of uncertainty in this regard. Increases of 20% and more over baseline cost assumptions tended to eliminate the selection of hydrogen FCV passenger cars. This equates to a price premium relative to conventional gasoline ICE of 38% or greater.
- Price premium is not generally an impediment in this market if a high status product is being sold, but hydrogen FCVs are competing in the space with hybrid vehicles that offer a much higher range, higher in fact than conventional technology, as well as good fuel economy. While FCVs have substantially higher range than Battery Electric Vehicles (BEV), their current range is well below that of conventional technology.
- Penetration of FCV into the minibus and large bus fleet in the model was less sensitive to system changes because these vehicles have much higher mileage, increasing the system cost impact of fuel savings, and longer operating life which discounts the investment cost premium over a longer period. This speaks to the suitability of these vehicles for launching a hydrogen transport economy. They have the added advantage of regular routes and central fuelling, which mitigates the very great challenge of rolling out hydrogen distribution and dispensing infrastructure in a mature and highly distributed transport economy. There are a number of fuel cell buses already running, around the world.
- A caveat to the above is that a public bus may undergo many engine rebuilds, and even engine replacements during its life. In SATIM this is reflected in the operation and maintenance costs assigned to a technology, but while these are quite accurate for plant refits for, say, power stations they have not been thoroughly researched for buses. A final cost benefit decision of fuel cell versus diesel traction for buses would therefore require a more detailed model taking maintenance into account, and the results of this study should be considered indicative.

Bibliography

Anandarajah, G., McDowall, W., & Ekins, P. (2013). Decarbonising road transport with hydrogen and electricity: Long-term global technology learning scenarios. *International Journal of Hydrogen Energy* xxx, 1–14.

Cengel, Y., & Boles, M. (1998). *Thermodynamics An Engineering Approach* Third Edition. International Edition: McGraw-Hill.

Dodds, P. E. & McDowall, W. (2012). *A review of hydrogen production technologies for energy system models – UKSHEC Working Paper No. 6*. London: UCL Energy Institute, University College London.

Dodds, P. E. & McDowall, W. (2012 b). *A review of hydrogen delivery technologies for energy system models – UKSHEC Working Paper No. 7*. London: UCL Energy Institute, University College London.

DOE. (2009). *Energy Balances 2006 V1 – Microsoft Excel File*. Retrieved from Department of Energy: http://www.energy.gov.za/files/media/media_energy_balances.html

DOT. (2005). K*ey Results of the National Household Travel Survey – The First South African National Household Travel Survey 2003*. Pretoria: Department of Transport, Republic of South Africa.

EIA. (2008, August). *The Impact of Increased Use of Hydrogen on Petroleum Consumption and Carbon Dioxide Emissions – 2. Hydrogen Economy Systems and Technology Review*. US Energy Information Administration.

EIA. (2013). *U.S. Energy Information Adminstration Annual Energy Outlook 2013* Early Release Overview.

EPA. (2012). *www.fueleconomy.gov – The Official US Government Source for Fuel Economy Information*. Retrieved February 2013, from United States Environmental Protection Agency: http://www.fueleconomy.gov

ERC. (2013, April). *Assumptions and Methodologies in the South African TIMES (SATIM) Energy Model*. Retrieved from http://www.erc.uct.ac.za/Research/Otherdocs/Satim/SATIM%20Methodology-v2.1.pdf

ESKOM. (2013). *Multi Year Price Determination 3 (MYPD3)*. Retrieved from http://www.eskom.co.za/c/345/mypd3/

European Commission. (2012). *Reducing CO_2 emissions from passenger cars*. Retrieved from European Commission Climate Action: http://ec.europa.eu/clima/policies/transport/vehicles/cars/index_en.htm

Gupta, R. (2009). *Hydrogen Fuel Production, Transport and Storage*. Boca Raton: CRC Press Taylor and Francis Group.

Heywood, J. (1988). *Internal Combustion Engine Fundamentals*. Singapore: McGraw-Hill.

IEA. (2005). *Energy Statistics Manual*. Paris: International Energy Agency.

James, B., & Kalinoski, J. (2009). *Mass Production Cost Estimation for Direct H2 PEM Fuel Cell Systems for Automotive Applications: 2008 Update*. Arlington, Virginia: Directed Technologies Inc.

Johnson Matthey Plc. 2(012.) *Publications: Platinum Today, The world's leading authority on platinum group metals*. [Online] Available at: http://www.platinum.matthey.com/publications [Accessed 25 February 2013].

Lefèvre, M., Proietti, E., & Jaouen, F. (2009). Iron-Based Catalysts with Improved Oxygen

Reduction Activity in Polymer Electrolyte Fuel Cells, 324, 71. *Science*, 324, 71–74.

McDowall, W. (2012). Endogenous Technology Learning for Hydrogen and Fuel Cell Technology in UKSHEC II: Literature Review, Research Questions & Data. *UKSHEC Working paper No. 8.*

McDowall, W. & Dodds, P. (2012). *A review of low-carbon vehicle and hydrogen end-use data for energy system models – UKSHEC Working Paper No. 5.* London: UCL Energy Institute, University College London.

Melaina, M., Steward, D., Penev, M., McQueen, S., Jaffe, S. & Talon, C. (2012). *Hydrogen Infrastructure Market Readiness: Opportunities and Potential for Near-term Cost Reductions – Proceedings of the Hydrogen Infrastructure Market Readiness Workshop and Summary of Feedback Provided through the Hydrogen Station Cost Calculator.* National Renewable Energy Laboratory (NREL) a national laboratory of the U.S. Department of Energy, Office of Energy.

Merven, B., Stone, A., Hughes, A. & Cohen, B. (2012). Quantifying the energy needs of the transport sector for South Africa: A bottom-up model. *ERC Working Paper.* Energy Research Centre (ERC), University of Cape Town.

National Academy of Sciences. (2004). *The Hydrogen Economy: Opportunities, Costs, Barriers and R&D Needs.* Committee on Alternatives and Strategies for Future Hydrogen Production and Use, National Research Council.

Rand, D. & Dell, R. (2008). *Hydrogen Energy Challenges and Prospects.* Cambridge, UK: The Royal Society of Chemistry.

Schafer, A. & Victor, D. (2000). The Future Mobility of the World Population. *Transportation Research*, 34(Part A), 171–205.

Schootsa, K., Feriolia, F., Kramer, G., & van der Zwaan, B. (2008). Learning Curves for Hydrogen Production Technology: An Assessment of Observed Cost Reductions. *International Journal of Hydrogen Energy*, 33, 2630–2645.

Schwoon, M. (2008). Learning by doing, learning spillovers and the diffusion of fuel cell vehicles. *Simulation Modelling Practice and Theory*, 16, 1463–1476.

Simbeck, D. & Chang, E. (2002). *Hydrogen Supply: Cost Estimate for Hydrogen Pathways Scoping Analysis.* NREL/SR-540-32525.

Sirosh, N. (2002). *Hydrogen Composite Tank Program.* Proceeding of the 2002 US DOE Hydrogen Program Review.

Sterman, J. (2008). *Why Bad Things Happen to Good Technologies.* Retrieved from MIT: http://video.mit.edu/watch/why-bad-things-happen-to-good-technologies-9329/

Topinka, J. (2003, September). Knock Behaviour of a Lean-Burn, Hydrogen-Enhanced Engine Concept. *Dissertation in Partial Fulfillment of the Requirements of the Degree of Master of Science in Mechanical Engineering.* Massachusetts Institute of Technology (MIT).

US DOE. (2013). *Progress and Accomplishments in Hydrogen Fuel Cells.* Retrieved from US Department of Energy, Energy Efficiency and Renewable Energy, Fuel Cell Technologies Office: http://www1.eere.energy.gov/hydrogenandfuelcells/pdfs/accomplishments.pdf

Wipke, K., Sprik, S., Kurtz, J., Ramsden, T., Ainscough, C. & Saur, G. (2012). *National Fuel Cell Electric Vehicle Learning Demonstration Final Report.* National Renewable Energy Laboratory (NREL) a a national laboratory of the U.S. Department of Energy, Office of Energy.

CHAPTER 6

GEOPOLITICAL IMPLICATIONS OF A GLOBAL HYDROGEN ECONOMY FOR SOUTHERN AFRICA

Fátima Ferraz

1 INTRODUCTION

Although geopolitical discourse has evolved as a discipline, there is no normative definition. Instead, geopolitics is best understood in its historical and discursive context of use. Geopolitics addresses the socio-technical landscape in which local, regional and global factors can be assessed as a complex adaptive system and trends can be identified. Such a system is multi-dimensional and includes different actors, elements and locations, in both spatial and conceptual terms. Geopolitics is used as a tool with which to gain an improved insight that contextualises an activity (platinum group metals mining and beneficiation regime) within the bigger picture (the landscape of the evolving world order). Geopolitics is closely associated with ideas of the economy, defence, politics and information.

The current geopolitical discourse can be characterised as a mix between 'new world order' and 'environmental' geopolitics (Table 1; see Appendix A1). At present, there is much discussion within new world order geopolitics on, among other things, the role of globalisation particularly in view of the global financial Crash of 2008 (Veseth, 2010) and global energy supplies. This has led to a questioning of the assumptions that geopolitics scholars bring to the discipline in the form of 'critical geopolitics'. As an example of the questioning of assumptions, the long-term trends in energy supplies, for instance, and the dominant practices of our industrial system, which relies on energy, are under scrutiny. Further, there are the routines adopted by people and organisations in relation to energy – what are the assumptions on which they are based?

Discourse	Key intellectuals	Dominant lexicon
Imperialist geopolitics	Alfred Mahan Friedrich Ratzel Halford Mackinder Karl Haushofer Nicholas Spykman	Sea Power Lebensraum Landpower / Heartland Landpower / Heartland Rimlands
Cold War geopolitics	George Kennan Soviet & Western political & military leaders	Containmen First/Second/Third World countries as satellites & dominoes Western vs Eastern bloc
New world order geopolitics	Mikahil Gorbachev Francis Fukuyama Edward Luttwak George Bush Leaders of G7, IMF, WTO Strategic planners in the Pentagon & NATO Samuel Huntington	New political thinking The end of history Statist geo-economics US led new world order Transnational liberalism / neoliberalism Rogue states, nuclear outlaws & terrorists Clash of civilisations
Environmental geopolitics	World Commission on Environment & Development Al Gore Robert Kaplan Thomas Homer-Dixon Michael Renner	Sustainable development Strategic environmental initiative Coming anarchy Environmental scarcity Environmental security

Table 1: Discourses of geopolitics (adapted after Ó Tuathail, 1998:5)

In South Africa, with its developmental state agenda – potential extensive state activism, regulation and planning – a strategy which looks to the needs of the country and its people, its resources and their long term use, is necessary. As Smil clarifies '[e]nergy flows and conversions sustain and delimit the lives of all organisms, and hence also of such superorganisms as societies and civilisations. No human action can take place without harnessing and transforming energies' (Smil, 2008, p. 75). The optimisation of a nation's energy conversion can be seen as an indicator of its progress. However, the hydrogen economy must be viewed in terms of 'global energy systems' (Smil, 2008). Hydrogen is but one component of the energy mix – it is useful to critically re-evaluate a hydrogen economy for southern Africa from the perspective of platinum group metals (PGM) usage. PGM are not a source of primary energy. Instead, their contribution is indirect via components of, for example, auto-catalysts and fuel cells. As elaborated in Chapter 1, three types of fuel cells – alkaline fuel cells (AFC), phosphoric acid fuel cells (PAFC), and proton exchange membrane fuel cells (PEMFC) – utilise PGM in the energy-generation system. While to date, the hydrogen economy has not shown itself to be the most intensive user of PGM, this is expected to change over the coming decades. The observations in this chapter are tentative: they do not constitute a strategy; but may be used to inform such.

1.1 OBJECTIVE

The objective of this chapter is to provide suggestions which might inform a strategy for southern Africa to adopt in relation to the geopolitics of a hydrogen economy which would be based on PGM, classified as a critical resource material. The suggestions provided are based on the understanding of strategy as long-term planning, including research, practical interventions, and close attention to plausible and speculative risks. The risks associated with the use and displacement of PGM are outlined against the background of the spatial distribution of PGM. PGM are an abundant resource in South Africa. The geopolitical dynamics related to access to PGM or the lack of access to them by various actors, are in a flux in the second decade of the twenty-first century. In the larger picture, the dynamics are related to a coming energy transition, great power dynamics, and the future of globalisation (Karlin, 2009).

1.2 DISCOURSE ON GEOPOLITICS

Although geopolitical discourse has evolved as a discipline, there is no normative definition. Instead, geopolitics is best understood in its historical and discursive context of use. Table 1 provides a summary of the discourse on geopolitics.

1.3 STRUCTURE

The chapter is divided into five sections. The first section clarifies what is understood as the geopolitics of a hydrogen economy for southern Africa. The second section outlines the socio-technical landscape of PGM in terms of where they are found worldwide, thus their spatial perspective. The dominant actors (non-human agents and human agents) which are part of the socio-technical landscape of the PGM are described. In the third section, geopolitical factors in the socio-technical regime or routines of handling PGM are considered. These current geopolitical ways of handling PGM are the factors that make them significant for South Africa in terms of developing a strategy for the future. Examples from other socio-technical regimes are considered. The fourth section is a description of the dominant practices in the South African socio-technical landscape in relation to PGM. The emphasis in this penultimate section is on access to PGM, a possible hydrogen economy, electricity supply, and restrictions to market entry. Finally, the fifth section offers a resumé of geopolitical discourse in use about a hydrogen economy and thence of PGM in southern Africa. In this section are the suggestions for a geostrategic role for PGM in southern Africa.

1.4 Clarifying the terminology and language

The geopolitical terminology is elucidated using language from transition theory and concepts from complex adaptive systems. We use terminology from transition theory on the understanding that many transitions would be needed to accomplish a hydrogen economy as our current energy economy is a fossil fuel based one (Figure 1). A hydrogen-dominated economy is understood as requiring production of large volumes of hydrogen with limited, or without, fossil energy. As elaborated in earlier chapters, the realisation of, or transition to, a hydrogen economy will be a long drawn-out process. Van den Bergh and Kemp (2006) explain transition theory as based on insights from evolutionary

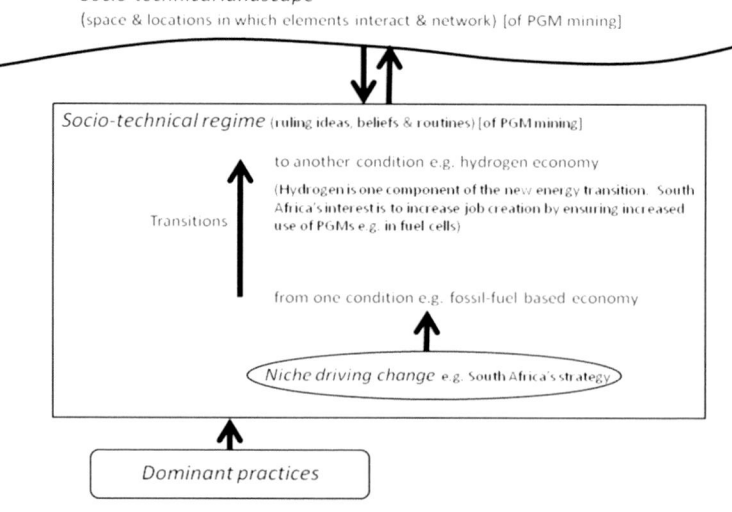

Figure 1: Terminology used in this chapter

economics, scholars of Science and Technology Studies (STS) and Innovation Studies which have developed a 'quasi-evolutionary' approach to studying technological change. In clarifying our use of the language and terminology both about transition theory and geopolitics we attempt to look at the complexities of the geopolitical implications of a global hydrogen economy for southern Africa from a fresh perspective.

The language used is driven by the fact that 'conventional geopolitical vocabularies that are the legacy of the Cold War, and the related spatial assumptions of the nation-state' (Dalby, 1998:307)' might not serve as the best descriptors for current geopolitics. The tendency to come to outdated conclusions because of using terms and language loosely can be a trap which hinders strategy development. After wars such as the Gulf War (1990–1991) or the Lebanon War (2006), or events such as the global economic crisis (2007–2011), or trends reflected in the surge of developing economies in the global balance of power, the world order is in serious flux, so is the discourse of geopolitics.

Geopolitical dynamics is the overarching term used in this paper for the changing network of relationships and alliances between actors involved in PGM and a potential hydrogen economy: including, among many others, governments, industry, cartels, countries, financial markets, financiers, regulators, labour and technological experts, and so on. This set of networked elements can be referred to as a geopolitical socio-technical landscape. We use the term landscape to describe the space and locations in which the various, exogenous, networked elements act. It is within the socio-technical landscape – in which local, regional and global factors can be assessed as a complex adaptive system – that trends can be identified, bearing in mind the economy, defence, politics, and information. The landscape of PGM and how they are and might be used to develop a hydrogen economy is the structure that gives stability to our understanding. Physically PGM are mined and extracted from a number of ore bodies although the Bushveld Complex in South Africa, the Great Dyke geological feature running through the heart of Zimbabwe (http://www.platinum.matthey.com/about-pgm/production/zimbabwe) and the Norilsk-Talnakh orebody in Russia dominate in respect of resource and reserves. In addition, PGM are also recycled from products such as auto-catalysts in many places outside the regions where they are mined.

The socio-technical regime, or 'the ruling ideas, beliefs and routines', concerns the value, current and potential uses which people have in relation to PGM. The socio-technical regime is deeply affected by geopolitical factors. But it is within a regime that it becomes possible to make transitions from one set of ideas and beliefs within the landscape, and then subsequently change the dominant practices. Or, as Schot and Geels (2008, p. 538) say, within a niche is where 'radical, path-breaking innovations can take place ... where rules, institutions and motives are different from the regime; these are "protected spaces" where "nurturing and experimentation with the co-evolution of technology, user practices, and regulatory structures" take place'. In the case of this research, it would be the regime of PGM as part of a potential hydrogen economy which undergoes the transition into a new regime.

The 'dominant and logical practices' which all actors follow in the landscape of PGM arise out of an understanding of both the landscape and, especially, the regime. To bring about a transition in a regime would lead to changes in dominant and logical practices. An example of a dominant practice in the socio-technical regime of PGM is when PGM prices rise, or users recycle or substitute for the PGM. The geopolitical socio-technical regime of PGM thus occurs within the socio-technical landscape of the global market with its shifting interests and alliances. Ecological sustainability, resource security, and resource scarcity represent some of the drivers that can further influence the regime. Conflicting interests amongst those currently extracting the metals, the governments receiving fiscal benefits from those operations, the labour which works on the mines, and those involved in research and other PGM value-add activities, speaks to the need for actors to enter into a dialogue. To do so effectively, and to have the power to negotiate strategically for long-term improvements, requires a broad understanding of the geopolitical dynamics.

Geopolitical dynamics make strategies for different practices difficult to develop. How to make counter-intuitive deductions is an important step in developing strategy. Assumptions need to be clearly stated so that conclusions drawn subsequently are not fallacious nor based on outdated scenarios. For example, assumptions underpinning deductions about the hydrogen economy in relation to PGM in South Africa cannot be based on the 'Cold War' world order, nor based solely on the beginnings of what is termed globalisation of the 'New World Order'. Relevant to this is the supposed shift of power to transnational corporations and increasingly to Asia and the rest of the developing world. And it cannot be assumed that because a natural resource is located within a sovereign territory, the locus of power over that natural resource lies with the nation-state.

1.4.1 GLOBALISATION AS USED IN THE DOCUMENT

Globalisation is different from globalism, its accompanying ideology. Globalisation is seen as an 'integrative process, leading to convergence, efficiency and more harmony' (Conversi, 2010, p. 40). The ideology underpins and legitimates the expansion of the transnational corporations with minimal restraints, except in competition among themselves.

In this document, globalisation is understood as the practice of corporate power that is largely hegemonic in nature, boundary-less in terms of territory, and expansionist in view. It is driven by capitalism and tends to rationalise and to encourage efficiencies on all levels (Veseth, 2010).

2 GEOPOLITICAL SOCIO-TECHNICAL LANDSCAPE OF PGM

The geopolitical socio-technical landscape refers to the space and locations in which the various networked elements act. The changing network of relationships and alliances between actors involved in PGM, and a potential hydrogen economy include among many others, governments, industry, cartels, countries, financial markets, financiers, regulators, labour and technological experts. Trends follow a pattern which can be incremental and gradual or marked by discontinuities and instability. There are various risks involved: the risk of long-term trends not being recognised in time; the inability to predict which trends will become embedded in society; and their unknowable effects on human society (Karlin, 2009). The landscape of PGM and how they are, and might be, used to develop a hydrogen economy, is the structure that gives our understanding stability. PGM are mined and extracted from two different sources – directly through mining, and indirectly from recycling.

The reformed neoclassical economic paradigm of the market economy and globalisation uses economic growth to drive development. Such growth is seen as an enabling mechanism for poverty reduction. Yet at the same time, globalisation and the market economy approach of 'core business focus' and 'cost-cutting' has led to mass unemployment. Alternatives to this approach are being sought to provide more socio-economic opportunities for people.

One area which is being considered in South Africa as a potential opportunity for increased development and socio-economic opportunities is that of the dislocation between extraction and processing centres of natural resources. The supply of PGM is dominantly from southern Africa (with more than 80 per cent of PGM reserves) and current markets and processing centres predominantly in Europe, North America and Asia (see Table 2).

Demand	2007	2008	2009	2010	2011
Europe	2800	2610	1830	2155	2350
Japan	1315	1735	1050	1155	1210
China	1540	1410	2165	2040	2075
North America	1525	1145	860	1400	1125
ROW	1090	1090	890	1155	1320
Supply					
South Africa	5070	4515	4635	4635	4775
Russia	915	805	785	825	825
North America	325	325	260	200	360
Zimbabwe	170	180	230	280	335
Others	120	115	115	110	100

Table 2: Distribution and quantum (in 000s of ounces) of demand and supply of PGM around the world

2.1   SPATIAL DISTRIBUTION OF PGM

Geologically, platinum group elements are associated with layered igneous intrusions. South Africa is the world's foremost primary producer of platinum group elements from mineralised horizons of the Bushveld Complex. The process of primary beneficiation of such mineralised material to PGM is complex although most of this production occurs in South Africa. Platinum is a rare metal and it is the chief metal of a group of metals termed PGM that includes palladium, rhodium, ruthenium, osmium and iridium. However, platinum possesses characteristics that make it a useful industrial metal despite its extraction being costly. In Africa, both South Africa and Zimbabwe contain significant PGM resources. Other primary producers of PGM include: in Eurasia – Norilsk Nickel in Siberia, Russia; in North America – Lac des Iles in Canada, and Stillwater igneous complex in southern Montana in the United States. Occurrences of PGM are also noted in Antarctica but these are not mined. A tabulation of dominant commodity produced at active mine properties and the associated ownership structure is available in Mudd (2010) (see also Appendices A2 & A3).

2.1.1 EurAsia
Russia

Russia is an important primary producer of PGM. Norilsk Nickel is the world's largest producer of nickel and palladium and one of the leading producers of platinum and copper from the Norilsk- Talnakh ore-body. It also produces various by-products, such as cobalt, rhodium, silver, gold, iridium, ruthenium, selenium, tellurium and sulphur. The PGM are extracted from operations located above the Polar Circle, on the 69th parallel on the Taimyr Peninsula in the Krasnoyarsky Region.

Figure 2: The locations of the Russian platinum mines (Mining-technology.com)

There are seven mines extracting ore from the Oktyabrsky, Talnakh, and Norilsk-1 deposits (Norilsk, 2013); these are listed in Appendix A2. Of significance is that the PGM are not the primarily desired commodity; instead, nickel is. This fact is relevant when evaluating a PGM alliance – PGM in Russia are extracted as one of a series of rich by-products albeit in an inhospitable region. Demand for nickel and copper remains high which ensures the income stream. In contrast, in South Africa, the PGM are being extracted as the primary saleable product.

2.1.2 Southern Africa
South Africa

Platinum mining occurs across a geographically extensive area in three provinces – North-West, Limpopo and Mpumalanga. Platinum-rich lithologies are associated with the layered Bushveld Complex which exposed at current levels of erosion, consists of the western (Rustenburg area), eastern (Lebowa area) and northern (Potgietersrus area) limbs (Figure 2). The Bushveld Complex comprises three ore bodies: the Merensky Reef, the Upper

Group 2 (UG2) chromitite, which together can be traced on surface for 300 km in two separate arcs, and the Platreef, which extends for over 30 km. The complex is some 7–9 km thick with mining already taking place at 2 km depth in some mines.

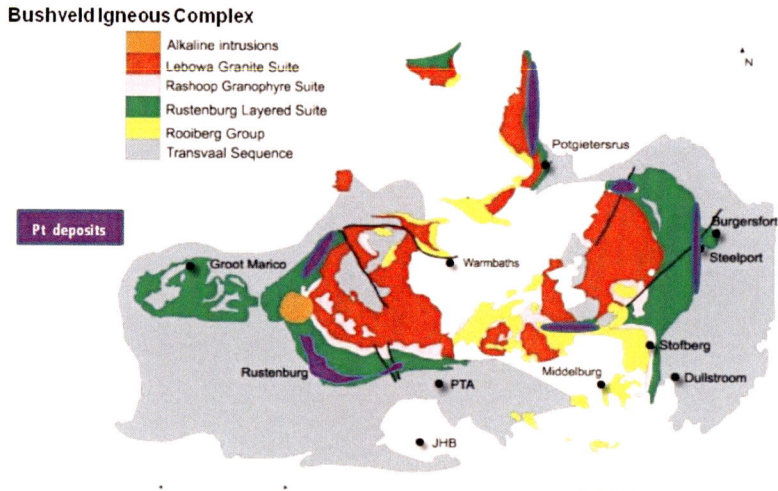

Figure 3: Simplified geological map of the Bushveld Complex (Kinnard, 2000)

The lateral continuity of the layers, a characteristic of this ore-body, justifies exploration of adjacent areas, evaluation of the economic feasibility for a new mine, mine planning and operation. Current mines are likely to be exploiting the most favourable sections of reefs. The major platinum mining companies hold most of the mineral rights to adjacent areas (Cawthorn, 1999; Cawthorn, *et al.*, 2002; Schouwstra & Kinioch, 2000). It is inevitable that the potential for new mining ventures will be at least evaluated and possibly realised beyond existing mine lease boundaries provided the demand for PGM remains economically feasible. A map showing the distribution of producing, developing and potential platinum-group metal mines is illustrated in Figure 4.

Communities aggregate in the vicinity of the producing, developing and potential platinum-group metal mines. Although the human activity of platinum mining reflects in multiple dimensions, it is most clearly affected in the technological, economic and social dimensions. It is the technological dimension – as an actor – with its many underlying system networks that interacts with socio-economic elements. This is exemplified by the extension of platinum exploration into areas adjacent to current mines and the

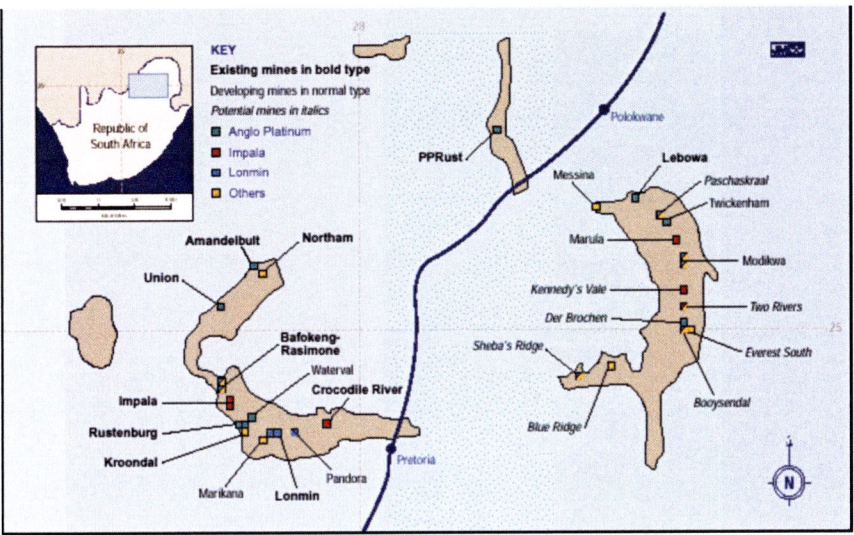

Figure 4: The Bushveld Complex, showing producing, developing and potential platinum-group metal mines (after Platinum 2003 in DME 2003)

economic evaluation of these for the feasibility of commissioning and operating new mines (Figure 3). The mine life cycle moves from exploration to development, to production, to closure, and to rehabilitation of the mine site. In the exploration stage, prospecting teams are relatively small and serve to collate technical information that confirms the feasibility of the development. In cycling through each stage, the technological dimension is inextricably linked to the economic dimension in a symbiotic relationship in which the monetary aspect of the economic dimension dominates. Two studies have been carried out – as part of the overall PGM research project – to contextualise and understand the socio-economic impacts on such mining communities. The first study focused on the theoretical foundation of sustainability, sustainable development and mining (Ferraz, 2012) and the second, on a sustainability orientated model beyond mining to address socio-economic needs (Ferraz, 2012).

Zimbabwe

The Great Dyke of Zimbabwe constitutes the world's second largest reserve of platinum group elements after the Bushveld Complex in South Africa. The Great Dyke is linear in shape and trends over 550 km north-north-east at a maximum width of about 11 km across the Zimbabwe Craton. The Great Dyke is a layered series comprising a lower ultramafic sequence and an upper mafic sequence. Economic concentrations of platinum group elements, nickel and copper are found in the main sulphide zone (MSZ) within the ultramafic sequence. PGM are currently mined at the Hartley, Ngezi, Unki and Mimosa mines (Figure 5).

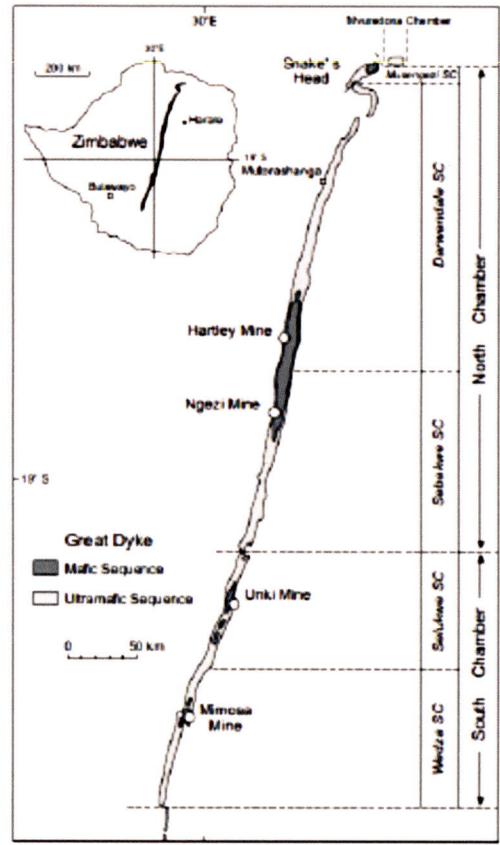

Figure 5: The Great Dyke showing producing and potential PGM mines and a simplified geological map with its subdivision into chambers and sub-chambers (SC) (After Oberthür, *et al.*, 2012)

2.1.3 NORTH AMERICA

The Lac des Iles Mine is located within the Lac de Iles layered gabbroic igneous complex in north-western Ontario and is the largest palladium-rich ore-body in Canada. This ore-body is one of only two primary palladium producers in the world and is currently mined, both opencast and underground, by North American Palladium (NAP, 2013).

The Stillwater igneous complex in southern Montana in the United States is a layered mafic intrusion. Chromium reserves are associated with the complex and these have been mined in the past. More recently, though, palladium and other platinum group elements have been identified and are currently being mined from the J-M Reef. This reef occurs within the lower banded zone of the complex and is extracted at both Stillwater and East Boulder mines by the Stillwater Mining Company (Stillwater, 2013).

2.1.4 OTHER: PGM AS BY-PRODUCTS IN SOUTHERN AFRICA

Smelters in South Africa are geared up for platiniferous ores from blends of the UG2, Merensky and Platreef horizons. Mining operations that extract chromium and nickel as a primary activity may also contain PGM of sufficient grade to be beneficiated as a by-product. The extracted volume of PGM though is relatively small in South Africa. An example of this is Nkomati (see Appendix A3).

Linkages from the platinum sector to the other mineral sectors, including chromium, are important. Chromium is often a by-product of the PGM generating mines. Chromium is considered an infrastructure mineral and is used in the manufacture of steel.

2.2 ACTORS IN THE LANDSCAPE OF PGM

There are many actors in the socio-technical landscape of PGM. The actors vary according to which stage of the PGM life cycle is in sight: such 'geopolitical dynamics' happen in a global market of shifting interests and alliances.

After the Cold War, the nature of geopolitical discourse revolved around nation states as key players. This is no longer possible in quite the same way in view of the emergence of powerful transnational corporations. The landscape in which transnational corporations operate is not bounded by geographical and spatial constraints. Whereas previously geopolitics was defined geographically, space is no longer the only factor to take into account. For example, the geology shows where the primary resource is located, however, this is not necessarily the locus of power.

Geopolitical knowledge is constructed from positions and locations of political, economic, and cultural power and privilege. Hence the histories of geopolitics have focused on states and elites (Routledge, 1998). This cannot be the only consideration in South Africa in view of high levels of poverty, inequality and economic marginalisation.

The materials stewardship life cycle is divided into process stewardship and product stewardship (ICMM, 2007; Figure 6). From a mining perspective it is important to understand the material flows and life cycle benefits; to build and strengthen relationships with other players in the value chain; and to optimise production and provide inputs to support decision-making. Process stewardship includes exploration, mining, concentration, smelting and refining. This value-add chain is effectively primary beneficiation. Product stewardship involves further beneficiation steps through fabrication, design and manufacture. Thus process and product stewardship can be used to ensure the supply of scarce metals through improved efficiencies, product development and optimisation, recycling and substitution.

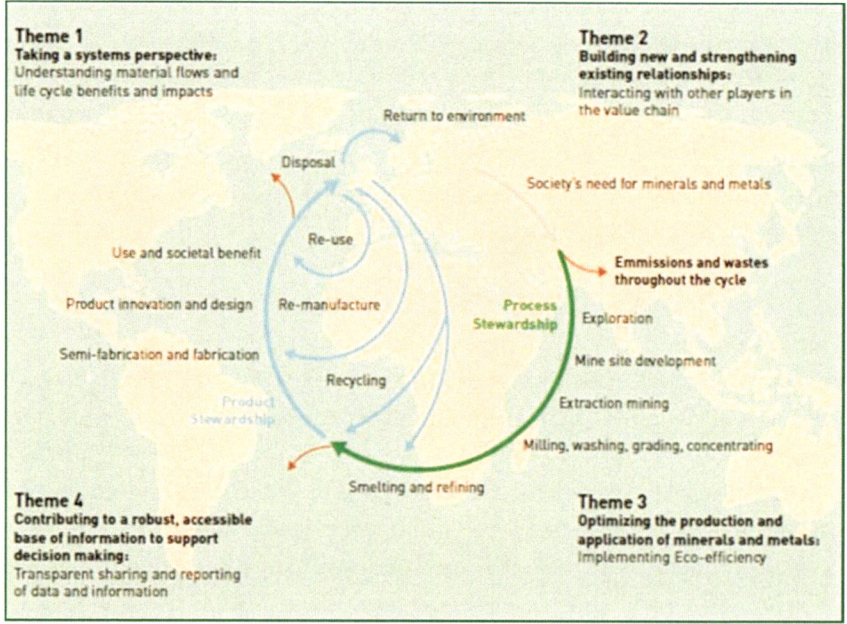

Figure 6:   Materials stewardship life cycle (ICMM, 2007)

2.2.1   THE EXTRACTIVE METALS LANDSCAPE – PROCESS STEWARDSHIP

A PRIMARY BENEFICIATION PERSPECTIVE

South Africa is the dominant producer of PGM in terms of volume although costs of extraction are high. This is the result of the complexity of mining the ore-bodies underground, from the intricacies of the smelting extractive processes and high labour costs. South Africa has a number of integrated platinum producers, operating from mining through to refining, who account for the bulk of production – including Anglo American Platinum Corporation, Impala Platinum Holdings and Lonmin Plc. Other operators such as Northam Platinum Limited, African Rainbow Minerals and the Royal Bafokeng Platinum do not own precious metal refining facilities and instead, have agreements with refineries for this process. At present, PGM ore produced in Zimbabwe is transported to South Africa for refining. There is interest in Zimbabwe to refine locally instead of continuing with the status quo. In Russia, the dominant producer of PGM, Norilsk Nickel, is an integrated platinum producer with an agreement with the precious metals refinery – Krasnoyarsk Precious Metals Plant. The Norilsk Nickel group is also involved in the marketing and the sale of base and precious metals products.

There are producing, developing and potential PGM mines. Of these a non-exhaustive tabulation of the 22 platinum companies listed on the stock exchanges of the world is presented in Appendix A3. Each of the PGM producer mining companies is an actor in the geopolitical socio-technical landscape. The bulk producers of PGM though are Anglo American Platinum Corporation, Impala Platinum Holdings, Lonmin Plc and Norilsk Nickel – the four dominant production actors (Figure 7).

Another significant factor in primary production or mining of PGM is regulatory policy. This is especially true in South Africa with its history of labour activism and post-apartheid redress policies. Regulatory policy, largely for national welfare benefits, often leads to an increased financial cost becoming one contributing factor governing the price of PGM. For example, compliance with more stringent laws governing aspects such as the awarding of mining licenses, improved environmental management and improved community inputs has been associated with higher expenditure in South Africa. Compliance with global protocols on 'eco-protectionism' to address, for example, climate change concerns, also increases the financial cost of PGM. Government in each of the producer countries is also an important actor from a different perspective. PGM may be declared a critical and strategic resource – as in the United States. Additionally, PGM mining contributes revenues to the government fiscus which can be made available for development – as in South Africa. Labour, too, is an important actor, especially in South Africa with high levels of trade union organisation and labour activism. Financial markets, financiers and technological expertise are further, critical actors.

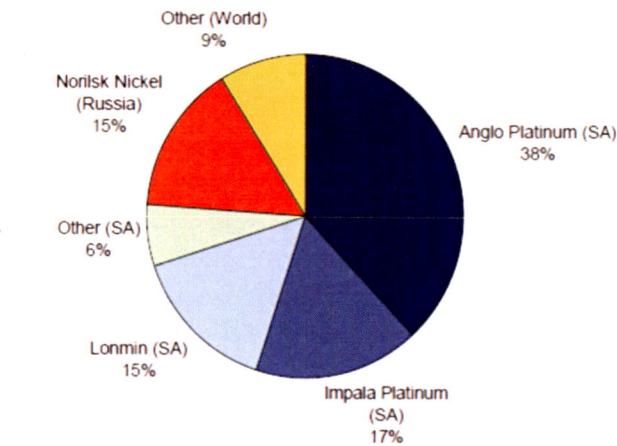

Figure 7: Bulk producers of PGM (After Anglo Platinum, 2004)

2.2.2 THE MANUFACTURING LANDSCAPE – PRODUCT STEWARDSHIP

A SECONDARY BENEFICIATION PERSPECTIVE

The actors involved in secondary beneficiation of PGM, downstream value-add processes, differ from those in primary beneficiation. PGM are used in different applications including jewellery manufacturing; power and energy applications (e.g. fuel cell technology[1]); transportation (e.g. auto catalysts); petroleum refining; machine components (e.g. in computer hardware); chemicals and glass.

The countries which are major consumers of platinum are the European Union nations, the United States, Japan and China. These are the dominant consumer actors. Fluctuations within these dominant consumer economies would impact the rate of consumption of PGM. This will reflect in either an increase or decrease of the market price of the primary commodities, namely PGM. Other factors that will impact the price of the PGM include fluctuations in the prices of other precious metals on global markets, and the size of the existing material stockpile. Here again, labour, financial markets, financiers, regulatory policy, and technological expertise are important actors.

RECYCLING

Recycling is one example of a dominant practice in the regime of PGM

---

1. This market currently accounts for less than 1% of PGM uses.

mining. The high cost of PGM induces increased recycling efforts. While this may not necessarily reflect direct (nor the only) causality, the correlation between platinum prices and recycling rates of the metal in the catalytic converter, industrial and jewellery markets, is illustrated in Figure 8. Recovery rates of PGM of more than 95% are possible with the rate of recycling highest in industry applications (80–90%), followed by the catalytic converter industry (50–55%), and lowest in the electronics sector (0–5%). (IRP, 2011.) More than 2,500 tonnes or 10 years supply of platinum is available in catalytic converters installed in vehicles around the world (Hageluken, 2007). This automotive fleet is growing and Belgian company, Umicore, refer to recycling of this platinum as 'mining on wheels'. Platinum recycling is therefore the fastest growing source of supply of the mineral (Whitburn, 2012).

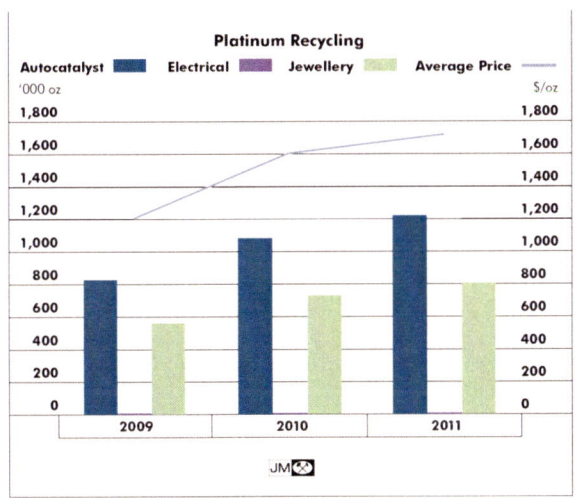

Figure 8: Platinum recycling and platinum prices for the years 2009, 2010, and 2011 (Butler, 2012)

The economic rationale for recycling platinum is driven by headgrades in excess of 2000g/ton in automotive catalytic converters which exceeds that of mining headgrades (of approximately 10g/ton) by orders of magnitude. Recycling thus has some effect in countering the high price of the metal and ensures security of supply to address demand (Figure 9). It can also be argued that recycling – and reduction in PGM-loading in some applications – will also ensure longevity in the various uses of PGM and thus in the exploitation of the reserves.

Assuming platinum fuel cells are adopted beyond niche markets and demand for the mineral increases, platinum prices will rise giving the recycling industry further impetus. Currently, the amount of platinum recycled is estimated at around 25% of global production from mines. However, improved recycling efficiencies resulting from technological innovations will increase this percentage.

Platinum recycling is an area of real growth potential especially if there is additional uptake of platinum into fuel cells. Recycling, however, competes directly with the extractive platinum mining industry for the same market. Market leader in PGM research and development, Johnson Matthey Plc reports also demonstrate that recycling is increasingly becoming a source of platinum. Recycling PGM is undertaken in many places including outside the regions where it is currently mined. This suggests a shifting landscape in the supply of PGM from where it has historically been sourced to a more evenly distributed geography via recycling. The role of recycling needs to be considered in any PGM strategy.

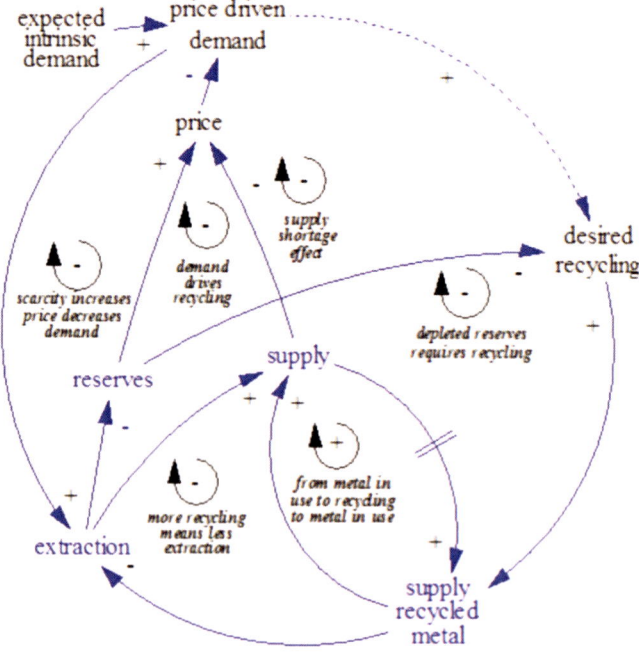

Figure 9: Demand and supply for recycled metal

SUBSTITUTION

Substitution is a further example of a dominant practice in the regime of PGM mining with high costs of PGM driving substitution strategies to ensure security of supply. Several countries have led initiatives on PGM substitution, the most recent of which is led by the European Union. The European Union Parliament Resolution 31 of 13 September 2011 'regrets that substitution and re-use are not sufficiently addressed in the [European Commission] Communication; recalls that substitution, particularly for critical[2] raw materials (CRM) including PGM, and rare earth elements (REE), is of great relevance and can offer efficient solutions to supply and environmental risks when possible; calls on the Commission, therefore, to ramp up its work in this field by leveraging research and innovation funding through the possible development of a substitution R&D programme in the forthcoming research framework programme, supporting demo-plants; encourages the Commission and the Member States to consider setting substitution targets while taking into account relevant impact assessment' (European Parliament, 2013).

In addition to the European Union, the USA and Japan have also undertaken much substitution research. The focus of research on fuel cells in Japan is based on the development of oxide-based cathodes involving a consortium of research organisations including Sumitomo. Non-platinum material is being tested for stability and high catalytic activity from Group 4 and Group 5 oxide-based compounds such as tantalum and zirconium. Current research on Zr-CNO and Ta-CNO fuel cell cathodes indicates that a significant cost saving is possible compared to using platinum (Ota, 2013).

3 Geopolitical Factors in the Socio-technical Regime of PGM

The 'socio-technical regime', or otherwise, 'the ruling ideas, beliefs and routines' which people have in relation to PGM about their value, as well as current and potential uses is affected by geopolitical factors. One of these factors that impacts on the socio-technical regime is the nature of international relations (New World Order); another is the developmental state; and another the hydrogen economy as a fledgling part of the energy transition in which the world is engaged.

According to transition theory it is within a regime that it is possible to

---

2. Critical raw materials (CRM) are those the risk of which is significant in terms of security of supply. A disruption of supply can have disruptive consequences for materials used in the defence, communications, and energy industries.

make transitions from one set of ideas and beliefs within the landscape to another; and that change will affect the dominant and logical practices of the new socio-political regime. In this instance, it would be the regime of PGM as part of a potential hydrogen economy which would be a new regime.

The geopolitical factors influencing the socio-political regime relate to how the PGM resources are extracted and by whom, and with which affiliations and agendas; additionally, how safety, health and the environmental questions are managed in addition to profit considerations. Specifically, geopolitical factors influence how the PGM resource has been organised to benefit people and transnational companies and organisations. Resources are considered from a politico-economic perspective when they are exploited. A significant part of the geopolitical discourse also relates to information. How can the economy be arranged to maximise benefits? A significant part of the geopolitical discourse also relates to information. There are two choices for information in its use and dissemination – it can be used as information, and as disinformation. The defence of a socio-political regime can take different forms, such as regulatory (for example, the China embargo on the export of REE resources) and military (for example, the US military intervention in Iraq to secure oil resources).

In geopolitical discourse another important way of weighing up the implications of the regime is to consider risk. Risk is the probability or threat of loss that is caused by external or internal vulnerabilities and that may be avoided through pre-emptive action. Politics is about risk – who can assess, who can predict, who can prevent? As much as risk is part of the socio-technical regime in the PGM mining sector, it also includes a consideration of benefits. Although the socio-technical landscape can have many niches, it is difficult to transition from one niche to another, unless 'rules, institutions and motives are different from the regime[3]' (Geels & Schot, 2008, p. 538). The niche has to be a 'protected space' in which 'nurturing and experimentation with the co-evolution of technology, user practices, and regulatory structures' (Geels & Schot, 2008, p. 538) are permitted to take place. Because there are significant PGM resources (see Appendix A3), the system is stable in this sense. To be strategic, though, requires a move out of what is the established and known, to undertake plausible and even speculative risk. There are many factors impacting the socio-technical regime, and because processes are in flux particularly at present, any situational analysis is complex.

---

3. This refers to the 'socio-technical regime', or 'the ruling ideas, beliefs and routines' around the value, current and potential uses which people have in relation to PGM.

The central question in this section is to understand the ruling ideas around the value, current and potential uses for PGM and where these are moving to. This is driven in particular through the developmental agenda of government and its efforts at job creation. Socio-technical regimes work in landscapes and work through alliances. To establish such alliances requires information; an assessment of geopolitical dynamics and trends, including the energy transition, the shifting balance of power, globalisation, and risks (Smil, 2008).

3.1 ENERGY TRANSITION

The socio-technical regime or the ruling ideas around the value, current and potential uses for PGM in relation to the energy transition is a further factor to discuss in geopolitical terms.

According to Smil (2008), ours remains a predominantly fossil-fuel based economy dependent on oil, coal and gas. The technological base of this fossil-fuel based civilisation requires large-scale uninterrupted supplies of energy to exist. The sources of energy remain constant for long periods on account of the difficulty of substitution, which involves discarding old infrastructure and building anew. Any transition to non-fossil fuel sources will be a long, drawn-out process, particularly in view of the scale of the shift required.

Reliable and affordable energy is a fundamental driver of the market economy. However, an energy crisis is currently developing across the world. The basis of this crisis is sustainability in terms of energy reserves, energy utilisation and environmental protection. 'Global consumption of coal (supplemented by small amounts of crude oil and natural gas) surpassed that of phytomass (wood, charcoal, crop residues) during the 1890s. Coal's share in the world's total primary energy supply (TPES), excluding the phytomass, stood at 95% in 1900; it slid below 50% during the early 1960s, but in 2005 it was still at about 28%. Crude oil accounted for 4% of global TPES in 1900, 27% in 1950, and about 46% in 1975. By 2005 it was about 36%, while the importance of natural gas kept rising; natural gas supplied nearly 24% of global TPES in 2005' (Smil, 2008, p. 75), illustrated in Figure 10.

One future energy scenario is the widespread use of hydrogen as an energy carrier – the basis of the 'hydrogen economy'[4]. A hydrogen economy requires the development and application of a range of different technologies, some similar to, and some quite different from, those in common use in current energy systems. Much improvement of the technologies will be required

---

4.   The term 'hydrogen economy' was first used by John Bockris in the 1970s (Bockris, 2011).

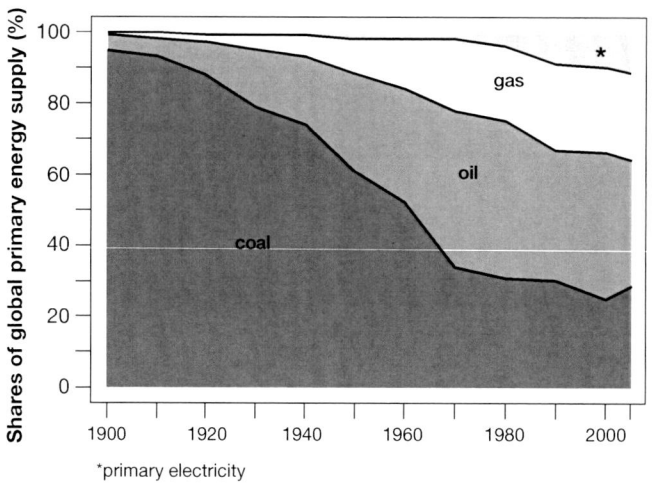

Figure 10: Shares of fossil fuels and primary electricity in global energy consumption, 1900–2000 (After Smil 2008:76)

before they become competitive with alternative means of delivering the same energy service (Ekins & Hughes, 2009). Hence the hydrogen economy is unlikely to dominate globally in the near future (Ekins & Hughes, 2009; Smil, 2008). It must be borne in mind that, ideally, a hydrogen-dominated economy is understood as requiring the production of large volumes of hydrogen without fossil energy. Currently, take-off steps or transitions to a hydrogen economy have initially to be powered by fossil fuels. This is illustrated globally with the predominant source of hydrogen being natural gas, a fossil fuel derivable from coal. Inputs from platinum in fuel cells – as a secondary energy source – into a hydrogen economy are expected to be rather small.

3.2 POLITICAL POWERS DYNAMIC

To understand geopolitical factors and their influence on the socio-technical regime, the shifting dynamic in global power is an important fact, particularly in relation to PGM and how they are and might be used to develop a hydrogen economy.

3.2.1 RUSSIA

South Africa recently concluded a bilateral agreement with Russia, under which the countries agreed to co-operate on PGM initiatives to create a suite of interventions necessary to stabilise the platinum industry (Esterhuizen, 2013). From a political powers perspective, Russia is a resurgent power although long-term prospects are uncertain (Smil, 2008). Russia is an energy superpower (especially gas) and its interest, for its own benefit, in stabilising the platinum industry is a factor to be considered in a hydrogen economy. Of further interest, Russia is a member of BRICS – the association of emerging economies: Brazil, Russia, India, China and South Africa.

Russia has comparative advantage. In the first instance, its PGM mines are primary producers of nickel with PGM as a by-product commodity and are therefore not dependent on the sale of PGM for continued mining. Second, palladium is the dominant PGM produced; this substitutes for platinum (in gasoline engines) when prices of the latter are high. In the third instance, Russia has long-term experience in the manipulation of markets in which it has pre-eminent endowments, as with natural gas in Europe. The Russia-South Africa partnership announced at the recent BRICS summit (Durban) is consequently an intriguing subject for analysis, especially if South Africa is interested in stabilising platinum markets and allaying security of supply concerns. It should be noted that, even during the Cold War, there seems to have been an arrangement between De Beers and the Soviet Union government to sell diamonds through the Central Selling Organisation, De Beers' marketing arm, so as to manage the markets for these precious stones. (Kretschmer, 1998).

3.2.2 China

China does not have significant resources or reserves of PGM. However, there is direct Chinese investment in PGM mining in South Africa through Wesizwe Platinum's Bakubung mine which aims to start production in 2018. This is China's first direct investment in the sector. The role of China is important. Paladini and George (2011) assert that 'state competition for natural resources has resulted in increased international tensions on strategic materials. China is at the centre of this scramble, and there is quite an amount of attention devoted to China's economic growth as well as its possible imperialistic aspirations in today's world. Resource-seeking Chinese corporations aggressively try to secure markets and resources. They generally compete against multinational corporations both from the OECD countries and from other emerging players' (Paladini & George, 2011). China is transitioning to a position of global economic predominance, seeking resources around the world. In its behaviours, it reflects business-state symbiosis (Karlin, 2009). The interest from China in resources is for its own benefit, and its interest in PGM is a factor to be considered in any strategy driving a hydrogen economy with PGM. China is also a member of BRICS.

3.2.3 INDIA

India, another member of BRICS, often competes with multinational corporations both from the Organisation for Economic Co-operation and Development (OECD) countries and from other emerging players such as China. For example, PetroChina is involved in the acquisition of resources abroad; this has put the parastatal on a collision course with Indian companies involved in the same activity (Paladini & George, 2011). India may have interests similar to China in resources and recycling, hence a possible interest in PGM.

3.2.4 ZIMBABWE

Zimbabwe, in contrast to Russia, China and India, is not a member of BRICS, but is a member of the Southern African Development Community (SADC) – and a neighbour to South Africa. Zimbabwean PGM ore is mined and extracted from the Great Dyke which contains good resources and reserves of platinum. Zimbabwe, too, would be interested, for its own benefit, to stabilise the platinum industry. Of note, in both Zimbabwe and South Africa, transnational corporations and other private sector mining companies dominate the PGM extraction industry and not government. In Zimbabwe, government owns 51% of each mine through its indigenisation programme, but despite this control over the supply side, the demand side is fully controlled externally.

3.2.5 OTHER

While an understanding of the dynamic of the emergent powers' interests is critical, the largest importers and processors of PGM are located in Europe, North America and Japan. Most of the transnational companies involved in PGM mining in South Africa and Zimbabwe are headquartered in these countries. Countries in Europe, such as Belgium and Germany, do lead in PGM recycling and research into PGM substitution. Similarly, Japan is a leader in PGM product development, recycling and research into PGM substitution.

The global leader, the United States, has undertaken much work on the hydrogen economy including research and product development. Companies such as Johnson Matthey Plc are leaders in the development and manufacturing of products containing PGM such as fuel cells. For instance, in addition to the concerns of the European Union referred to earlier, the US government has been encouraged to consider factors that would affect security of supply of PGM, including: 'potential for political instability in major platinum producing countries; control of platinum production by a limited number of companies in the major producing countries; future growth/decline in the world economy; potential for significant increases in platinum demand from new applications other than fuel cells; ability of fuel cell technology developers to overcome technical hurdles; ability of the fuelling infrastructure to support commercialisation of fuel cell vehicles' (TIAX, 2003).

Therefore, the interests of the EU, the US, Japan and, as intimated in earlier chapters, Canada and South Korea in the PGM industry need to be taken into account in a hydrogen economy strategy.

3.3 GLOBALISATION

Corporatism, the driver of globalisation, with its 'built in' characteristics (boundary-less, open-endedness, and expansionist), relies upon a structure of national interdependence, thus rendering the system vulnerable to disruption.

The destabilising force rises from the growing inequality between nations and, more importantly, across the populations within nations (Karlin 2009).

Governments can introduce legislation about the licensing of PGM resources and their extraction; however, the ownership of the extracted PGM resources lies in the hands of private sector mining companies. The extracted PGM are sold or advance-sold on the international market. Access to raw PGM comes through purchase agreements with the individual mining companies and metal exchanges. The ownership of the PGM consumption lies also in the private sector, currently dominated by the automotive catalytic converter industry. Such arrangements create complications for the emergence of serious PGM beneficiation, underpinned by mining cluster development based on downstream and upstream activities.

These geopolitical factors or ruling ideas highlight the nature of the current PGM industry in view of globalisation. They need to be considered strategically if a role for PGM is to grow as part of the energy transition, possibly to a hydrogen economy, from which South Africa can draw maximum benefit.

3.4   GOVERNANCE AND TRANSPARENCY

Despite mandatory corporate reporting, the perception remains that the extractive sector – oil, gas, and mining – needs closer investigation. In the PGM sector, for example, the practice of transfer pricing[5] by the transnational corporations raises questions about accountability and governance.

Organisations such as Global Witness continue to monitor and campaign internationally against natural resource-related conflict and corruption, and associated environmental and human rights abuses (Global Witness, 2013). The Extractive Industries Transparency Initiative, EITI, has been established as a global standard to promote revenue transparency and accountability in the sector (EITI, 2013). It is a methodology for monitoring and reconciling company payments and government revenues from oil, gas and mining at the country level based on two components:

- Transparency: Oil, gas and mining companies disclose their payments to the government, and the government discloses its receipts. The figures are reconciled and published in annual EITI Reports alongside contextual information about the extractive sector.
- Accountability: A multi-stakeholder group with representatives from government, companies and civil society is established to oversee the process and communicate the findings of the EITI Report.

3.5   RISKS

Risk is the probability or threat of loss that is caused by external or internal vulnerabilities and that may be avoided through pre-emptive action. The significance for South Africa of PGM in the current geopolitical socio-technical regime lies in its classification as a critical raw material (CRM) (Figure 11). A recent review of several studies which evaluated CRM across different geographies, sectors and time horizons, reveals concerns about ready access to PGM and REEs (Erdmann & Graedel, 2011). Disruptions can influence security of supply to the detriment of production levels in various industries as noted previously, such as the defence, communications, and energy industries.

Countries resort to different activities as a result of security of supply concerns for resources deemed CRM. Governments from countries with large endowments of such resources have also sought to take advantage of

---

5. 'Transfer price is the price at which goods or services are sold between divisions of a company, or between companies in the same group (Domfeh, 2011).'

Legend:

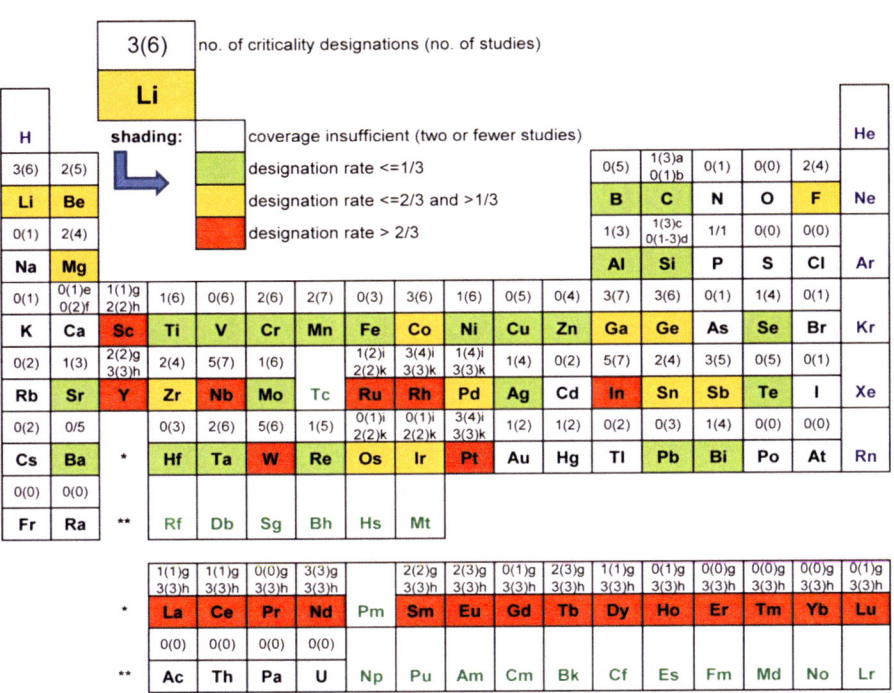

Figure 11: Frequencies of criticality designations and of coverage (in brackets) for materials addressed in several studies (cited in the reference). In several cases, minerals or metal groups were evaluated instead of individual elements. For PGM, i as element, k as component of PGM (Erdmann & Graedel, 2011)

them, including through cartels such as the Organisation of Petroleum Exporting Countries (OPEC) in the case of oil (Khusanjanova, 2011), and China's restrictions on the export of REE (Jepson, 2012). Economic entities from demand centres have resorted to different measures to minimise and mitigate security of supply concerns. These measures include development assistance in exchange for access (Campos & Vines, 2007), war or threats thereof (Le Billon, 2004), stockpiling (US DoD, 2013), recycling (Graedel, *et al.*, 2011) and substitution (Graedel, 2002).

If the need for PGM were to increase substantially, additional concerns about availability would be raised (Alonso, *et al.*, 2012; Yang, 2009), accelerating efforts to substitute or to access the mineral from other sources, including space (*The Economist*, 2013)[6].

---

6. It has been suggested that 'a single 500-metre metal-rich asteroid might contain the equivalent of all the platinum-group metals mined to date'. Such a scale of supply may lead to a slump in Pt prices that may undermine the business case for space mining.

SA AND THE GLOBAL HYDROGEN ECONOMY

In order effectively to assess the potential significance of PGM for South Africa, the risks need to be adequately estimated. This involves understanding fully the complexity of human systems and their discontinuities. The method suggested by Smil (2008) is to analyse key variables categorised as a) catastrophes, b) powerful trends (the effects of globalisation), and c) the shifting balance of power (the marginalisation of Japan, an unstable Islamic region, Russia's partial resurgence, the uncertain rise of China, and an increasingly faltering United States). Catastrophic events are further classified into a) known catastrophic risks, for example earthquakes, asteroid strikes and super-eruptions; b) plausible catastrophic risks, such as nuclear wars and pandemics; and, c) speculative risks – mining platinum from asteroids and the seabed.

	Plausible risk	Speculative risk
Supply		
Constraints from resource nationalism	√	
Infrastructure in relation to electricity, railways, water, poor education & skills	√	
Labour & governance in relation to human rights, labour costs, strikes, regulation, corruption, transparency & reporting	√	
Demand		
Constraints from countries, e.g. China, India		√
Technology, e.g. increased uses for metals, e.g. hi-tech components, batteries ...		√

Table 3: Supply and demand constraints versus plausible and speculative risk (Modified after Burgess, 2011)

There is a need to ensure security of supply: that means the evaluation of risk in conjunction with an evaluation of the persistence of the consumption culture. Factors informing the assessment of risk of access to South African PGM resources can be considered as either plausible or speculative risk, in relation to supply and demand constraints (Table 3).

The geopolitical framework is also important in relation to access. The framework contextualises the interest in local value-add beneficiation initiatives. In terms of supply, if a product is deemed valuable enough, it can lead to conflict and the development of a black market as in the case of blood diamonds. Carbon taxes can act as impediments to market entry and they may affect supply. The practice of transfer pricing, for example, by the transnational corporations, raises questions about accountability and governance in the PGM sector. PGM producers sell their products two to three years in advance although there are also spot sales. Approximately 12% of PGM production is sold into South Africa. Countries have their own raw material resource strategies to ensure security of supply through policy instruments. Such economic policy instruments can be used to give impetus to developmental goals. For these instruments to be effective new agreements with PGM producers may have to be entered into and bilateral agreements renegotiated. The potential for local beneficiation of PGM and the growth of an export market would be encouraged. In this way beneficiation becomes a tool for economic upliftment. Economic upliftment, though, would have to be managed to ensure developmental interests are addressed. The market for PGM recycling and substitution goes some way to address and ensure security of supply. Temporal risk, also associated with both supply and demand, is a consideration of how long the consumption culture, the need for PGM, will persist. The introduction of a PGM commodities' exchange would help to mitigate this temporal risk as well as other challenges outline above.

3.5.1 STABILISING COMMODITY MARKET VOLATILITY

One mechanism to stabilise the volatility of the commodity market is a PGM commodity exchange located in South Africa (Abedian, 2013). South Africa

1	A commodity exchange is a market mechanism, and as such it is structured and managed by a private sector entity.
2	A commodity exchange has both a 'physical' and a 'financial' component. Both are complex, yet well-known operations. Both generate considerable economic activity, and generate well-paid jobs, and make sustainable contributions to GDP.
3	A successful exchange requires a binding participation by the majority, if not all the producers of the commodity concerned. The fewer the producers, the easier it is to structure and operationalise the commodity exchange.
4	In effect, a commodity exchange is a co-ordinated market structure based on the voluntary participation by all the key producers.
5	The producers may or may not be investors/shareholders in a commodity exchange.
6	Government is a major stakeholder, but not necessarily a shareholder of a commodity exchange.
7	A well-functioning exchange helps smooth the demand-supply balances over the business cycle.

Table 4: Key attributes of a commodity exchange (After Abedian, 2013)

is considered to have a strategic advantage through its dominant position in commodities such as platinum, chrome, and manganese – a market share of 70–90% of the known reserves of these commodities. Consequently, global commodity cycles will dominate for the next few decades with attendant cyclicality – of boom-bust cycles – if left solely to market forces. As a dominant supplier, South Africa (and all its producers) needs to ensure continuity of supply. Any actual or potential disruption of supply will rebound negatively on South Africa's geo-political interests, and also encourage the search for substitutes and hence erode considerable value. To avoid excessive market volatility, commodity exchanges offer an effective framework for long-term profit maximisation and volatility minimisation. Canada's Potash Exchange is a case in point. The key attributes of a commodity exchange are listed in Table 4.

4 The Dominant Practices and the PGM Sector

The 'dominant practices' which all actors follow in the landscape of PGM are fuelled by the ideas and beliefs, and the hegemonic status of a socio-technical regime. In other words, the ideas and beliefs which drive dominant practices result in the hegemonic status of a socio-technical regime. Within the regime a transition would lead to changes in dominant and logical practices. If we review one of the probable dominant practices in the socio-technical regime of PGM – when PGM prices rise, for instance, users recycle or substitute for the PGM – the question of how a transition could be managed in the practice requires innovative thought, approaches and actions in a wide variety of areas in the economy and in the technology sphere. Recycling and substitution initiatives in Europe exemplify an earlier transition from PGM accessibility to a lack of easy access. The transition strategy Europe has sought to use is recycling and substitution, as the geopolitical socio-technical regime of PGM is affected by the socio-technical landscape of the global market with its shifting interests and alliances. Ecological sustainability, resource security and resource scarcity represent some of the pressures that can further influence the regime. Conflicting interests amongst those currently extracting the metals; the governments receiving fiscal benefits from those operations; the labour which works on the mines; and those involved in research, PGM value-add activities, results in a need for actors to engage in a dialogue. To do so effectively and to have the power to negotiate strategically for long-term improvements, a broad understanding of geopolitical dynamics is required.

The global economic outlook is precarious: the world is in its sixth year of a global downturn, which is impacting on both developed and developing nations. Now, instead of a 'collapse of markets', it is a 'collapse of countries'. Consequently, reserve banks which traditionally have been responsible for financial stability have additional concerns (Marcus, 2013). In addition to high unemployment, environmental conditions have also not improved. 'Core business' with 'focus' and short-termism have been the biggest restrictions to development of new opportunities. As an alternative, the Green Economy is not much viable because it has high costs attached to it; it is subsidised and involves tax rebates with a focus on cutting costs. The dominant institutional practices associated with PGM mining are regulated by organisational and regulatory frameworks. The dominant technological practices relate to PGM extraction: the physical activity of mining. Informational practices relate to capacitating actors. There is, though, a question of how to create a capacitated state and how regulation can be informed. The dominant demographic practices in relation to PGM mining are classified into three distinct groupings – government, platinum mining companies (including their employees) and communities neighbouring mining operations. The dominant social practices are associated with the actors that benefit, or not, from PGM mining.

The private sector represents the proportion of the national economy owned and resourced by private enterprise. It includes the personal sector (households) and corporate sector (firms) and is responsible for allocating most of the resources within an economy. A transnational corporation differs from a multinational corporation in that it does not identify itself with one national home. Whilst multinational corporations are national companies with foreign subsidiaries, transnational corporations have operations in many countries, sustaining high levels of local responsiveness.

In contrast to the private sector, government's mandate is different. Government comprises a body of people that sets and administers public policy, and exercises executive, political and sovereign power through customs, institutions and laws within a state. Public sector is that part of a national economy supported financially by government (financed through taxation and levies), which provides basic goods or services that either are not, or cannot be, provided by the private sector. It comprises national and sub-national governments, public corporations, and quasi-autonomous non-government organisations. Public sector is one of the largest sectors of any economy. Multi-lateral organisations are international groups or agencies comprised of member states (Ferraz, 2012).

The central issue in this section is how the envisaged (HySA, 2013) hydrogen strategy for South Africa can evolve in terms of the dominant practices of PGM activities. South Africa's goal is to address the need to increase work opportunities for its people. To do this, South Africa will have to develop capacity in PGM beneficiation (including recycling) and research, and to negotiate with potential collaborators to realise socio-economic opportunities within multiple niches. Niche innovation occurs in three areas: the articulation (and adjustment of expectations), the building of networks and addition of more actors, and learning and articulation processes on various dimensions such as technical design, infrastructural needs, organisational issues and business models (Geels, 2011).

4.1   Access to PGM

Access has to be understood from the perspective of South Africa as predominantly a supplier of raw PGM. The PGM extractive industry is dominated by private sector transnational corporations. As a consequence, supply of PGM is primarily controlled by the private sector. The list of mines and ownership structures is outlined in Appendices A2 and A3.

Besides maximising shareholder value, there is need for greater focus on the rehabilitation of, and catering for, aggregated communities and long alliances to develop uses for PGM[7]. An example is the set-up of HySA Systems as an industry, technology and product development centre to develop, build, commission and validate prototype systems (HySA, 2013). Transnational corporations involved in PGM mining such as Anglo American Plc have added interests in PGM recycling to their business portfolio (Creamer, 2012a). This is but one element of the niches which are needed to initiate a serious transition from a fossil-fuel energy economy to a hydrogen energy economy – or at least, to increase the demand for PGM especially within South Africa.

The key impetus for transitions in PGM activity is the interactions between supply and demand that reflect changes in consumption behaviour. For example, a high growth rate of middle class sections of society in an individual country such as China represents an area of significant growth potential for the automotive industry market. Accordingly, if Chinese legislation requires motor vehicles to have auto catalysts (which contain

---

7. There have been some initiatives in collaboration with the Department of Science and Technology.

PGM), this is a growth area for the PGM industry. The automotive industry would require access to raw PGM also as the supply from recycling activities would be insufficient.

PGM is recycled in different geographies located away from where it was mined. Consequently, recycling has become an important source of PGM, competing for the same markets as the raw product. It is therefore important to avoid the territorial trap where boundaries are confused with barriers and flows, and linkages are obscured by the assumption that autonomous states are the only actors of importance (Dalby, 1998). Economic advantages today are in the hands of those *without* direct territorial access to the mineral resource, i.e. no PGM deposits. The benefits emerge from *negotiated access and strategic collaboration* of the market economy. This reflects a transition from territory-based to non-territory based loci of power driven primarily by globalisation. 'Sovereignty, therefore, operates through a number of spatial modalities: territorial, spatial-interactional, and place-based' (Agnew, 2009, p. 21).

4.1.1 Role of Transnational Corporations

Transnational corporations strive to minimise barriers to trade as well as investment and revenue flows. In the absence of appropriate regulation and partnerships among the various economic role-players, this can result in popular resentment and social instability.

It is consequently incumbent on government to design differentiated and strategic policies toward mutually-beneficial partnerships. This allows host countries to intelligently use transnational corporations for their long-term development plans (Chang, 2003, p. 247–269).

4.1.2 Restrictions to markets

The changing geopolitical environment is characterised by growing protectionism. An example of such protectionism is the use of technical infrastructure such as standards, quality assurance and metrology in global trade by industrialised and advanced developing countries as *Technical Barriers to Trade and Non-Tariff Barriers* that make it increasingly difficult to access their markets (DTI, 2010). These countries have put in place demanding standards generally related to health and safety, termed 'eco-protectionism' under the guise of addressing climate change concerns, particularly from advanced countries. Some countries are considering the imposition of 'border adjustment taxes' on goods imported with greater carbon emissions than similar products produced domestically and subject to carbon emission limits.

The response to these restrictions requires South African Technical Infrastructure policies and institutions to re-orientate themselves to play a strategic industrial policy role. Key institutions include the South African National Accreditation System, the National Regulator for Compulsory Specifications, the South African Bureau of Standards and the National Metrology Institute of South Africa. The strategic industrial policy role has to address how to disallow unsafe and poor-quality imports, and how to ensure access to increasingly demanding export markets.

Proposed restrictions affect both the extractive and manufacturing landscape, particularly in the management and mitigation of process-generated air emissions. The extractive and manufacturing landscape represents a significant component of South Africa's industrial space. Primary beneficiation activities are associated with PGM mining in the extractive socio-technical landscape. Any secondary beneficiation activities that involve adding value to PGM occurs in the manufacturing component of the landscape.

In the extractive metals landscape, process stewardship, and air emissions in particular, have many restrictions. While the environmental attributes of platinum-based fuel cells are relatively positive, such is not necessarily the case for the PGM producers. Air emissions from the PGM mines in South

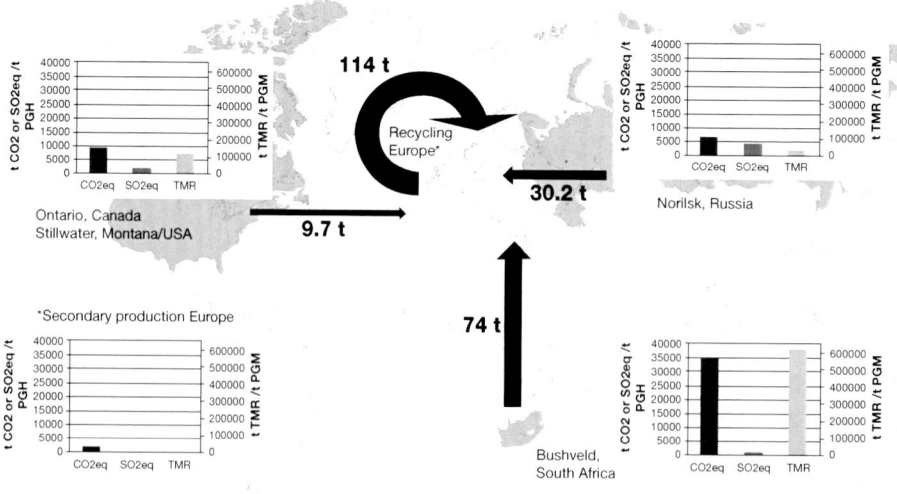

Figure 12: PGM flows to Europe with emissions, and total material requirement (TMR) per tonne PGM produced (Saurat & Bringezu, 2008)

Africa have comparatively low sulphur oxides but somewhat higher greenhouse gases such as carbon dioxide (Figure 12, Saurat and Bringezu, 2008). The amount of greenhouse gases would increase with escalating demand if the commercialisation of PGM fuel cells were to be realised.

The implications of additional restrictions to access the market for raw PGM derived from South African operations would be multifaceted. This would represent an additional cost to the already expensive extractive processes. The high greenhouse gas figures for PGM production are primarily the result of the use of fossil-fuel energy – electricity derived from coal-fired power plants and using oil – in the mining and metallurgical plants. One way to reduce the effect of burning coal and its impact on the atmosphere is to adjust the energy mix used to power PGM mining in South Africa. Alternate power sources need to be assessed collectively by the PGM mining sector. Diversifying the energy mix is one area in which transitions are considered necessary to contribute to the growth in the demand for PGM. Affordability and security of supply of the requisite energy are important considerations to ensure PGM prices do not become an obstacle to the uptake of fuel cells. Incidentally, many of operations in the PGM mines could themselves be powered by fuel cells.

5 Some considerations in a geostrategic approach

The geopolitical implications of a global hydrogen economy for southern Africa have been placed in the context of the need for reliable and affordable energy. Affordable energy is a fundamental prerequisite for the market economy. A major effort in countries globally is to secure their own energy supplies. In South Africa the dominant form of energy is fossil-fuel based, that is coal and oil; and this is unlikely to change rapidly in the near future. Speculation about the potential for a hydrogen economy to maximise PGM inputs is being considered as though the locus of power is in the abundance of the raw materials. But, as this chapter has sought to show, the complexity of global dynamics calls for rigorous strategising at many levels, not only the geological, if South Africa is to take advantage of the fledgling hydrogen economy.

The significance of PGM for South Africa has to be considered from the perspective of how PGM address the country's requirements. The niches in which transitions or change can happen have to be identified. The prospect of a global hydrogen economy is insufficient to create value – measured as significant job creation – today.

Collaboration between relevant actors from government, industry and society is important if job creation and development is to be the focus. For South Africa to facilitate this and close the knowledge gap, alliances with relevant institutions and countries are critical. Alliances from South Africa's perspective must focus on driving its own agenda. Partnerships would assist with growing a broader PGM industrial base: for instance, with Japan, South Korea and Canada to access their PGM research; with the Belgians and Germans for PGM recycling; with the transnational corporations for PGM extraction; with Russia for technical and economic information on PGM. The socio-technical regime (or routines) of PGM mining are not the places for innovative thinking, because extractive processes are very effectively carried out. These routines lead us to think that business-as-usual is possible in perpetuity but such an observation is based on incorrect assumptions and conclusions.

Diversifying the energy mix used in PGM mining is one area in which transitions are considered necessary. Although inputs from PGM in fuel cells – as a secondary energy source – into a hydrogen economy are expected to be small, the retention of the PGM mining industry and the introduction of PGM beneficiation industries are a significant opportunity for South Africa to meet some of its job creation goals. Affordability and security of supply of the requisite energy are important considerations to ensure PGM prices do not become an obstacle to the uptake of PGM-rich products such as fuel cells. Importantly, the market for PGM products also has to be created within South Africa itself.

South Africa has defined a vision for a hydrogen economy in view of its supply dominance over the raw PGM product and the fact that catalytic properties of PGM remain superior despite investigations for substitutes. Also, there have been investments in R&D initiatives in PGM-related studies both in and outside South Africa which have identified some business opportunities but these have yet to be fully realised. Is the vision of a hydrogen economy enunciated by the Department of Science and Technology 'to create knowledge and human resource capacity that will develop high value commercial activities in hydrogen and fuel cell technologies utilising local resources and existing know-how', broad enough, given the observations made above?

Strategies which are decided on in advance are intended strategies, while those that are put into operation, are deliberate strategies. However, the suggestions presented here are for an intended strategy to be realised for South Africa. That would mean taking on the idea of a transition to more diversified PGM activities.

5.1 Suggestions for a geostrategic role for PGM

The suggestions for a geostrategic role for PGM have been informed by the trends in the socio-technical landscape. These trends are long-term. What appears plain is that there has been a change from one geopolitical discourse to another in the New World Order and that discourse needs to inform the intended strategies for South Africa about the role of PGM.

For there to be a turnaround, or development in the socio-technical regime, a comprehensive review of what is current knowledge is required. To access this knowledge, alliances with those who possess it must be set up. At the same time, niches would need to be identified in protected space; it would be possible for innovations to be developed and piloted.

In developing a strategy for PGM, the global reality in which nation states are not the leaders of geopolitical dynamics has to be kept in mind. This should include how corporations respond to the short-term ebbs and flows of the market, and how such responses should relate to long-term strategic considerations.

PGM are classified as a critical raw material, the risk of which, in terms of security of supply, is significant. Such disruptions of supply could impact materials used in the defence, communications, and energy industries. Although there is a strategic advantage to PGM recycling, there is still a need for newly-mined PGM to meet the demand, even though technological advances have lowered the quantity of PGM needed in products.

5.1.1 Room for PGM alliances

The geopolitical dynamic should be used to clarify the role of diverse alliances and the impact of such on policy. Some alliances in this dynamic are discussed and suggested from the perspective of optimising the outcome for South Africa.

In the socio-technical landscape (see Figure 1) in which PGM mining occurs globally, regional and local factors need to be considered. The recently concluded bilateral agreement between South Africa and Russia will mean that the countries will agree to co-operate on PGM initiatives to create a suite of interventions necessary to stabilise the platinum industry (Esterhuizen, 2013). Although Russia is an energy superpower (especially as a gas supplier) its interest, for its own benefit, in stabilising the platinum industry is a factor to be considered in a hydrogen economy.

In South Africa, an alternative conversation is also taking place about curbing the volatility of the PGM market. Economists have proposed a PGM exchange to stabilise prices (Abedian, 2013; Creamer, 2012c). Although initial thoughts of an organisational structure along the lines of OPEC may have been considered, it is suggested that this style would be inopportune for three reasons: first, the PGM mineral assets, although located largely in South Africa, Zimbabwe and Russia, are owned predominantly by private sector entities; second, transnational corporations are active in both the recycling of, and substitution for, PGM activities; the presence of the private sector requires a different form of engagement as the geopolitical considerations of these differ from those of governments; and third, a cartel among private entities would not necessarily benefit society as a whole. Thus, as elaborated in the last chapter, suggestions on the form that a 'platinum exchange' can take should bear in mind these realities.

The niche driving change in this regime of PGM mining is the role for PGM in a global energy transition from a fossil fuel-based to a new energy mix that in the future could be dominated by hydrogen. However, there is no certainty that this will happen. The prospect of displacement of PGM from the automotive sector and other markets must not be ruled out . For example, the global recession makes it possible to ignore environmental considerations, thus reducing demand (Clifford & Martin, 2011). This is supported by significant amounts of research and development to identify not only applications, but also appropriate substitute materials for PGM, as in Japan (Ota, 2013).

5.2   SUMMARY

The suggestions made for inclusion in a geostrategic approach for PGM revolve around efficient PGM mining, processing and recycling:

- In the mining of PGM, labour, supply and demand, and price need to be stabilised. Agreements have to be reached between the actors: government, the PGM sector and labour in South Africa.
- A vehicle for stabilising the PGM price such as a commodity exchange should be considered by the actors: the government and the PGM sector in South Africa as well as economic entities, both locally and internationally. Global alliances should be considered from this perspective.

- The efficiency of PGM mining must continue to improve to optimise mining and lower costs. This may mean the increased use of mechanised mining. Job opportunities could be created in other fields along the PGM value chain.
- South Africa should also develop skills and other capacities for PGM recycling. This would grow work opportunities in specific fields, including research and development in the business of recycling. Collaboration with relevant actors in this field is suggested. In the long-term, greater demand for PGM as social dynamics change in countries such as China and India, and as the hydrogen economy takes off, would ensure expanding markets for PGM from mining. At the same time, recycling would also ensure that the rate of depletion of the PGM endowment is mitigated.
- Through research and development, South Africa should continue to develop more uses for PGM beyond auto-catalysts, fuel cells and jewellery. New uses would grow work opportunities in specific fields. Alliances with relevant actors are suggested.
- The need to introduce a new energy mix that includes PGM components within South Africa must be addressed. This involves establishing a hydrogen economy infrastructure in the country to create demand for fuel cells that use PGM. Further, the capacity to produce PGM-based catalysts must be expanded.

Bibliography

Abedian, I. (2013). The case for a commodity exchange for SA's platinum, chrome and manganese, 15 May 2013, IDC Seminar on Beneficiation, South Africa.

Agnew, J. (2009). *Globalisation and Sovereignty*. New York: Rowman & Littlefield.

Aizhu, C. (2013, January 23). *Reuters*. Retrieved April 18 2013, from Beijing to slap tougher emission standards on vehicles: Xinhua: http://www.reuters.com/article/2013/01/23/us-beijing-emission-idUSBRE90M10R20130123

Ali, S. & Dadush, U. (2012, May 16). *Foreign Policy*. Retrieved April 18 2013, from The global middle class is bigger than we thought: http://www.foreignpolicy.com/articles/2012/05/16/the_global_middle_class_is_bigger_than_we_thought

Alonso, E., Field, F. R. & Kirchain, R. E. (2012). Platinum availability for future automotive technologies. *Environmental Science & Technology*, 46(23), 12986–12993.

Anglo Platinum. (2004). Corporate Review. http://www.angloplatinum.com [Accessed 5 October 2004].

Anglo American. (2013). Operations Overview. http://www.angloplatinum.com/business/operations/overview.asp [Accessed 25 June 2013].

Arkhipov, I. & Wild, F. (2013, March 27). *Russia, South Africa seek to create OPEC-style Platinum Bloc*. Retrieved 3 April 2013, from Bloomberg: http://www.bloomberg.com/news/2013-03-26/russia-south-africa-seek-to-create-opec-style-platinum-bloc.html

Bewag Aktiengesellschaft. (2001). Energy moving into the future, the fuel cell: A technical report, www.fuelcellpark.com [Accessed 22 April 2013].

Bockris, J. (2011). Hydrogen. *Materials*, 4:2073–2091.

Brecher, J. & Costello, T. (1994). Reversing the Race to the Bottom. *The Geopolitics Reader*. Edited by G. Ó Tuathail, S. Dalby, P. Routledge, London: Routledge, 299–304.

Burgess, S. (2011). *The effect of China's scramble for resources and african resource nationalism on the supply of strategic southern african minerals. What can the United States do?* INSS.

Burgis, T. (2009, January 25). *Financial Times*. Retrieved April 11, 2013, from Slowdown brings relief to South Africa's Eskom: http://www.ft.com/intl/cms/s/0/f4bbe266-eafa-11dd-bb6e-0000779fd2ac.html#axzz2QAVqQwZQ

Butler, J. (2012). *Platinum 2012*. Hertfordshire: Johnson Matthey Plc.

Campbell, K. (2013, April 5). *Engineering News*. Retrieved April 11 2013, from SA looks nearly set to start new nuclear power station build programme : http://www.engineeringnews.co.za/article/sa-looks-nearly-set-to-start-new-nuclear-power-station-building-programme-2013-04-05

Campos, I. & Vines, A. (2007). *Angola and China: A pragmatic partnership*. London: Chatham House.

Cawthorn, R. G. (1999). The platinum and palladium resources of the Bushveld Complex. *South African Journal of Science*, 95:481–489.

Cawthorn, R. G., Merkle, R. K. W. & Viljoen, M. J. (2002). Platinum-group element deposits in the Bushveld Complex, South Africa. *In The Geology, Geochemistry, Mineralogy and Mineral Beneficiation of Platinum-Group Elements*. Edited by L. J. Cabri. Canadian Institute of Mining, Metallurgy and Petroleum, Special Volume 54:389–429. Canada: Marc Veilleux Imprimeur.

Chang, H. J. (2003). *Globalisation, Economic Development and the Roles of the State*. London: Zed Books.

Chinguno, C. (2013). Unpacking the Marikana massacre. Corporate Strategy and Industrial Development, University of the Witwatersrand. http://www.polity.org.za/article/unpacking-the-marikana-massacre-february-2013-2013-02-13 [Accessed 22 February 2013].

Clifford, S. & Martin, A. (2011, April 21). *The New York Times*. Retrieved April 9 2013, from As consumers cut spending, 'green' products lose allure: http://www.nytimes.com/2011/04/22/business/energy-environment/22green.html?pagewanted=all&_r=0

COM. (2011). Facts and Figures, Chamber of Mines of South Africa. http://www.bullion.org.za/content/?pid=71&pagename=Facts+and+Figures [Accessed on 19 September 2012].

Conversi, D. (2009). Globalization, ethnic conflict and nationalism. *Handbook of Globalization Studies*. Edited by B. Turner, London: Routledge, 346–366.

Conversi, D. (2010). The limits of cultural globalisation? *Journal of Critical Globalisation Studies*, 3:36–59.

Creamer, M. (2012a). Amplats initiates platinum recycling project. http://www.miningweekly.com/article/amplats-pgm-recycling-2012-02-13 [Accessed 24 April 2013.]

Creamer, M. (2012b). Anglo reviewing stoppage-hit platinum business – Carroll. http://www.miningweekly.com, 2012-02-17 [Accessed 30 April 2012].

Creamer, M. (2012c, July 14). *Miningweekly.com*. Retrieved April 08 2013, from Amplats' Nqwababa rejects platinum exchange, backs consolidation: http://www.miningweekly.com/article/amplats-nqwababa-rejects-platinum-exchange-backs-consolidation-2012-07-24

Creamer, M. (2013). Eastplats puts Crocodile River mine on care and maintenance. http://www.miningweekly.com/article/eastplats-puts-crocodile-river-mine-on-care-and-maintenance-2013-06-25 Accessed 25 June 2013.

Dalby, S. (1998). Conclusions: Geopolitics, knowledge and power at the end of the century. *The Geopolitics Reader*. Edited by G. Ó Tuathail, S. Dalby, P. Routledge, London: Routledge, 305–312.

DME. (2003). Platinum-group Metal Mines in South Africa 2003. Directorate: Mineral Economics, Department of Minerals and Energy, D6/2003. http://www.dme.gov.za [Accessed on 17 September 2012].

Dodds, K. (2007). *Geopolitics – A very short introduction*. New York: Oxford University Press.

DTI. (2010). 2010/11-2012/13: Industrial Policy Action Plan, February 2010. Economic Sectors and Employment Cluster, Department of Trade and Industry. http://www.info.gov.za/view/DownloadFileAction?id=117330 [Accessed on 10 September 2010].

Ealey, L. A. & Mercer, G. A. (2002, August). Tomorrow's cars, today's engines. *The McKinsey Quarterly*, 40–53.

Eberhard, A. (2013, March 19). *BusinessDay BDlive*. Retrieved April 11, 2013, from Nuclear power is neither necessary nor cost-effective: http://www.bdlive.co.za/opinion/2013/03/19/nuclear-power-is-neither-necessary-nor-cost-effective

Eggert, R. G., Carpenter, A. S., Freiman, S. W., Graedel, T. E., Meyer, D. A., McNulty, T. P., *et al.* (2008). *Minerals, critical minerals, and the U.S. economy.* National Academies Press.

EITI. (2013). Homepage. Extractive Industries Transparency Initiative. http://eiti.org [Accessed on 11 July 2013].

Ekins, P. & Hughes, N. (2009). The prospects for a hydrogen economy (1): hydrogen futures. *Technology Analysis & Strategic Management,* 21:783–803.

Emerson, J. W., Esty, D. C., Hsu, A., Levy, M. A., de Sherbinin, A., Mara, V., *et al.* (2012). *EPI 2012: Environmental Performance Index and Pilot Trend Environmental Perfomance Index.* New Haven: Yale University.

Energy Regulator. (2008). *Inquiry into the national electricity supply shortage and load shedding.* National Energy Regulator of South Africa.

Erdmann, L. & Graedel, T. E. (2011). Criticality of non-fuel minerals: a review of major approaches and analyses. *Environmental Science & Technology,* 45, 7620–7630.

Esterhuizen, I. (2013). Shabangu announces gold, platinum rescue plan. Mining Weekly. http://www.miningweekly.com/article/shabangu-announces-gold-platinum-rescue-plan-2013-05-28 [Accessed on 30 May 2013].

European Parliament. (2013). An effective raw materials strategy for Europe. *Official Journal of the European Union,* 56, 21–37.

Fakir, S. (2011, August 5). *Engineering News.* Retrieved April 22 2013, from There are still many unknowns about shale gas extraction in the Karoo: http://www.engineeringnews. co.za/article/there-are-still-many-unknowns-about-shale-gas-exctraction-in-the-karoo-2011-08-05

Ferraz, M. F. F. (2012). Sustainability, Sustainable Development and Mining: Theoretical Foundation, Work Package 01. MISTRA, Project Report. MR005: South Africa and the Global Hydrogen Economy: The Strategic Role of Platinum Group Metals.

Ferraz, M. F. F. (2013). Mining, Sustainable Development and Sustainability: Sustainability Orientated Model, Work Package 02. MISTRA, Project Report. MR005: South Africa and the Global Hydrogen Economy: The Strategic Role of Platinum Group Metals. GCIS. (2012, September 7). *Government Communications.* Retrieved April 11 2013, from Post-Cabinet Lekgotla media statement: http://www.gcis.gov.za/content/ newsroom/media-releases/cabstatements/post-cabinet-lekgotla-media-statement-phumla-williams-acting-cabinet-spokesperson

Geels, F. W. (2011). The multi-level perspective on sustainability transitions: Responses to seven criticisms. *Environmental Innovation and Societal Transitions,* 1:24–40.

Global Witness. (2013). Homepage, Global Witness. http://www.globalwitness.org [Accessed on 11 July 2013].

Graedel, T. E. (2002). Material substitution: a resource supply perspective. *Resources, Conservation and Recycling,* 107–115.

Graedel, T. E., Allwood, J., Birat, J. P., Buchert, M. & Hageluken, C. (2011). *What do we know about metal recycling rates?* Lincoln: USGS.

Grant, T. Ed (2004a). Gold Fields Ltd – History of Gencor Ltd. http://www.enotes.com/company-histories/gold-fields-ltd/history-gencor-ltd [Accessed 16 November 2011].

Grant, T. Ed (2004b). Gold Fields Ltd – Gold Fields and Gencor Unite in 1998. http://www.enotes.com/company-histories/gold-fields-ltd/gold-fields-gencor-unite-1998 [Accessed 16 November 2011].

Green, D. (2013, February 21). *WardsAuto*. Retrieved April 16, 2013, from Experts say Beijing's new auto emissions standards to ripple through China: http://wardsauto.com/asia-pacific/experts-say-beijing-s-new-auto-emissions-standards-ripple-through-china

GRI. (2011). Global Reporting Initiative. Homepage. http://www.globalreporting.org/Home [Accessed 15 November 2011].

Hageluken, C. (2007). Closing the loop – Recycling of automotive catalysts. *Metall-Forschung*, 24–39.

Hageluken, C. (2012, January). Recycling the Platinum Group Metals: A European perspective. *Platinum Metals Review*, 29–35.

Halme, K., Piirainen, K. A., Vekinis, G., Sievers, E. U. & Viljamaa, K. (2012). *Substitutionability of critical raw materials*. Brussels: European Parliament.

HySA. (2013). HySA Systems. http://www.hysasystems.org/wp-content/uploads/hysa/images/2012/07/Welcome-to-HySA.pdf [Accessed 25 May 2013].

ICMM. (2007). Materials Stewardship, Eco-efficiency and Product Policy. www.icmm.com [Accessed on 7 May 2008].

IMF. (2011). International Monetary Fund. Glossary page. http://www.imf.org/external/np/exr/glossary/showTerm.asp#91 [Accessed on 10 June 2011].

IRP. (2011). *Recycling rates of metals – a status report*. International Resource Panel. Paris: United Nations Environment Programme.

Jaffe, A. M. (2010, May 10). Shale gas will rock the world. *The Wall Street Journal*.

Jepson, N. (2012). *A 21st Century scramble: South Africa, China and the rare earth metals industry*. South African Institute of International Afffairs.

Karlin, A. (2009). Review of 'Global Catastrophes and Trends' (V. Smil). http://www.vaclavsmil.com/wp-content/uploads/docs/smil-bookreview-global-catastrophes-20090928-karlin.pdf [Accessed 24 May 2013].

Khusanjanova, J. (2011). OPEC's benefit for the member countries. *Research in world economy*, 14–23.

Kinnard, J. A. (2000). The Bushveld Large Igneous Province. http://www.largeigneousprovinces.org/sites/default/files/BushveldLIP.pdf [Accessed 10 September 2012].

Kretschmer, T. (1998). De Beers and beyond: The history of the international diamond cartel. http://pages.stern.nyu.edu/~lcabral/teaching/debeers3.pdf [Accessed on 30 April 2013].

Lapper, R. (2008, December 5). *Financial Times*. Retrieved April 11 2013, from Eskom cancels plan to build nuclear plant: http://www.ft.com/intl/cms/s/0/f89a4712-c2fd-11dd-a5ae-000077b07658.html#axzz2QAVqQwZQ

Lapper, R. (2010, May 25). *Financial Times*. Retrieved April 18 2013, from China seals African platinum deal: http://www.ft.com/intl/cms/s/0/6c44a70e-67f6-11df-af6c-00144feab49a.html#axzz2QpKQaIJn

Le Billon, P. (2004). The geopolitical economy of 'resource wars'. *Geopolitics*, 1–28.

Loder, A. (2013, February 13). *Bloomberg*. Retrieved April 8 2013, from Fracking threatens OPEC as U.S. output at 20-year high: http://www.bloomberg.com/news/2013-02-13/fracking-threatens-opec-as-u-s-output-at-20-year-high.html

Marcus, G. (2013). Welcome and introduction. Comments presented at Growth, Job

Creation and Financial Stability Workshop, 2 May 2013, South African Reserve Bank Conference Centre, Pretoria, South Africa.

Mauro, P., Chamon, M. & Okawa, Y. (2008). Mass car ownership in the emerging market giants. *Economic Policy*, 245–296.

Mbeki, M. (2011). Advocates for Change: How to Overcome Africa's Challenges. Paper presented as public lecture at Department of Politics, Faculty of Humanities, 20 September 2011, University of Johannesburg, South Africa.

McCutcheon, R., Mishal, B. M., Scamuffa, A., Haffner, S., Simone, D., Portnoy, M., *et al.* (2011). *Shale Gas: A renaissance in US manufacturing?* pwc.

Medlock III, K. B., Jaffe, A. M. & Hartley, P. R. (2011). *Shale gas and U.S. National Security.* Houston: James A. Baker III Institute for Public Policy, Rice University.

Mudd, G. M. (2010). Platinum group metals: a unique case study in the sustainability of mineral resources. *The 4th International Platinum Conference, Platinum in transition 'Boom or Bust'*, the Southern African Institute of Mining and Metallurgy.

NAP. (2013). Overview Lac de Iles, North American Palladium. http://www.nap.com/operations/lac-des-iles-mine/default.aspx [Accessed 30 April 2013].

NIC. (2012). *Global trends 2030: Alternative Worlds.* National Intelligence Council.

Norilsk. (2013). About Norilsk Nickel. http://www.nornik.ru/en/about/ [Accessed 25 April 2013].

NPC. (2011). *National Development Plan.* The Presidency – Republic of South Africa.

Ó Tuathail, G. (1998). Introduction: Thinking critically about geopolitics. *The Geopolitics Reader*. Edited by G. Ó Tuathail, S. Dalby, P. Routledge, London: Routledge, 1–14.

Oberthür, T., Melcher, F., Buchholz, P. & Locmelis, M. (2012). The oxidised ores of the Main Sulphide Zone, Great Dyke, Zimbabwe: Turning resources into minable reserves – mineralogy is the key. *Platinum 2012*, Southern African Institute of Mining and Metallurgy, 647–672. http://www.saimm.co.za/Conferences/Pt2012/647–672_Oberthur.pdf [Accessed 25 April 2013].

Ota, K. (2013). Trends in fuel cell development in Japan. Paper presented as public lecture at Platinum Valley Development Initiative, 10 April 2013, IDC Conference Centre, Johannesburg, South Africa.

Paladini, S. & George, S. (2011). Chinese corporations and the scramble for resources. A study in the oil business. *L'Espace Politique*, 15:3. http://espacepolitique.revues.org/index2151.html [Accessed 25 May 2013].

Powertech. (2013). *A review of South Africa's electricity sector.* Creamer Media's Research Channel.

REACH. (2013). European Chemicals Agency. Homepage. http://echa.europa.eu/web/guest/regulations/reach [Accessed 30 April 2013].

Richter, A., Burrows, J. P., Nub, H., Granier, C. & Niemeier, U. (2005). Increase in tropospheric nitrogen dioxide over China observed from space. Nature, 129–132.

Routledge, P. (1998). Introduction: Anti-geopolitics. *The Geopolitics Reader.* Edited by G. Ó Tuathail, S. Dalby, P. Routledge, London: Routledge, 245–255.

Saurat, M. & Bringezu, S. (2008). Platinum Group Metal Flows of Europe, Part I. *Journal of Industrial Ecology*, 12(5/6), 754–767.

Schot, J. & Geels, F. W. (2008). Strategic niche management and sustainable innovation journeys: theory, finding, research agenda, and policy. *Technology Analysis & Strategic Management*, 20:537–554.

Schouwstra, R. P. & Kinioch, E. D. (1999). A short geological review of the Bushveld Complex. *Platinum Metals Review*, 44(1):33–39.

Silberglitt, R., Bartis, J. T., Chow, B. G., An, D. L. & Kyle, B. (2013). *Critical materials – present danger to U.S. manufacturing*. RAND Corporation.

Smil, V. (1994). Some contrarian notes on environmental threats to national security, from Canadian Foreign Policy. *The Geopolitics Reader*. Edited by G. Ó Tuathail, S. Dalby, P. Routledge, London: Routledge, 212–215.

Smil, V. (2008). *Global Catastrophes and Trends: The Next 50 Years*. London: MIT.

Stillwater. (2013). Stillwater Mining Company. http://www.stillwatermining.com

The Economist. (2013, March 9). *The Economist*. Retrieved April 13 2013, from Fool's Platinum: http://www.economist.com/news/technology-quarterly/21572924-asteroid-mining-two-start-ups-see-promise-extracting-valuable-resources

TIAX. (2003). Platinum availability and economics for PEMFC commercialization. http://www1.eere.energy.gov/hydrogenandfuelcells/pdfs/tiax_platinum.pdf [Accessed on 30 April 2013[.

Timmons, H. & Vyawahare, M. (2012, February 1). *The New York Times Interational Herald Tribune*. Retrieved April 18 2013, from India's Air the World's Unhealthiest, Study Says: http://india.blogs.nytimes.com/2012/02/01/indias-air-the-worlds-unhealthiest-study-says/

Trembath, A., Jenkins, J., Nordhaus, T. & Shellenberger, M. (2012). *Where the shale gas revolution came from*. Breakthrough Institute.

Umicore. (2013). Homepage. http://www.umicore.com/en/ [Accessed 28 April 2013[.

US DoD. (2013). *Strategic and critical materials – 2013 report on stockpile requirements*. United States Department of Defence.

US EIA. (2011). *World shale gas resources: an initial assessment of 14 regions outside of the United States*. Washington DC: US Department of Energy .

Van den Bergh, J. C. J. M. & Kemp, R. (2006). Economics and transitions: lessons from economic sub-disciplines. http://kemp.unu-merit.nl/docs/Economics%20%20transitions%20-%20vandenBergh%20%20Kemp%20-%20%20Final.pdf [Accessed 28 March 2013].

Veseth, M. (2010). *Globaloney 2: The Crash of 2008 and the Future of Globalisation*. New York: Rowman & Littlefield.

Wager, P. A., Lang, D. J., Wittmer, D., Bleischwitz, R. & Hageluken, C. (2012). Towards a more sustainable use of scarce metals. *GAIA*, 300–309.

Wald, M. L. (2004). Questions about a Hydrogen Economy. *Scientific American*, 290:66–73.

WBCSD. (2011). World Business Council for Sustainable Development. Homepage. www.wbcsd.org [Accessed on 9 May 2011].

Whitburn, P. (2012, April). The goose that laid the platinum egg. *REVIEW*, 10–17.

Wong, E. (2013, March 21). As pollution worsens in China, solutions succumb to infighting. *The New York Times*.

Wu, Y., Wang, R., Zhou, B. L., Fu, L., He, K. & Hao, J. (2010). On-road vehicle emission control in Beijing: Past, present, and future. *Environmental Science and Technoloy*, 45, 147–153.

Yang, C. J. (2009). An impending platinum crisis and its implications for the future of the automobile. *Energy Policy*, 1805–1808.

Chapter 7

Beyond Mining: Sustainability and Sustainable Development
Theoretical Foundation

Fátima Ferraz

1 Introduction

Mining and its beneficiation activities can be described as 'pivotal and growing instruments of human advancement, but also concentrated sites of poverty and dysfunctionalities' (Bugliarello, 2011:3). Growth, such as mining enables, also leads to sustainability failures as the mining activities impact on natural and human resources (Ashton, *et al.*, 2001; BMF 2008, 2011 & 2012). Understanding how mine life cycle interactions evolve is important for responding to emerging sustainability challenges – and to balance growth with the need for continued socio-economic opportunities. The challenges which emerge in this attempt at balancing can be considered in seven dimensions: ideological, institutional, economic, demographic and social, informational and technological dimensions. In mining traditionally social (labour), water, energy, and infrastructure challenges are acknowledged.

An understanding of sustainability, sustainable development and the mining life cycle, but for numerous hidden assumptions, needs to be made overt. Decisions based on a more holistic and integrated approach which takes account of current trends in thinking about sustainability will have more value. A further aid to decision making and risk reduction is to conceptualise the innumerable interlocking systems and networks which are a feature of the greater system in which mining is only one small system. Taking account of the systems and networks allows a form of management of the transition processes that drive much needed change which is fundamentally different from either what government or the unions call for.

The emerging and fundamental challenges in the mining life cycle are, however, inextricably interconnected in multiple systems. This is because the mining life cycle is embedded in and makes use of the resources of the biosphere (which includes the environment). But the interconnectedness of all that happens makes discussion about meeting challenges difficult. What is most important? What can be left till later? What has to change right away? An effective response to emerging and interconnected challenges '…will determine the future trajectory of human settlements and, ultimately, of our species. This will require comprehensive paradigms of urban dynamics and a new vision of engineering' (Bugliarello, 2011:3). Consensus amongst prominent thinkers suggests that a holistic, integrated approach would serve those who are charged with planning and executing human settlements in any setting, from the outset (Maiello, *et al.*, 2011; Bugliarello, 2011).

Any approach to planning and the subsequent execution of interventions to address the socio-economic challenges among others facing the mining sector, requires an understanding of the socio-technical regime in which this industry operates. This chapter provides a description of the socio-economic and socio-political landscape in which mining operates in South Africa. The chapter is a consideration of the social and environmental impact of platinum mining as it currently operates. The theoretical foundation starts with an examination of the evolution of diverse concepts of sustainability, and the expression of sustainable development as it has been interpreted globally, and specifically within the mining sector in South Africa. There is a brief outline of mineral beneficiation. The relationship between current practice in mining and beneficiation as it is construed from a neoclassical economic paradigm is offered. There is a discussion about globalisation in relation to neoclassical economics. A description of regulatory structures underpinning the concepts of sustainability in mining is also presented.

The chapter draws from current ideas across several disciplines. In this sense, it is transdisciplinary. Transdisciplinarity is understood as intellectual space in which the manifold links among isolated issues can be explored, the space where issues are rethought, alternatives reconsidered and interrelations revealed (Nicolescu, 2010:19–38).

The five areas identified – situational analysis, beneficiation including economics and globalisation, sustainability and sustainable development and regulatory structures – are interconnected and key to understanding the landscape in which the mining sector, government, and communities function.

1.1   General Background

Directed transitions in beneficiation, it is suggested, can ultimately make a feasible contribution to sustainability. The systemic transformations required to make secondary beneficiation (upstream and downstream) benefit people, the economy and the environment can be done gradually. In the economic landscape of mining, a neoclassical landscape, with its 'dominant practices, rules and ensuing logic of appropriateness' (Kemp & Rotmans, 2004:140) some actors have not allocated value to non-monetary components but to a high-profit margin with growth as a focus. Other actors have regularly pointed out the unequal distribution of wealth. The mining industry has avoided giving value to non-monetary components as the instability in the sector that might follow such a decision could affect profit and growth. Sustainable development initiatives are treated from a single profit maximisation perspective *in mining*. This aligns with the nature of business which is to simplify systems and dimensionality as Magala (2010 asserts: 'The institutional reduction of complexity follows a reflection on the "core competence" and the raison d'être of an institution' (pp. 147–149). The core competence of mining is primary beneficiation involving extraction, smelting and refining processes.

Sustainable development activities are undertaken in two ways generally, through corporate social responsibility programmes and sustainable development projects. Projects in sustainable development, especially if grounded in science and technology, often come to summary ends, or are truncated before they could be said to have achieved sustainability objectives, outside of the narrow scope of economic and technological dimensions. In such interventions, urgent questions requiring technological solutions for the mines' benefit are resolved; however, social questions are largely left to simmer. Certain environmental questions are addressed but in a limited way (BMF, n.d.). From the mining sector's perspective, social and environmental questions relate primarily to post-closure liabilities. Nevertheless, there are many derelict and ownerless mines and mine-polluted areas throughout the country that have become the responsibility of government to rehabilitate, at considerable (unplanned) cost to the fiscus. As a result, current legislation places an onus on mine owners/operators to obtain closure certificates for mines once their extractive capacity has been reached[1]. Other questions – anticipating community needs once a mine has closed, or after organisational restructuring and downsizing in the transnational corporations has created unpredictable conditions; or, imagining what the needs would be of migrants

---

1. Time delays in awarding of, for example, alternate use permits, of mine closure certificates and finalising social and labour plans, can be attributed to each actor – regulator, mining company, community.

who had come into the area looking for opportunities – are ignored or addressed inadequately.

Sustainable development (as currently implemented within the mining sector) was and is not persistent; sustainability is conceived as a short-term solution to long-term abuses of the biosphere; reporting structures on how sustainability is achieved continue to pass through operational departments whose core concern is not sustainability. There are few career development paths open to sustainability practitioners in mining companies. Budgets are designed according to mining constraints without consideration of broader obligations. The relationship between people and the biosphere in which we all, including mines, exist is approached in an apparently *ad hoc* way in mining. It might be argued that this is a false claim – the dominant practices of the mining sector are well regulated and monitored – but their practices are shown to be *ad hoc* when a broader definition of sustainability is used than the approach current in the industry.

The approach in the mining industry to sustainability, as in many other industries, is reductionist. The oversimplification of sustainability into three dimensions – *economic, social* and *environmental* that industry terms the *triple bottom line* – does not allow the complex, interrelated dimensions of the biosphere to be considered. The reason for the reductionist view is so that primacy can be given to economic considerations. This is no surprise in the global economic landscape in which industry operates. Second, disparate elements of the biosphere, each of which is its own complex adaptive system, are approached as if they were of the same value, when plainly human and environmental capital are not interchangeable. Meyer (2005) notes that 'these values ought to be conceptualised in a manner inclusive of material and non-material connections to the environment, both of which are threatened by practices of a deregulated, neo-liberal, global economy' (p. 83). Third, economic and social dimensions are human constructs. By including the environmental dimension in the *triple bottom line* concept to support sustainability, an important fact is obscured. This fact is that the environment refers to the biosphere, the ground in which human constructs are built. The fallacy of the *triple bottom line* or *weak sustainability* argument becomes clear in the implementation of sustainability in many sustainable development projects in the mines. When the biosphere is considered as a complex adaptive system, many more complex relationships are revealed between the ongoing damage to the environment through mining, the deleterious effects of ongoing growth, and the intimate connection between social well-being and a healthy environment. Systems as complex as these can be made easier for discussion if we speak about seven dimensions (albeit in a reductionist way) namely: ideological, institutional, economic, social, demographic, informational and technological (after Gell-Mann, 1994; after Speth, 1992). The need for changes in these dimensions from current practice to better practice can be articulated more clearly in terms of the dimensions.

My first consideration in understanding sustainable development as conceived in mining terms was to look into the activity of *beneficiation* over the last hundred or so years. 'The dominant practices, rules and ensuing logic of appropriateness' which Kemp and Rotmans (2004:140) speak about would reveal themselves and it would be clear how little or how much mining had changed in the face of the sustainability debate. European and American money funded the earliest mine extraction processes in South Africa. The Chamber of Mines of South Africa was formed in 1887 to protect vested interests. The close relationship between the government of the day and the mine owners is well recorded (Stephens, 2003; van Onselen, 1982; Van-Helten, 1997). Beneficiation is best understood in terms of the economic and ideological paradigm in which the business of mining was conducted in the South African colony. Despite the historical moves from being a colony to a democracy, some of the patterns ensuing from pre-sustainability thinking about *beneficiation* versus *waste* continue to be entrenched in the early twenty-first century. When mining emerged in the British ruled Cape Colony in the nineteenth century (1875), practices appropriate to the period were already in place. As early as the 1870s diamond mining was well-established in Kimberley (Worger, 1987) and as a non-agricultural and labour-intensive activity it provided the template for what followed in other mining sectors. Duncan (1995, p. 19) sums up those labour practices, '…using blacks for spade and shovel work under white supervisors, recruiting migrant labour for fixed contracts, and compounding African workers to maximise control and productivity.' The master-servant relationship was a well-established norm for a colony. The colony sent gold and profits back to Imperial Britain. It was incumbent on colonists to understand the concept of 'service' to the empire, as colonists; Williams (1958, p. 39) succinctly says of the colonist: 'He must subordinate his … [own interests] to a larger good, which is called the Queen's peace, or national security, or law and order, or the public weal. This has been the charter of many thousands of devoted lives, and it is necessary to respect it'. The commitment to service of a larger good metamorphosed into a commitment to serve the stakeholders of the mines, even when South Africa was no longer a colony.

In a context such as existed in South Africa during the colonial period, the consideration of environmental and human capital was not important. Some of the colonial attitudes – to approach environmental legislative compliance only in its narrowest sense, and the well-being of workers from a position which underlines the privilege of the old 'master' in relation to his 'servants', have remained. Old concepts used to interpret new ideas, technologies and societal expectations are the norm amongst businesses and organisations. Geels (2004, p. 42) speaking about transitions, nevertheless optimistically reminds us that, 'The existing regime should not just be analysed as a barrier. Ongoing processes in the regime can also provide opportunities for novelties to link up with'. It is this optimism which underlies the development of a sustainability-orientated model, the transitions to an ecologically focused or institutional economic paradigm are hindered by the slow rate of change in the industry.

South Africa only in the 1990s began to legislate for sustainability, reflecting the worldwide recognition of a looming environmental crisis. The crisis was described in terms of climate change, poor air quality, and the loss of natural resources. There was no overt statement about how far and how immediately development and growth would impact the biosphere and compound the quality of life of the communities around the mines. The Brundtland definition of sustainable development, formulated as 'development that meets the needs of the present without compromising the ability of future generations to meet their own needs' (Brundtland, 1987) had emerged in Europe as a response to the environmental crisis, and South Africa responded rapidly.

To implement sustainable development is complex. The understanding of the transnational mining corporations about what to do with the sustainability agenda was important as they and government would institute the appropriate actions. Changes in society's expectations of industry would play a role over the century or so that mining had occurred. Industry's response came finally in the 1990s in the form of *Corporate Social Responsibility* reporting. This practice was introduced at the same time as transnational corporations continued to shed jobs and close operations often in one area (specifically South Africa), while opening new operations in other countries where financial benefits were more favourable. This practice allowed business to focus on maximising economic profit. An increasing alienation between transnational corporations and South African society was reflected in increasing levels of nationalist sentiment. Consequently stricter regulatory policies were introduced to minimise the impact on local landscapes where such actors operate, pollute, disengage or close.

Interventions in the mines in sustainable development are institutionalised responses to the larger call for sustainability to be on their agendas. As a result, the understanding of sustainability is reduced to *triple bottom line* efforts. Making the distinction between sustainability and sustainable development is a necessary next step in this chapter. My understanding that a global consciousness is the basis of thoughts on sustainability led me to further understandings – by virtue of its inclusivity, sustainability is transdisciplinary and systemic. Sustainability can be described in terms of a complex adaptive system in which the needs of the present do not compromise the ability of future generations to meet their own needs. Extinction of future and present biodiversity will result from the excessive exploitation of resources in the biosphere, at many levels and in many dimensions. So, a more effective way of implementing sufficiency and sustainability would be to invoke the need for *systemic* adaptations and transitions through the establishment of an industry based on secondary beneficiation. Such a proposal can be referred to as a *system innovation*, which includes, but is wider than, product and process innovation (Geels, 2005, p. 3). Once the idea of *sustainability* includes multiple interactions of human activity, well-being, and the biosphere, it is useful to speak of transitions in seven dimensions.

The brief review that follows of the primary activities of the mining industry indicates why the industry is slow to embrace a larger perspective on sustainability.

2 Beneficiation

Beneficiation in the mining industry as is currently practised is described so that it can be assessed in relation to the economic model in which transnational corporations have been established, and operate.

Beneficiation from a technical perspective in mining is understood as the series of processes which allow extracted ore to be separated into mineral or metal, and gangue material. The mineral and metal products can be processed further or used, while the gangue material is stockpiled as waste, or for possible processing should it become economically viable at a later stage. The value chain from physical mining processes, through metallurgical extractive processes with their resulting mineral and metal products is termed *primary beneficiation*. Beneficiation may be described principally in terms of technological and economic dimensions, although institutional and ideological dimensions are significant in as much as they are the dimensions in which the mines have well-developed arrangements to support their key activities.

The technological dimension of the mining industry, as a result of the nature of the activity, has evolved rapidly and progressively. Ore destined to undergo pyrometallurgical processes is physically concentrated through milling and flotation beforehand. Ore destined for hydrometallurgical extraction is broken or crushed before being leached (Davenport, *et al.*, 2002, pp. 31–54; Merkle & McKenzie, 2002). Minerals, such as copper and platinum, for example, are too dilute in the ore-bearing rock to be directly smelted (0.5 to 2% for copper, and greater than 2 grams per tonne for platinum group elements, Barnes & Maier, 2002, p. 432). To heat and melt the large quantities of rock to extract these small quantities of the desired mineral and metal from the ore requires a great deal of energy and furnace capacity. There is great pressure on the mines to undertake cleaner production and use less energy. Additional demands are also made on the mining industry to reduce the deleterious impact of exploration, extractive, smelting and refining activities on the social and environmental (including biodiversity, air and water quality) spheres.

2.1 CLEANER PRODUCTION

The commonly used definition and established view of cleaner production was developed by the United Nations Environment Programme (UNEP) in 1990 and states that 'cleaner production is the continuous application of an integrated preventative environmental strategy, to processes, products and services, to increase efficiency and reduce risks to humans and the environment' (UNEP, 2011d). Such practices focus on the idea *reduce, reuse, recycle* to conserve raw materials and energy, eliminate toxic inputs and reduce toxic outputs reflecting a design philosophy for its processes. The United Nations Industrial Development Organisation (UNIDO) is the agency mandated to focus on sustainable industrial development. It is also responsible for the roll-out of National Cleaner Production Centres to encourage the use of cleaner production methodologies in developing countries.

For *reduce, reuse, recycle* to be implemented in a country on a systematic basis, requires a regulator. The imperative to change towards cleaner production, in particular for the mining sector, has had to be driven by legal mandates. Regulations have been put in place during the period in which mines have operated in South Africa, in keeping with evolving ideas about processes, procedures and the impact on the economy, the workers and the environment.[2] The mines require some form of inducement to offset existing investments in capital intensive infrastructure. Continuous improvement strategies, which focus on incremental adjustments, are sought after, particularly in cases in which the same inputs result in more output, a reduced amount of waste or lower energy use per unit of output. With this changed perspective, waste is viewed as a by-product of the process, and with cleaner production methodologies, management looks to gain financially from the value of this waste.

The International Council on Mining and Metals (ICMM) has taken a systems approach to materials stewardship. Materials are considered from the angle of resources, processes and product life cycles. This perspective is outlined as the *minerals cycle* (ICMM, 2007, Figure 6, Chapter 6). The starting premise is that it is critical to understand the materials flows and life cycle benefits; to build and strengthen relationships with other players in the value chain; and to optimise production and provide inputs to support decision-making.

The concept of life cycle assessments of materials from cradle-to-grave and cradle-to-cradle is a good mechanism with which to assess the implementation of sustainability in transnational corporations. McDonough and Braungart (2002) and Benyus (1997) have argued persuasively for the need to reassess design. They raise questions about how products are made to address cradle-to-cradle environmental concerns. The transnational corporations in the mining sector readily and willingly implement cleaner production methodologies provided that they result in process and production efficiencies, and that these efficiencies can be quantified financially. The same principles of cleaner production can also be used to evaluate the sustainability of post-mining economic activities. The difficulty is quantifying such activity in clear financial terms

The minerals cycle is divided into two main flows, *process stewardship* and *product stewardship* (see Figure 6, Chapter 6). Process stewardship includes exploration, mining, concentration, smelting and refining. This value-add chain is effectively primary beneficiation. Primary beneficiation is characterised by capital-intensive plants; these plants generally have low employment levels engaged in the production of mass intermediate products.

---

2.   These regulations are mainly aimed at changing processes to improved levels of efficiencies.

This is the case in South Africa, where such intermediate products account for a significant part of the national mineral revenue. Product stewardship, on the other hand, involves transforming the feedstock downstream or upstream from the process stage in fabrication, in design and in manufacture.

Beneficiation or value-add of both upstream by-products, such as wastes, and downstream by-products, such as the readily saleable product, can be seen as an extension of cleaner production. Residues or wastes are co-generated as by-products of industrial, mining, agricultural, municipal and other processes. These resultant materials may be organic, inorganic, non-hazardous but in some instances, hazardous in nature. Increasingly, the large volumes of residues or wastes generated through such processes require long-term environmental management. Advances in materials technologies have opened up numerous economic opportunities to *reduce, reuse and recycle* these materials or natural resources.

An illustration of the case for beneficiation of residues, in this instance materials research and development, is presented by Pappu, *et al.* (2007). India, for example, produces approximately 960 Megatonnes per year of waste, including inorganic mining waste residues; this could represent a significant resource available for further exploitation and enterprise development were product stewardship to be considered differently than it is now. There is a worldwide shift towards the recycling of metals, although economies of scale still favour large primary mining and smelting complexes over smaller recyclers (Ayres, 1997:145). The relationship between environmental performance and business performance can be undertaken by assessing low and high cost activities in terms of their financial contribution to the business performance. In one case, Zeng, *et al.* (2010) found an overall financial, positive impact from cleaner production in the business performance of industrial firms with low-cost activities (exemplified by improving working conditions to reduce waste), and they found that high-cost activities make a greater contribution to non-financial performance (such as using energy efficient and clean technologies). Some transnational corporations in the mining sector have begun, through commitment to materials stewardship, to contribute to the creation of societal value while minimising the impact on humans and the environment.

At present, industry sells its primary beneficiated commodities through existing economic market mechanisms such as the London Metals Exchange. The mining sector, with its focus on the primary beneficiation of minerals generates millions of tonnes of waste material. Such mining waste residue may comprise slag, tailings, slimes, brines and waste rock. Solid materials of this nature are typically stockpiled: a process that requires long-term management even after the mining operation has closed. The disposed minerals may carry some intrinsic value, including minor remnants of the target commodity, and other commodities that were not of primary interest. For example, after the collapse of the nuclear energy industry in the late 1970s, uranium in gold mining ores was no longer extracted and instead was disposed of to the tailings storage facilities (WNA, 2011). Secondary beneficiation of mining waste residues that are stockpiled opens opportunities of a different nature to the established market mechanisms.

The routine extractive processes of mining do affect the environment and the complex social setting associated with those operations. The industry is guided by the need to give effect to regulatory and industry specific guidelines. The role of (mining) beneficiation in generating value for local economic development to benefit communities has grown with the increasing awareness of sustainable development. Examples of positive outcomes from this awareness by the industry can be shown in relation to safety, conservation and air quality. One area of improvement has been safety as reflected in the drive to improve the levels of workplace safety on the mines in South Africa. Safety is a challenge for the mining industry but through the introduction of safety initiatives, the industry has reduced the number of fatalities due to workplace accidents in mines – a target of zero fatalities and injuries has been set. Industry's performance since 1996 shows a reduction in the fatality frequency (Figure 1), defined as fatalities per million hours worked. The Chamber of Mines in South Africa set an objective of 20 per cent annual improvement; since 2003 when safety milestones were agreed to for 2013, there has been a 54 per cent reduction in fatalities (COM, 2006, p. 63; 2011, p. 68).

The Diamond Route is an example of how the mining industry in South Africa is making a contribution to sustainable development, in this case through conservation. Established by the Oppenheimer family, De Beers (Ltd) and its broad-based black economic empowerment consortium partner, the Ponahalo Group, the route extends over nine mine properties in five provinces. It seeks the preservation of diverse habitats and species protection and to enhance environmental awareness while contributing to social development through evolving opportunities in education and tourism (Diamond Route, 2011). Education and tourism initiatives generate revenue from opening former mining properties to historic, cultural and conservation visitors while contributing to conservation.

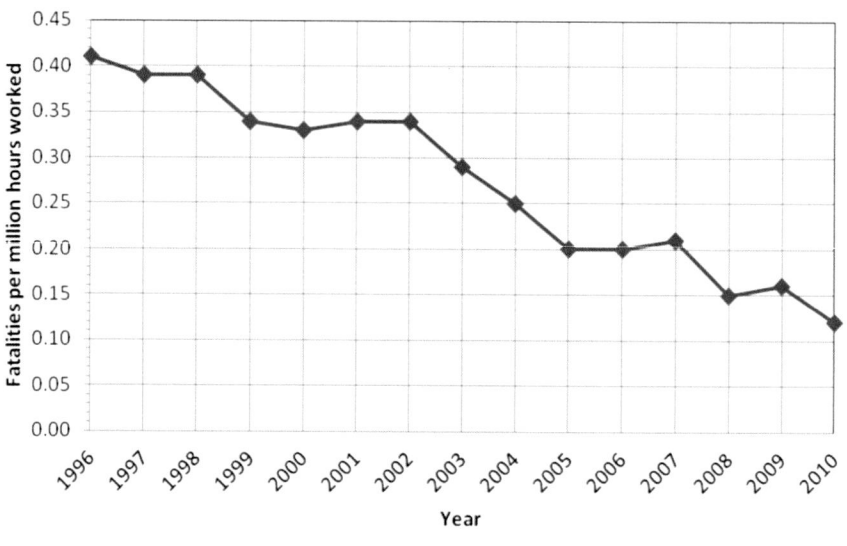

Figure 1: Improvements in South Africa's mining industry's safety performance (Modified after COM, 2006, p. 63; 2009, p. 10; 2011, p. 68)

A further example relates to air quality. In support of the Department of Minerals and Energy (DME)[3], and as a contribution to corporate social responsibility, Anglo Coal sponsored the trial roll-out of the top-down fire lighting methodology for *imbaula* (coal-burning braziers), renamed in a vernacular language as *Basa njengo Magogo* (literally, make fire like the granny), which results in lower smoke emissions at no additional cost to the user. The trial roll-out in a coal-burning township was planned involving awareness and education campaigns, drawing in schools and households in addition to the characterisation of the township's air quality (Nuwarinda, 2007, p. 4). Although the uptake of the methodology was high initially, a measure of the retention of such uptake together with repeat air quality studies is needed to determine the degree of actual improvement in smoke emissions particulate matter compared to the controlled laboratory results (which showed a greater than 80 per cent reduction in smoke emissions).

All the efforts by the mines to move to cleaner production are conceptualised from a narrow view of sustainability.

---

3. The Department of Minerals and Energy was reorganised into two departments, the Department of Mineral Resources and the Department of Energy, in 2008.

2.2 Ideological driver behind beneficiation as a function of the economy

An outline of the economic dimension in which mining is carried out globally and locally contextualises beneficiation as it is currently conceived. To understand the economic theory behind the praxis of the mines, an outline of the development of economic theory is helpful.

The economy affects everyone. It is a system of production, distribution, and consumption of goods and services that a society uses to allocate scarce resources to satisfy their unlimited wants and needs through supply and demand. The essential task of an economy is to transform resources into useful goods and services (the act of production), then to distribute or allocate these products to useful ends (the act of consumption). Virtually all economies accomplish this task through a combination of decisions made through voluntary market exchanges and involuntary government rules and regulations. A concise guide to the economy of South Africa is available from Roux (2002), while the fundamentals of economics, particularly from a neoclassical paradigm, are available in compilations such as that by McConnell (1984).

Beneficiation activities as exemplified by primary and secondary value-add processes in the mining and industrial sectors have as an end goal the commoditisation of products for the market. Commodities can be understood as 'true commodities ... objects produced for sale on the market. Fictitious commodities – labour, land and money – are obviously not produced for sale, cannot be wholly subsumed to a market and hence are not properly understood as commodities' (Meyer, 2005, p. 88). 'To fully commoditise labour and land, then, would amount to treating humans and nature as fully reducible to the quantities of labour activity or property that are exchanged in the market' (Meyer, 2005, p. 89). This can be seen as a series of adaptations toward the goal of economic liberalisation. In terms of sustainability efforts, the careful distinction between *true* and *fictitious commodities* needs to be made.

Economics is a particular view of how goods are traded, and until the environmental crisis became apparent, classical economics was dominant with its understanding of the role of the individual as a unit of labour. Economic paradigms affect humans as quality of life is regularly measured in terms of monetary value with little or no consideration of human satisfaction. The neoclassical economic model continues to measure the individual similarly. However, the neoclassical paradigm has become dominant and brings the tools of conventional economic analysis such as price mechanisms and market forces, to development. Consequently, governments and international agencies make use of economic analyses in development policies (Weiss, 2002). Such development policies 'allow markets to function freely, but also to control inflation, primarily by monetary policy and in general to reduce substantially the share of the state in economic activity' (Weiss, 2002, p. 45).

In the late twentieth century, social and ecological modernisation of the neoclassical economic model has led to a reformed view that takes social, environmental and economic factors into account. Consequently, social and environmental conditions have improved in mining. But growth of the market economy, accompanied by increased globalisation, has magnified the awareness of social injustices. The role of transnational corporations and the planetary rules, as set by the International Monetary Fund and the World Bank, have resulted in global disempowerment (Conversi, 2009, p. 346; Korten, 2000).

In terms of mining and primary beneficiation, the chronological development of the economic paradigm shows mining to be ideologically aligned with classical and then later with neoclassical profit maximisation approaches (Moodie & Ndatshe, 1994). Economics can thus be viewed as a system which adapts in response to crises such as wars, power shifts and environmental crises. A rapid review of economic approaches is given to aid understanding of the perspective adopted in this study, i.e. that mining has always been embedded in a particular economic paradigm – first *neoclassical*, then *reformed neoclassical*.

Of the many schools of economic theories, an economic policy which fosters development that is environmentally and socially benign would seem to be a useful approach to sustainability. The relationship between economic theories and sustainability can be broadly categorised into three positions: a *status quo* position, a *reformist* position, and a *radical* or *transformational* position (Sedlacko & Gjoksi, 2009; Söderbaum, 2008). The status quo position defines neoclassical economists and organisations. Further, the status quo position holds that economic growth is to the ultimate benefit of all with emphasis on the material component of well-being; no inconsistency is seen between economic growth and environmental degradation. The reformist position, usually held by international agencies and governments, looks to political consensus and involvement of several actors (including transnational corporations). This position is illustrated by the technology-orientated *Green Economy* and low-carbon economy discourse leading to regulatory and policy measures. The radical or transformational position holds that significant

Economic theory		Examples of principal	Capital	Fundamental principle of value authors	Normative value	
	Classical economics	Malthus, Ricardo, (Smith)	Limits on natural capital (land), human-made and financial capital	Scarcity of production factor (especially labour, land and human-made capital)	Productive efficiency	Monetary
Classical economics	Neoclassical economics	Smith, Marshall, Jevons, Menger, Walras	Substitution possible between human-made, financial, human capital	Real exchange value	Economic efficiency in exchange	
	Environmental economics	Pigou, Piers	Substitution possible between human-made, financial, human and natural capital	Imputed exchange value	Economic efficiency in exchange inclusive of environmental externalities	Monetary
	Ecological economics	Daly, Costanza	Substitution between human-made, financial, and human and social capital on the one side and natural capital on the other is not possible	Economic rent of natural and environmental resources	Sustainability	Monetary but no trade-offs
Institutional economics	Neo-institutional economics	Coase, Eggertson	Entitlements to human-made, financial, human, social and natural capital	Real or imputed exchange value with inclusion of transaction costs	Economic efficiency in exchange inclusive of transaction cost	Monetary and non-monetary
	Evolutionary economics	Veblen, Schumpeter	Human-made, financial, human, social and natural capital used in the evolutionary process of production	Instrumental principles of social value, focusing on the continuity of human life	Continuity (survival or systems norm)	Monetary and non-monetary

Table 1: Economic theories and fundamental principles determining value
(Modified after Blignaut & de Wit, 2004b:60–61)

changes in social organisation are required to move societies away from the current pathway. Further, this position is generally held by environmental scientists and grass-roots movements. One overview of environmental economics is available in the compilation by Gilpin (2000) while a reflection on economic growth is described by Blignaut (2004). A tabulation of a selection of economic theories and fundamental principles is presented in Table 1 (Blignaut & de Wit, 2004b:60–61).

Economic paradigms reflecting the radical or transformational position in addition to reformist and status quo positions are schematically illustrated in Figure 2. This schematised representation of the ecological crisis and continuing global injustice shows how sub-specialities of economics based on the work of the original theorists in classical economics have interacted.

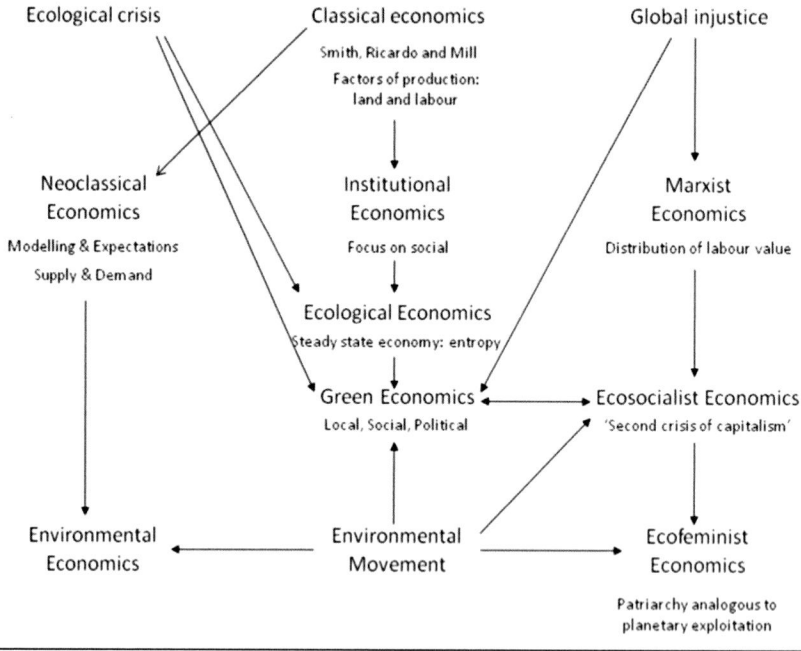

Figure 2: Economic paradigms and the environment (Cato, 2011:11)

Classical economics theory was the dominant paradigm for the system of production, distribution, and consumption of goods and services in South Africa during the eighteenth and nineteenth centuries. The classical economic paradigm affected the miners in the eighteenth and nineteenth centuries. These mine workers had poor working conditions and no benefits, as individuals were not considered beyond their value as 'units of labour' (Duncan, 1995, p. 37). In that colonial era, allocation of scarce resources, such as skilled labour and availability of finance (Katz, 1994) was to satisfy not only the local economy of the colony but also the needs of the colonial power through supply and demand. Malthus (1766–1834), Ricardo (1772–1823) and Smith (1806–1873), who were among the principal developers of the classical economic paradigm, referred to *political economics* and not simply *economics* (Söderbaum, 2008a:4). Their focus was on factors of production and wealth. Theories of trade between nations, concerns about the availability of natural resources (land) for food production because of population growth and the advantages of specialisation or division of labour were what they concentrated on. Mining in the Barberton Mountain Land during the gold rush of 1875–1883 (UNESCO, 2011), the diamond rush at the Big Hole in the Kimberley Diamond Fields in 1871–1914, and the discovery of the Witwatersrand Goldfields in 1886 coincided with the period of classical economics as the dominant paradigm (Gray, 1936). With the development of neoclassical economics, the mining companies adjusted their economic systems accordingly.

Neoclassical economics emerged in the late nineteenth century. From the start 'reference was made to "economics" rather than "political economics"' (Söderbaum, 2008a, p. 5). To achieve objectivity and be as scientific as possible, emphasis was placed on specialisation, on modelling, on supply and demand. Neoclassical economists do not place a specific value on the environment because the belief is that the market will naturally resolve all problems. Neoclassical economists argue natural and man-made capital can stand in place of each other. 'the objective of the economic laws is to achieve a balance, this should not be equated with a fixed system – rather the economy is expected to grow relentlessly' (Cato, 2011:36). The market system is deemed to be efficient and hence to be superior to other forms of economic organisation, and does not concern itself with the allocative fairness of the share. Neoclassical economics deals with environmental problems through externalities, or as Cato (2011:39) has it, 'a consequence of economic activity that does not impinge on the person or business conducting that activity'. Accordingly, the mining sector involved as it is in primary beneficiation continued with its focus on the maximisation of wealth and growth of market share.

The shifts in emphasis from classical/neoclassical economics to *environmental economics* are significant. Environmental economists suggest the market economy is used as an allocation mechanism and the limits of the planet *are* taken into account. However, they apply the market pricing techniques of neoclassical economics to value the environment. They argue that environmental management that takes account of ecological and social realities is a prerequisite for the creation of economic welfare. Environmental resource economics assigns a monetary value or an externality cost to a resource. The preferred definition of sustainability in environmental economics is that of weak sustainability (technology will lead to manufactured capital substituting for natural capital) (Blignaut & de Wit, 2004a & 2004b; Cato, 2011:52–68). However, in terms of the primary beneficiation policies of the mining sector, implementing a *weak sustainability* paradigm allows the sector to continue to focus on the maximisation of wealth.

In *ecological economics* both natural and man-made capital are necessary and the belief is that loss of one cannot be made up for by the other. In this instance, the definition of sustainability is that of *strong* sustainability in which man-made capital cannot substitute for natural capital. Ecological economics is concerned with ensuring the size of the economy does not expand beyond what the planet can sustain. Ecological economists are also committed to social justice. Once these matters have been addressed then there is concurrence with neoclassical economics in the sense that the market should be allowed to allocate resources. There is no fundamental restructuring of the economic philosophy envisaged (Cato, 2011, p.68–84). Proponents of ecological economics such as Costanza and Daly (1987), Daly and Cobb (1989), and Costanza (1989) envisage this as an interdisciplinary and holistic way in which to consider linkages between ecological and economic systems to a balanced and steady-state economy. The implementation of ecological economics *sensu stricto* on mining in its business of primary beneficiation would restrict the process of wealth creation both in terms of monetary value and market share.

Approaches that build on neoliberal philosophies (including neoclassical, environmental and ecological economics) support monetary valuations of the environment. (Ecological economics *sensu stricto* however, does not allow for trade-offs in environmental quality and thus this approach implies lower profits and less economic growth.) Other more radical economic views critique capitalism as an ideology and its neoclassical economic system. These include the green economists, institutional economists and the anti-capitalists with a pro-environmental bias. Both the green and institutional economists believe that fundamental change is required in terms of the way that the economy is structured, while the anti-capitalists argue that the capitalist system of thinking prevents progress in analysis and prescription (Cato, 2009 & 2011; Goodwin, 2008 & 2010; Speth, 2008).

Green economics, proposed as an alternate system to market economics, is concerned with social justice and uses a bottom-up approach through environmental movements to achieve a sustainable economy. Its objective is to move from economic growth to a balanced economy that allows for diversity in place of uniformity in the global marketplace. The green economics paradigm envisages the economy as operating within society and the whole of society as embedded within the natural world, the *Nested model* of sustainable development (Figure 4). Green economics recognises that the political nature of the economy and the environmental problems faced as a result of economic activity 'cannot be solved without fundamental political changes' (Cato, 2011:85). A comprehensive discussion on green economics is available in Cato (2009 & 2011). The primary objective for the Green Economy is the meeting of needs rather than the generation of profit.[4] Although the Green Economy is defined as low-carbon, resource-efficient and socially-inclusive (UNEP, 2011b, p. 2; 2011c) through its promotion of a green consumerism, it reverts to market economics in implementation with its predominantly financial metrics and its quest for economies of scale among its characteristics. An example of this is the derivation of primary energy from renewable resources. The pursuit of solar power through the installation of large-scale undertakings is often favoured over small-scale distributed photo voltaic systems reducing the number of available opportunities to a few large organisations. From a mining perspective, the implementation of the Green Economy will not have a significantly deleterious effect on its primary beneficiation activities. Instead, efficiencies across the mining process may reflect ultimately as profit.

Indicator	Negative impact on quality of life
Technology	Reduced skill in work, lower wage levels, increased stress
Health	Pollution and inequality cause physical and mental disease
Crime	Growth generates envy and higher crime levels
Community	Intensified work patterns undermine relationships
Inequality	The proceeds of growth are unfairly distributed

Table 2: The negative consequences of economic growth for quality of life
Source: R. Douthwaite (1992) *The Growth Illusion*, Totnes: Green Books, in Cato 2009:10

4. In the market economy growth is measured one-dimensionally, in purely (financial) economic terms; there is no reflection of quality of life indicators (Table 2). Cato (2009:9) says of this phenomenon: 'In capitalist ideology it does not matter that economic growth is destructive and does not increase human well-being; it only matters that there is more money changing hands in the global market'.

In contrast, the underlying philosophy of *institutional* economics is social value theory (Blignaut & de Wit, 2004b, p. 59). Institutional economics looks to a 'coherent representation of economic processes within and as part of a complex social system and their interaction' (Kapp, 1976, p. 213), and focuses on the nature of relationships within such a complex system and their interconnectedness. A comparison between the neoclassical and an institutional conceptual framework reveals the differences in emphasis (Table 3).

View of:	Neoclassical economics	Institutional economics
History	Not very relevant	Evolutionary perspective, path-dependence
Individual	Economic man	Political economic person, PEP as actor
Organisation	Profit-maximising firm	Political economic organisation, PEO as actor
Economics	Ideologically closed idea of efficient resource allocation	Ideologically open ideas about efficiency and resource allocation
Decision-making	Optimisation	Matching, appropriateness, pattern recognition
Approach to decision-making & sustainability assessment	Cost-benefit analysis, CBA	Positional analysis, environmental impact assessment, etc.
Relationships between actors	Markets	Non-market and market
Market	Supply and demand for single commodities	Social (and power) relationship between market actors, fairness, multi-functionality, multiple commodities
Progress in society	Growth in gross domestic product, GDP	Ideologically open – interpretations of sustainable development among options

Table 3: A comparison between the neoclassical and an institutional conceptual framework
Source: Söderbaum, 2008a:45

A distinguishing characteristic of institutional economics is its inclusion of values and an ideological component. In this way, economic mankind – defined as consumers are wage earners in neoclassical economics – can be regarded as political economic persons whose roles are considered and whose ideological orientations are not limited to one specific ethic (Söderbaum, 2002, p. 1; 2008a, pp. 37–75). A comparison of the roles of different actor categories between neoclassical economics and institutional economics is presented in Table 4.[5]

	Neoclassical economics	Institutional economics
State	The State is the main actor; market-based instruments versus 'command and control'	The State is the main actor but there are other political actors / policy-makers as well who influence state regulation, and initiate, and implement institutional change processes.
Organisations / business	Profit maximisation; adaptation to new state regulation	PEO[1] assumptions; a company may act as policy-maker by changing its mission statement and behaviour, and adapting to new regulation. Institutional change processes may be initiated and influenced.
Organisations / non-business	(Not part of neoclassical theory)	PEO[1] assumptions; these organisations may act as policy-makers by changing their mission statements and behaviour, and adapting to new regulation. Institutional change processes may be initiated and influenced.
Individual	Consumer maximising utility; adaptation to new state regulation	PEP[2] assumptions; individuals may act as policy-makers by changing their ideological orientation and behaviour, and adapting to new regulation. Institutional change processes may be initiated and influenced.

Table 4: Sustainability politics: roles of different actor categories
Source: Söderbaum, 2008a:74
1. Political economic organisation, PEO
2 Political economic person, PEP

Another economic paradigm with a different underlying philosophy from that of neoclassical economics is sufficiency economics. The *sufficiency economy* is based on a philosophy that 'stresses the middle path as an overriding principle for appropriate conduct by the populace at all levels' (Piboolsravut, 2004). This philosophy is more likely to offer self-protection from internal and external influences through its emphasis on a stepped and

---

5. Söderbaum (2011: slide 3) supports 'an institutional version of sustainability economics understood as a readiness for critical reflection in relation to the mainstream and also in relation to the existing political economic system'

balanced approach to economic development.

For the purposes of economics, ideally sustainability can be understood to mean 'not only longevity, in the sense that an ecosystem will endure or the planet as a whole will endure, but also that the quality of the environment has not been degraded by our activities' (Cato, 2011, p. 8). However, economics has seen a near hegemonic domination by the neoclassical approach, a system of production, distribution and consumption of goods, defined through supply and demand using natural resource capital, manufactured capital and human capital. Neoclassical economics, as the dominant economic paradigm, has fuelled the current growth-based market economy, which in turn has evolved as the process of globalisation.

2.3   IDEOLOGICAL DRIVER: GLOBALISATION

South Africa became independent of the colonial power in 1910 and transitioned to a democracy free of the apartheid system in 1994. These major political transition points represent the move of the country into the global community and globalisation. However, the mining industry in South Africa has taken a series of more evolutionary steps to become globalised, as exemplified through mergers, joint ventures, relocation of headquarters and more. The evolution of the transnational corporations from their origins to current form shows this. In South Africa, for example, General Mining & Finance Corporation was established in 1895 and Union Corporation in 1897, both in the nineteenth century. The merger between these two corporations led to the formation of Gencor in 1980, which later purchased Billiton in 1994. In 1997 the business was restructured, with base metals and non-gold commodities reorganised into Billiton plc, with its head office in London (Grant, 2004a). Subsequently Billiton merged with BHP, an Australian firm, to form BHP Billiton in 2001, the current formation. The Gencor gold interests merged with Gold Fields in 1998, forming Gold Fields Ltd (Grant, 2004b). The dominance of the neoclassical economic paradigm becomes obvious in that the mining companies remain structured to maximise profit for the shareholders.

Globalisation for the mines represents the continuance of primary beneficiation as a generator of wealth in the same manner as in classical and neoclassical economic paradigms. Globalisation is defined by the International Monetary Fund as 'The process through which an increasingly free flow of ideas, people, goods, services, and capital leads to the integration of economies and societies. Major factors in the spread of globalisation have been increased trade liberalisation and advances in communication technology' (IMF, 2011). Global business refers to the globalisation of a neoclassical economic system premised on economic growth and a market economy in which goods and services are exchanged for one another, or for money, on the basis of their perceived worth.

The private sector is the part of the national economy made up of, and resources owned by, private enterprises and not by the government. It includes the personal sector (households) and corporate sector (firms) and is responsible for allocating most of the resources within an economy. A transnational corporation differs from a multinational corporation in that it does not identify itself with one national home. Whilst multinational corporations are national companies with foreign subsidiaries, transnational corporations have operations in many countries sustaining high levels of local responsiveness. (An example of a transnational corporation is Nestlé, which employs senior executives from many countries and tries to make decisions from a global perspective rather than from one centralised headquarters.)

In contrast to the private sector, government's mandate is different. Government comprises a body of people that sets and administers public policy, and exercises executive, political and sovereign power through customs, institutions and laws within a state. Public sector is that part of a national economy supported financially by government (financed through taxation and levies) which provides basic goods or services that either are not, or cannot be, provided by the private sector. It comprises of national and local governments, public corporations, and quasi-autonomous non-government organisations. The public sector is generally one of the largest sectors of any economy. Multilateral organisations are international groups or agencies comprised of actors from bodies such as the United Nations.

Globalisation has been seen as the vehicle with which to entrench the neoclassical market economy through the unfettered expansion of the 'capitalist mode of production, culminating in the corporatisation of the remaining public sector' (Conversi, 2010, p. 40). The response increasingly to the '...*demise of national sovereignty has directly or indirectly contributed to boost nationalism and other boundary-building practices*' (Conversi, 2009) is reflected in growing protectionism. An example of such protectionism is the use of technical infrastructure such as standards, quality assurance and metrology in global trade by industrialised and advanced developing countries as *Technical Barriers to Trade and Non-Tariff Barriers* that make it difficult to access their markets (DTI, 2010). Transnational corporations have been leaders in this process of globalisation – in this way growing the interdependence of the world on these corporations. However, strategic industrial policies remain the purview of government. It is consequently incumbent on government to design differentiated and strategic policies toward transnational corporations. This allows host countries to intelligently use transnational corporations for their long-term development plans (Chang, 2003:247–269). Globalisation has also led to an increase in the number of response-stimulus feedback loops. These generate events that increasingly cannot be predicted. Such an event, termed *black swan event* by Taleb (2007), is characterised as random, having an extreme impact and only retrospective predictability, as alluded to in Chapter 1 in relation to 'boom-bust' cycles.

3 Sustainability and Sustainable Development

In this section, an examination of the evolution of diverse concepts of sustainability and the expression of sustainable development as it has been interpreted globally and specifically within the mining sector in South Africa is explored.

3.1 'Development' added to 'sustainability'

Sustainable development is the mechanism by which sustainability, as the larger concept and ideal, is believed to be achieved. When sustainability is made operational through the idea of *sustainable development*, it is a 'contested concept' (Söderbaum, 2008, p. 1). How it comes to be contested is unsurprising as it is a catch-all phrase which has to do many duties. In linking the concept of sustainability to 'development', the global community after the 1980s, when they realised how urgent the matter was, did what is usual in such cases – choose the prevailing narrative for 'progress', namely 'development' and put the qualifier 'sustainable' in front of it.[6] How far 'development' runs against the grain of 'sustainability' has been at the heart of the contested concept since then. Although sustainable development can be viewed as the means to achieve the ideal of sustainability, the great range of matters in which 'development' traditionally is conceived makes it difficult.

The matters commonly understood as areas in which 'development' (in the

---

6. The idea of 'development' arises out of the international and economic development paradigm in which it is imagined that humanity has a 'standard of living' and everyone would like to move towards a better one; it can also be measured, hence the *Human Development Index* proposed by Mahbub ul Haq and Amartya Sen (HDR, 1990). The Human Development Index (HDI) is a comparative measure of life expectancy, literacy, education, and standards of living for countries worldwide. It is a standard means of measuring well-being, especially child welfare. It is used to distinguish whether the country is a developed, a developing or an under-developed country, and also to measure the impact of economic policies on quality of life

sense of 'progress') takes place are – land use, social science, international and regional, business and professional, science and technology. Pertinent here are *land use* and *social sciences.*

In land use, the term 'sustainable development' has come to include:

> **Green development** – consideration in implementation of community-wide or regional environmental implications, usually with a mixture of expected and unexpected consequences;
>
> **Land development** – the process of altering the landscape in any way with a mixture of expected and unexpected consequences;
>
> **Mixed-use development** – the practice of allowing more than one type of use with a mixture of expected and unexpected consequences; and
>
> **Urban planning or development** – the concept includes land use planning and transportation planning to improve lives of communities with a mixture of expected and unexpected consequences.

In the generally accepted understanding of 'development' these matters are also considered:

> **Community development** – actions taken to improve life for local communities;
>
> **Economic development** – the economic elements in social change;
>
> **Rural development** – actions and initiatives taken to improve the standard of living in non-urban neighborhoods, countrysides, and remote villages; and
>
> **Social development** – processes of change in societies.

Finally 'sustainable development' comes to be understood as a way to meet human needs, while preserving the environment, and can be described as a pattern of resources use, but not one which has a general consensus on its aims or its outcomes. Depending on the ideological outlook of those implementing 'sustainable development' there could be said to be several approaches. The institutionalisation of the approach to sustainable development was a convenient way to ensure that industry could continue without systemic changes (Geels, 2005).

3.2 Main approaches to sustainable development

The four main approaches to sustainable development are: *needs approach, limits-to-growth approach, capital-based approach* and *human development approach*. They are closely connected and overlap, but all of them serve the ends of the hegemonic idea that the economic dimension is the most important one for the global world.

The *needs approach* is one in which sustainable development is 'development that meets the needs of the present without compromising the ability of future generations to meet their own needs' (Brundtland, 1987). It introduces the concept of *needs that can be satisfied* through trade and economic growth and the idea of *limitations on the capacity* of the environment to meet present and future needs. In this approach human actors have the unquestioned privilege and advantage of having needs which must be satisfied. There is acknowledgement that the environment has massive limitations if it is exploited, but *trade and economic growth* are the drivers of the development.

The *limits-to-growth approach* is one in which sustainable development means 'improving the quality of human life while living within the carrying capacity of supporting eco-systems' (IUCN, *et al.*, 1991). The focus here is on the quality of life and a steady state economy that is in equilibrium, rather than growth orientated (Daly, 1974). It does not deplete the environment beyond its regenerative capacity nor pollute it beyond its absorptive capacity (Daly, 1993:815–816). The emphasis is on the quality of human living as something which can be upgraded. It is not, however, made clear how these difficult concepts can be brought into a discourse with each other.

The *capital-based approach* to sustainable development treats the environment as a form of capital. In this approach, natural capital is similar to other capital such as produced capital, human capital and social capital. However, natural capital increases through natural replenishment while consumption depletes stocks. Consequently, sustainable living requires achieving a balance between 'living off the interest on natural capital' without depleting the capital stock as postulated by Daly (1974). The terms used are from the economic domain but the idea remains that natural capital cannot be substituted for other types of capital – 'strong sustainability' (Kirsch, 2009).

If 'strong sustainability' is understood as not allowing the natural resource base to deteriorate and that the interdependence of human economies and the environment are not interchangeable, this definition of sustainable development as 'a dynamic process which enables all people to realise their potential and improve their quality of life in ways which simultaneously protect and enhance the Earth's life support systems' (LSX, 2011) begins to reflect an approach to the paradox of 'sustainable development' by including the idea of a dynamic process. In 'strong sustainability' weight is also given to social equity, human dignity and ecological integrity. The natural resource base provides the ultimate means, the basic life-support functions, on the path toward the ultimate end, that of human well-being (Figure 3). Science and technology is used to convert natural capital to intermediate means defined as built capital and human capital, the productive capacity of the economy. These in turn feed into the political economy to accomplish all higher purposes. The intermediate ends are goals that governments and economies are expected to deliver – outputs. However, there is no guarantee of the ultimate end – human well-being (Daly, 1974).

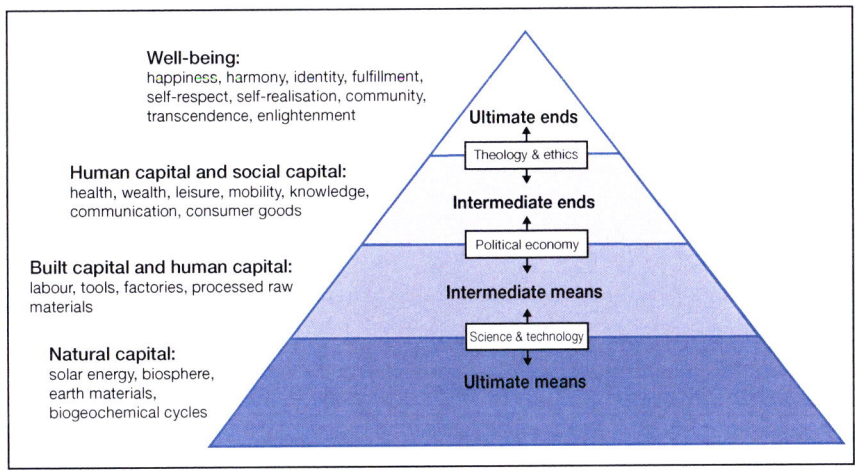

Figure 3: Daly's triangle of sustainability (In Meadows, 1998, after Daly, 1973)

Strong sustainability is further subdivided into two fields; one field entails that the value of natural capital be preserved, which for mining requires that primary beneficiation be compensated for by an investment in substitute shadow projects of equivalent value, and the other field calls for a portion of the total natural capital to be allocated to conservation so that its functions remain intact as critical natural capital (Sedlacko & Gjoksi, 2009). The counter argument runs that, although there is some biophysical degradation, regulated behavioural change will not guarantee sustainable development, nor result in an equitable distribution of resources. Instead, economic growth is the better indicator of development (Taylor, 2002:6).

Sustainable development, in which the natural resource base would be allowed to deteriorate as long as biological resources are maintained at a minimum critical level and the wealth generated by the exploitation of natural resources is preserved for future generations, is known as *weak sustainability*. This extends the neoclassical theory of economic growth to non-renewable resources as a factor of production, suggesting in this way that natural capital can be traded for other forms of capital if the total amount of capital grows (or at least stays the same) (Neumayer, 2003). Supporters of weak sustainabilit' in their development agenda argue that natural capital and manufactured capital are interchangeable and that sustainable development is achieved when the summed value of capital remains the same or increases. The mining industry subscribes to this definition of sustainable development, although aspects of strong sustainability are sometimes addressed. Such aspects of strong sustainability can be implemented in the mining industry through delineating areas of critical natural capital for conservation, such as the *Diamond Route* (Diamond Route, 2011); or by investment in substitute shadow projects such as independent power production from mining waste residue stockpiles of coal mine fines to generate electricity.

The *human development approach* defines sustainable development as being based on theories of social choice. Such a human development approach is understood in terms of human capabilities – what individual actors are able to do – and gives weight to functional capabilities rather than to resources. In this way, capabilities reflect multi-dimensionality, pluralism and democracy (Sen, 2005, pp. 157–160).

The sustainability paradigm is a contested concept as evidenced by the lack of a normative definition. A complicating factor remains, among others, clarity on the time frame to be assumed for sustainability, 'is a sustainable society one that endures for a decade, a human lifetime, or a thousand years?' (Worster, 1993, p. 134).

In order to make sustainable development something more than a shibboleth, some organisations have built their interpretation on a three-part division consisting of an environmental, an economic and a social dimension. This way of approaching the problem of sustainable developmen' is often called the *three pillars, triple bottom line*, or the *three Es* (where social is translated to equity/equality) of sustainable development (Cato, 2009, 2011; Daly 1974; Giddings, *et al.*, 2002).

While sustainable development has to be viewed from how it is currently understood, its close links to economics can be further explored.

3.3   Sustainable development and economics

Although the definitions of sustainable development are so numerous, the Swedish economist, Söderbaum (2008, p. 13–36), has categorised them into three ideological orientations that reflect the relationship to economics: (i) business-as-usual (BAU), based on neoclassical economics; (ii) social and ecological modernisation or reformist neoclassical economics; and (iii) a radical interpretation of sustainable development looking at alternative political economic models. In the business-as-usual model, the economy stands always at the heart of society, and independent of the environment, according to neoclassical thinking about economics. The *Three-ring* model (Giddings, *et al.*, 2002) (Figure 4) considers society, the environment and the economy as largely independent entities, with overlapping intersections in which activities with a mutual impact on each other happen. Each section of the model or each entity is given the same significance – the economy, society, the environment in which society exercises its economic interventions.

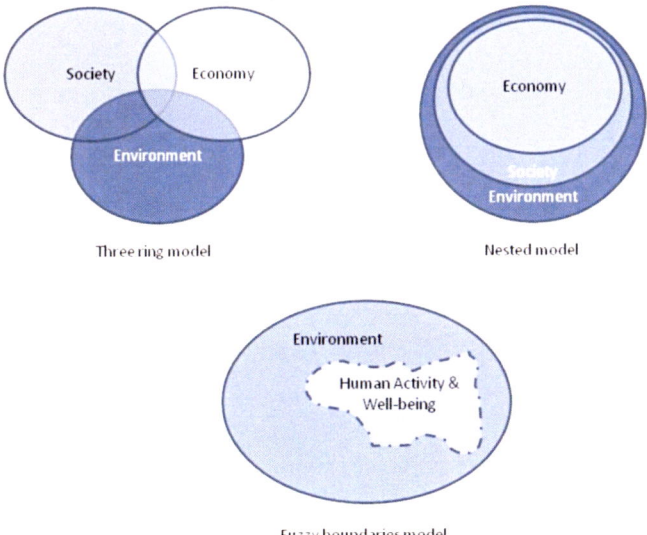

Figure 4: Three conceptual representations of sustainable development (Adapted from Giddings, *et al.*, 2002)

In response to the identified shortcomings of the ideological orientation reflected in neoclassical and adjusted neoclassical paradigms, the *Nested model* places the economy and society *within* the environment (Figure 4). Another example of a radical interpretation of sustainable development, the supporters of which acknowledge it requires a major shift in paradigm, ideology and institutional frameworks is represented by the *Fuzzy Boundaries* model (Figure 4). This model is conceived as a merging of society and the economy and an opening up to the environment through porous boundaries. This model emphasises the system interdependencies (Giddings, *et al.*, 2002). Such thinking is a result of theories moving toward a pluralistic and democratic philosophy. It is this last definition of sustainable development, based on a system of interactions between human actors, their well-being and the environment that is posited as ideal.

3.4 SUSTAINABILITY AND MINING

There have been many projects in sustainable development in the mining sector. Generally, such projects have not persisted much beyond the initial impetus. Of the many reasons which might be adduced for non-persistence, one is immediately apparent – in any sustainability efforts in the sector, it is their *economic* sustainability which is the key consideration. Dimensions other than economic are not given the same prominence, either in planning, execution or reporting, as is to be expected in the sector as it currently operates. This observation points to the core of the debate between the two competing approaches to sustainable development discussed in this study. There is, on the one hand, the dominant neoclassical economic model predicated on continuous growth and maximisation of shareholder return, in which sustainability projects are an add-on and are policed by the regulator. On the other hand, there is an economic model which favours the maintenance of a balance between technological development and profitability of organisations, and environmental preservation. In this second view, organisations such as transnational corporations, are politically economic actors (Söderbaum, 2008, p. 45). This is not a view the mines have taken up at all in South Africa.

To contextualise economic sustainability, the mines in South Africa are largely owned by private sector entities with a small state mining entity. Policies and practice in dealing with local socio-economic considerations for the transnationals are driven by business principles that reflect the praxis of economic globalisation. The growth of a global market economy has led to the emergence of a *global political space* and a *global civil society*; in this space and society the economic dimension is prominent. Thorn (2006, p. 249) describes some of the history of the trend, 'it is evident that the present mobilisation of a global civil society in relation to economic globalisation and supranational political institutions such as the World Trade Organization (WTO), the

International Monetary Fund (IMF), and the World Bank, has historical links to the post-war, transnational political culture of which the anti-apartheid movement formed an important part'. Nation-states are no longer independent hegemonic systems; instead there are now coalitions between countries and social movements (Thorn, 2006). Environmental and sustainability movements arise from such civil society or social movements and they could represent an 'energy-source which makes change possible' (Midgley, 2001, p. 186). It is through civil society initiatives that questions are being asked of industry, of government and of individuals about humanity's focus on the kind of materialism which benefits some sectors of humanity to the detriment of other interrelated systems in several dimensions.

3.5   CONCLUSION

Increasingly, discussions on sustainability and sustainable development highlight concepts of sustainability from a perspective in which human activity, as it occurs in and affects the biosphere, as a central consideration. This systemic view of human activity on the biosphere can be described in more dimensions that are also interconnected. In this way, a broader conception of sustainability as a complex adaptive system can be articulated. This is why I am positing that sustainability can be described in terms of *seven dimensions* – ideological, institutional, economic, social, demographic, informational and technological – in which transitions need to occur to move toward greater sustainability as defined by Gell-Mann (1994).

4   REGULATORY STRUCTURES IMPACTING MINING

4.1   SUSTAINABILITY FRAMEWORK WITHIN WHICH MINING OPERATES

The World Summit on Sustainable Development (WSSD) in 2002, through the Johannesburg Plan of Implementation, detailed key responsibilities for the business sector in order to bring about sustainability (DESA, 2004). This event changed the expectations of and the relationships between society, industry and the environment. Within the framework of the UN, the changed role identified for the private sector relates to its involvement in addressing the Millennium Development Goals, such as poverty eradication, primary education, reduction of child mortality, the ensuring of environmental sustainability, and the enabling of a global partnership for development (UN, 2001). Private sector organisations, under the leadership of the World

Business Council for Sustainable Development (WBSCD), have accepted the need for such a changed role. This is evidenced in the principles now incorporated into the business of major private sector entities that address sustainable development through capacity building, regulatory compliance and mandatory environmental impact assessments (GRI, 2011; WBCSD, 2010).

Transnational corporations from the mining sector are members of both the International Council on Mining and Metals (ICMM) and the WBSCS. The ICMM is a Chief Executive Officer-led industry group representing the mining sector (ICMM, 2011). The council addresses key priorities and emerging topics which affect the mining sector. This is done by making approved guidelines available, to promote good practice and improved performance.

The ICMM anticipates that by industry engaging in this way, governments and communities will see the mining sector as a preferred partner in the development of resources. As a guide, the Council has developed a sustainable development framework comprising principles, public reporting and independent assurance that all corporate member companies are required to implement. Companies report annually against these principles in accordance with guidelines set by the Global Reporting Initiative (GRI), while independent assurance is provided through third-party verification.

The WBCSD is a chief executive officer-led, global association of some 200 companies from different sectors dealing with business and sustainable development. This Council provides a platform for companies to explore sustainable development, share knowledge, experiences and best practices, and to advocate business positions on these concerns in a variety of forums, working with governments, non-governmental and intergovernmental organisations. Of note is that membership is by invitation of the executive committee to companies committed to sustainable development and to promoting the role of eco-efficiency, innovation and corporate social responsibility. Member companies publicly report on their environmental performance and aspire to widen their reporting to cover all three pillars of sustainable development – economic, social and environmental (WBCSD, 2011).

Transnational corporations in the mining sector report annually on results and on sustainability. Two separate reports are prepared: (i) the annual results report according to the host country guidelines and international standards such as the International Financial Reporting Standards (IFRS) and Generally

Accepted Accounting Principles (GAAP) guidelines; and (ii) the sustainability report compliant with Global Reporting Initiative (GRI) guidelines. Sustainability reporting conforms to national regulatory frameworks and with the WBCSD and the ICMM reporting guidelines and UN Global Compact commitments.

The mining sector transnational corporations all state a clear commitment to sustainable development within commonly agreed parameters. Sustainable development is defined as ensuring business remains viable and contributes lasting benefits to society through social, economic and environmental considerations, the *triple bottom line*. These global transnational corporations, tasked with maximising shareholder value, are often faced with challenges when evaluating the extent of their involvement in the broader framework of sustainable development. To justify involvement in non-core activities requires that the value-add for shareholders, employees, contractors, suppliers, customers, business partners and host communities be measureable by the business often as a reduction of long-term risks and liabilities.

4.2 Global Protocols and Directives

International agencies, such as the multilateral UN and its affiliates, and the private sector WBCSD, drive the public and private sector responses to global challenges, as evidenced through international agreements. A number of the international framework agreements and protocols that directly affect the resources industry are given below:

- Framework Convention on Climate Change, 1992 and the Kyoto Protocol, 1997;
- Universal Declaration on Human Rights, 1948; International Covenants on (i) Civil and Political Rights, (ii) Economic Social & Cultural Rights, 1966;
- UN Global Compact (Human Rights, Labour, Environment), 2000;
- Agenda 21 and the Rio Declaration on Environment and Development, 1992; and
- World Summit for Sustainable Development, Johannesburg, 2002.

Regulation may take the form of obligatory requirements in loan agreements, taxes such as incentives and penalties, market mechanisms and targets, both voluntary and involuntary.

4.3 COUNTRY-SPECIFIC REGULATORY STRUCTURES: SOUTH AFRICA

The Constitution of the Republic of South Africa of 1996 together with mining-specific and environmental legislation regulate the mining sector. Specific legislation includes:

- Sustainable development: Section 24 of The Constitution of the Republic of South Africa, 1996, is an affirmation of some globally-recognised principles of sustainable development that have emerged in various South African Acts.

- Mining, residue stockpiles and residue deposits: The Mineral and Petroleum Resources Development Act, No. 28 of 2002 (MPRDA); and the Mineral and Petroleum Resources Development Amendment Act, No. 49 of 2008;

- Prevention and remedying of environmental damage: The National Environmental Management Act, No. 107 of 1998 (NEMA); and the National Environmental Management Amendment Act, No. 62 of 2008;

- Atmospheric emissions: National Environmental Management: Air Quality Act, No. 39 of 2004;

- Disposal and treatment of gaseous, liquid and solid waste: The Environmental Conservation Act, No. 73 of 1989;

- Prevention and remedying of water pollution: The National Water Act, No. 36 of 1998; and the Water Services Act, No. 108 of 1997; and,

- Waste other than residue stockpiles and deposits: National Environmental Waste Management Act, No. 59 of 2008.

Significant changes have occurred in the legislative framework in South Africa that serve as regulatory drivers for instituting new policies and programmes, including public private partnerships. Liabilities and opportunities in terms of environmental management and regulatory structures are expanded upon online[7] together with the different forms of public private partnerships.

5 SITUATIONAL ANALYSIS POINTING THE WAY FORWARD TO RETHINKING SUSTAINABILITY ON THE MINES

The socio-technical landscape in which mining currently operates in South Africa reveals that mines value the technological and economic dimensions very highly. As a consequence less attention is accorded dimensions other than

---

7. www.mistra.org.za on the Knowledge Economy and Scientific Advancement (KESA) webpage.

those two, namely the ideological, institutional, social, demographic and informational (after Gell-Mann, 1994; after Speth 1992). The assumptions made about the ideological and institutional dimensions are addressed as a fait accompli in the implementation of sustainable development. Ideological issues are in line with globalisation, i.e. the neoclassical economic framework in which mining currently operates. The social and demographic dimensions are treated rather summarily as corporate social responsibility or as data for sustainability reports.

The situational analysis directing a rethinking of sustainability on the mines described as the socio-technical landscape can be viewed online[8]. The analysis is presented using the dimensions – ideological, institutional, economic, social, demographic, informational and technological – from a complex adaptive systems perspective, but embedded in the current economic paradigm in which the mining sector operates.

6 Sustainable development as managed transitions

For sustainability to become the focus of an agenda which includes support for communities, neighbouring mines, corporate social responsibility, and sustainable development projects cannot be the only commitments made by the mines. Nor is it sufficient to use innovation in the technological and economic dimensions as drivers for change when the systemic changes are kept *within mine boundaries*. What has become apparent is that the systemic changes in those two dimensions have benefits for people in some of the dimensions, but as the projects are not conceived in a holistic way they are not persistent. It is true that the nature of mining is geared to targeting financial sustainability and not to achieving developmental goals. Sustainability can better be served by moving *beyond mining*. Increasingly, the mines, particularly transnational corporations, evaluate the implementation of sustainability through the concept of life cycle assessments of materials from cradle-to-grave and cradle-to-cradle. An innovative mechanism with which to move sustainability of communities which have aggregated around mines is to set up a new collaborative model (Ferraz, 2012). Such a model would enable sustainability to inform all its goals. The approach is focused on sufficiency and sustainability to achieve improved levels of social well-being through encouraging public private engagement.

---

8. www.mistra.org.za on the Knowledge Economy and Scientific Advancement (KESA) webpage.

6.1 RATIONALE FOR A NEW WAY TO IMPLEMENT SUSTAINABILITY BEYOND MINING

The situational analysis has highlighted some key actors in the various dimensions that have been distinguished: the platinum mining industry, the government, and communities living in the vicinity of the mines. One pathway toward greater sustainability involves systemic innovation with an eye on the development of a new industrial sector in secondary beneficiation. Were secondary beneficiation activities to be re-envisioned as impacting all the dimensions of the biosphere, we would need to rethink change, transitions and adaptations. How could change come about?

To give effect to change, conceptualised as transitions with specific characteristics and linkages between processes at multiple levels, it is useful to realise that new ideas most often grow in technological niches (Geels, 2005, p. viii). Hence a proposal to chart a way forward towards sustainability as beginning in the technological niche of beneficiation has logic. That social and technical systems depend on each other in the production of whatever artefacts are developed because it is humans who create them, is obvious. Using slag, for instance, from the mining waste residues to build roads which can then allow access for people around the mines to different kinds of transportation, and can have an impact on communication, housing, energy supply, recreation, health care, is how a new development in the socio-technical niche can be said to be part of a dynamic system of continuous incremental adaptation. The incremental adaptations enable stability of the system. So, the trajectory of an idea is influenced by the *socio-technological regimes* in which it is embedded. The *socio-technical regime* consists of different actor groups and is a dynamic system of continuous incremental adaptation. Such incremental adaptation accounts for its stability. The *social technical landscape* encompasses the technological trajectories. Transitions can usefully be seen as multi-actor, multi-factor and not caused by change in one but the result of interactions between factors and multi-level processes. Multi-level processes occur between and within levels. Further, such levels may be located within societal groups and between societal groups. Changes in the interactions take time to develop. Within systems, innovation or change is often incremental, leading to trajectories in different dimensions. Although transitions cannot be managed, the trajectory of such transitions can be influenced (Berkhout, *et al.*, 2004, p. 48–75).

Bibliography

Alexander, P. (2010). Rebellion of the poor: South Africa's service delivery protests – a preliminary analysis. *Review of African Political Economy*, 37(123):25–40.

Ashton, P., Love, D., Mahachi, H. & Dirks, P. (2001). *An Overview of the Impact of Mining and Mineral Processing Operations on Water Resources and Water Quality in the Zambezi, Limpopo and Olifants Catchments in Southern Africa. Contract Report to the Mining, Minerals and Sustainable Development Project*, Southern Africa. ENV-P-C 2001-042, Pretoria, South Africa: CSIR.

Ayres, R. U. (1997). Metals recycling: economic and environmental implications. *Resources, Conservation and Recycling*, 21:145–173.

Baxter, R. (2011). The Vision Towards Competitive Growth and Meaningful Transformation of South Africa's Mining Sector. Presented to the *South Africa's Mining Industry Day at PDAC (The Prospectors and Developers Association of Canada)*, 7 March 2011, Toronto, Canada. http://www.bullion.org.za/documents/COM-presentation-to-MIGDETT-Roadshow-07-03-2011.pdf [Accessed on 15 November 2011].

Benyus, J. M. (1997). *Biomimicry Innovation Inspired by Nature*. NY, NY: Harper Collins.

Berkhout, F., Smith, A. & Stirling, A. (2004). Socio-technological regimes and transition contexts. *System Innovation and the Transition to Sustainability: Theory, Evidence and Policy*. Edited by B. Elzen, F. W. Geels & K. Green, Massachussetts: Edward Elgar, 48–75.

Blignaut, J. (2004). Reflecting on economic growth. *Sustainable Options: Economic Development Lessons from Applied Environmental Resource Economics in South Africa*. Edited by J. Blignaut & M. de Wit, Cape Town: UCT Press, 32–52.

Blignaut, J. & de Wit, M. (2004a). A perspective on the South African economy. *Sustainable Options: Economic Development Lessons from Applied Environmental Resource Economics in South Africa*. Edited by J. Blignaut & M. de Wit, Cape Town: UCT Press, 5–31.

Blignaut, J. & de Wit, M. (2004b). The economy of the environment. *Sustainable Options: Economic Development Lessons from Applied Environmental Resource Economics in South Africa*. Edited by J. Blignaut & M. de Wit, Cape Town: UCT Press, 53–81.

BMF. (n.d.). The Policy Gap. A Review of the Corporate Social Responsibility programmes of the platinum mining industry in the North West Province. http://www.bench-marks.org.za/ [Accessed 19 Sept 2012].

BMF. (2008). The Policy Gap Series. Corporate Social Responsibility and the Mining Sector in Southern Africa. A focus on Mining in Malawi, South Africa and Zambia. The Bench Marks Foundation. http://www.bench-marks.org.za/ [Accessed 19 Sept 2012].

BMF. (2011). Rustenburg Community Report, defending our Land, Environment and Human Rights. The Bench Marks Foundation. http://www.bench-marks.org.za/ [Accessed 19 Sept 2012].

BMF. (2012). Communities in the Platinum Mine fields. A review of Platinum Mining in the Bojanala District of the North West Province: A Participatory Action Research Approach. The Bench Marks Foundation. http://www.bench-marks.org.za/ [Accessed 19 Sept 2012].

Brundtland, G. H. (Editor) (1987). *Our Common Future*. Report of the World Commission on Environment and Development of 20 March 1987. http://www.un-documents.net/wced-ocf.htm [Accessed on 18 July 2011].

Bugliarello, G. (2011). Critical new bio-socio-technological challenges in urban sustainability. *Journal of Urban Technology*, 18(3):3–23.

Cato, M. S. (2009). *Green Economics. An Introduction to Theory, Policy and Practise*. London: Earthscan.

Cato, M. S. (2011). *Environment and the Economy*. Oxon, UK: Routledge.

Cawthorn, R. G. (1999). The platinum and palladium resources of the Bushveld Complex. *South African Journal of Science*, 95:481–489.

Cawthorn, R. G., Merkle, R. K. W. & Viljoen, M. J. (2002). Platinum-group element deposits in the Bushveld Complex, South Africa. In *The Geology, Geochemistry, Mineralogy and Mineral Beneficiation of Platinum-Group Elements*. Edited by L. J. Cabri. Canadian Institute of Mining, Metallurgy and Petroleum, Special Volume 54:389–429. Canada: Marc Veilleux Imprimeur.

Chang, H. J. (2003). *Globalisation, Economic Development and the Roles of the State*. London: Zed Books.

Cohen, J. & Stewart, I. (1994). *The Collapse of Chaos, Discovering Simplicity in a Complex World*. New York, NY: Penguin Books.

COM. (2006). Annual Report 2005–2006, Chamber of Mines of South Africa. http://www.bullion.org.za/documents/full%20text.pdf. [Accessed on 15 November 2011].

COM. (2009). Facts and Figures, Chamber of Mines of South Africa. http://www.bullion.org.za/Publications/Facts&Figures2009/F&F2009.pdf [Accessed on 15 November 2011].

COM. (2011a). Facts and Figures, Chamber of Mines of South Africa. http://www.bullion.org.za/content/?pid=71&pagename=Facts+and+Figures [Accessed on 19 September 2012].

COM. (2011b). Annual Report 2011, Chamber of Mines of South Africa. http://www.bullion.org.za/documents/AR_2011-small.pdf [Accessed on 15 November 2011].

Conversi, D. (2009). Globalization, ethnic conflict and nationalism. *Handbook of Globalization Studies*. Edited by B. Turner, London: Routledge, 346– 366.

Conversi, D. (2010). The limits of cultural globalisation? *Journal of Critical Globalisation Studies*, 3:36–59.

Costanza, R. (1989). What is ecological economics. *Ecological Economics*, 1:1–7.

Costanza, R. & Daly, H. E. (1987). Toward an ecological economics. *Ecological Modelling*, 38:1–7.

Creamer, T. (2011). Industry failures sow 'terrible' nationalisation seeds – Ramaphosa. Mining Weekly. http://www.miningweekly.com/article/industry-failures-sow-terrible-nationalisation-seeds-ramaphosa-2011-07-13-1 [Accessed 14 July 2011].

Daly, H. E. (1974). The economics of the steady state. *The American Economic Review*, 64(2):15–21.

Daly, H. E. (1993). Steady state economics: A new paradigm. *New Literary History*, 24:811–816.

Daly, H. E. & Cobb, J. B. (1989). *For the Common Good: Redirecting the Economy toward Community, the Environment, and a Sustainable Future.* Boston, MA: Beacon Press.

Davenport, W. G., King, M., Schlesinger, M. & Biswas, A. K. (2002). *Extractive Metallurgy of Copper.* 4th edition. Oxford, UK: Elsevier.

DEA. (2010). Green Economy Summit Report. Pretoria: Department of Environmental Affairs. http://www.sagreeneconomy.co.za [Accessed on 5 May 2011].

DESA. (2004). Johannesburg Declaration on Sustainable Development. United Nations Department of Economic and Social Affairs (DESA), Division for Sustainable Development. www.un.org/esa/sustdev/documents/WSSD_POI_PD/English/POI_PD.htm [Accessed on 5 July 2007].

Debswana. (2011). Debswana About us: Introduction. http://www.debswana.com/About%20Debswana/Pages/Introduction.aspx [Accessed 15 November 2011].

Diamond Route. (2011). Homepage. http://www.diamondroute.co.za/index.htm [Accessed 15 November 2011].

DME. (2003). Platinum-group Metal Mines in South Africa 2003. Directorate: Mineral Economics, Department of Minerals and Energy, D6/2003. http://www.dme.gov.za [Accessed on 17 September 2012].

DMR. (2011a). A Beneficiation Strategy for the Minerals Industry of South Africa. Department of Mineral Resources. http://www.dmr.gov.za/publications/viewdownload/162/617.html [Accessed on 9 August 2011].

DMR. (2011b). Strategic Plan: 2011–2014. Department of Mineral Resources. http://www.dmr.gov.za/publications/viewcategory/124-strategic-plan-201011--201213.html [Accessed on 10 August 2011].

DMR. (2011c). Scorecard for the Broad Based Socio-Economic Empowerment Charter for the South African Mining Industry, Government Gazette, 13 August 2004. http://www.dmr.gov.za/publications/viewcategory/24-mining-charter.html. [Accessed 15 November 2011].

DMR (2011d). An Overview of Current Platinum-Group Metals Projects and New Mine Developments in South Africa. Directorate: Mineral Economics, Department of Mineral Resources, R51-2011. http://www.dmr.gov.za [Accessed on 17 September 2012].

DTI. (2010). 2010/11-2012/13: Industrial Policy Action Plan, February 2010. Economic Sectors and Employment Cluster, Department of Trade and Industry. http://www.info.gov.za/view/DownloadFileAction?id=117330 [Accessed on 10 September 2010].

Duncan, D. (1995). *The Mills of God: The State and African Labour in South Africa 1918–1948.* Johannesburg: Witwatersrand University Press.

EDD. (2010). The New Growth Path: The Framework, Economic Development Department. http://www.pmg.org.za/node/24614 [Accessed 11 May 2011].

EDD. (2011). Comments in the State of the Nation Address debate, National Assembly, by Minister of Economic Development Ebrahim Patel, Economic Development Department. http://www.info.gov.za/speech/DynamicAction?pageid=461&sid=16545&tid=28906 [Accessed on 15 May 2011].

Ferraz. M. F. F. (2012). *Sustainability beyond Mining: Transformations in Systems for Secondary Beneficiation.* Unpublished Doctoral thesis. Johannesburg: University of Johannesburg.

Geels, F. W. (2004). Understanding system innovations: A critical literature review and a conceptual synthesis. *System Innovation and the Transition to Sustainability. System Innovation and the Transition to Sustainability: Theory, Evidence and Policy*. Edited by B. Elzen, F..W. Geels, K. Green, Massachussetts: Edward Elgar, 19–47.

Geels, F. W. (2005). *Technological Transitions and System Innovations. A Co-Evolutionary and Socio-Technical Analysis*. Massachussetts: Edward Elgar.

Gell-Mann, M. (1994). *The Quark and the Jaguar, Adventures in the Simple and the Complex*. London: Little, Brown and Company.

Gibson, J. T. R. & Comrie R. G. (1977). *South African Mercantile and Company Law*. 4th edition. Johannesburg: Juta.

Giddings, R., Hopwood, W. & O'Brien, G. (2002). Environment, economy and society: Fitting them together into sustainable development. *Sustainable Development*, 10:187–196. www.interscience.wiley.com. Accessed on 3 August 2011.

Gilpin, A. (2000). *Environmental Economics: A Critical Overview*. West Sussex, England: John Wiley.

Global Economic Prospects. (2009). *Forecast Update. March 30, 2009*. World Bank. DEC Prospects Group. http://siteresources.worldbank.org/INTGEP2009/Resources/5530448-1238466339289/GEP-Update-March30.pdf [Accessed 11 June 2011].

Goodwin, N. R.(2010). A new economics for the twenty-first century. *World Future Review*. http://www.wfs.org/Upload/PDFWFR/WFR_JunJul2010_Goodwin.pdf [Accessed on 17 August 2011].

Grant, T. Ed (2004a). Gold Fields Ltd – History of Gencor Ltd. http://www.enotes.com/company-histories/gold-fields-ltd/history-gencor-ltd [Accessed 16 November 2011.]

Grant, T. Ed (2004b). Gold Fields Ltd – Gold Fields and Gencor Unite in 1998. http://www.enotes.com/company-histories/gold-fields-ltd/gold-fields-gencor-unite-1998 [Accessed 16 November 2011].

Gray, J. (1936). Payable Gold: *An Intimate Record of the History of the Discovery of the Payable Witwatersrand Goldfields and of Johannesburg in 1886 and 1887*. South Africa: Central News Agency.

GRI. (2011). Global Reporting Initiative. Homepage. http://www.globalreporting.org/Home [Accessed 15 November 2011].

HDR. (1990). Defining and measuring human development. *Human Development Report*, http://hdr.undp.org/en/media/hdr_1990_en_chap1.pdf [Accessed 11 April 2012].

ICMM. (2007). Materials Stewardship, Eco-efficiency and Product Policy. www.icmm.com [Accessed on 7 May 2008].

ICMM. (2012a). Community development toolbox. http://www.icmm.com/page/1991/news-and-events/news/articles/community-development-toolkit-presented-at-bsr-conference-in-san-francisco [Accessed 10 September 2012].

ICMM. (2012b). Mine closure planning toolbox. https://www.icmm.com/page/9566/icmm-publishes-closure-toolkit [Accessed 10 September 2012].

ICMM. (2011c). International Council on Mining and Metals. Homepage. www.icmm.org [Accessed on 5 October 2011].

ILO. (2011). Global Employment Trends 2011: The challenge of a jobs recovery. International Labour Organisation. http://www.ilo.org/wcmsp5/groups/public/@dgreports/@dcomm/@publ/documents/publication/wcms_150440.pdf [Accessed on 2 December 2011].

IMF. (2011). International Monetary Fund. Glossary page. http://www.imf.org/external/np/exr/glossary/showTerm.asp#91 [Accessed on 10 June 2011].

IUCN, UNEP, WWF. (1991): Caring for the Earth: A Strategy for Sustainable Living. IUCN-The World Conservation Union; UNEP-United Nations Environment Programme; WWF-World Wide Fund For Nature. http://coombs.anu.edu.au/~vern/caring/care-earth1.txt [Accessed on 30 August 2011].

Jacobs, A. (1948). *South African Heritage: A Biography of H. J. van der Bijl.* Pretoria: Caxton.

Jerneck, A. & Olsson, L. (2011). Breaking out of sustainability impasses: How to apply frame analysis, reframing and transition theory to global health challenges. *Environmental Innovation and Societal Transitions,* 1:255–271.

Jooste, A., Kruger, E. & Kotzé, F. (2003). Standards and Trade in South Africa: Paving Pathways for Increased Market Access and Competitiveness. *Standards and Global Trade: A Voice for Africa.* Edited by J. A. Wilson & V. O. Abiola, Washington: World Bank, 235–370.

Kapp, K. W. (1976). The nature and significance of institutional economics. *Kyklos,* 29(2):209–232. http://www.kwilliam-kapp.de/documents/NatureandSignificance ORIGINAL.pdf [Accessed on 29 October 2011].

Katz, E. (1994). *The White Death, Silicosis on the Witwatersrand Gold Mines: 1886–1910.* Johannesburg: Witwatersrand University Press.

Kemp, R. & Rotmans, J. (2004). Managing the transition to sustainable mobility. *System Innovation and the Transition to Sustainability: Theory, Evidence and Policy.* Edited by B. Elzen, F. W. Geels, K. Green, Massachussetts: Edward Elgar, 137–167.

Kinnard, J. A. (2000). The Bushveld Large Igneous Province. http://www.largeigneousprovinces.org/sites/default/files/BushveldLIP.pdf [Accessed 10 September 2012].

Kirsch, S. (2009). Sustainable mining. *Dialectical Anthropology,* 34:87–93, DOI 10.1007/s10624-009-9113-x.

Korten, D. (2000). Creating a Post-Corporate World. Paper presented at Twentieth Annual E. F. Schumacher Lectures, October 2000, Salisbury, Connecticut. http://new economicsinstitute.org/publications/lectures/korten/david/creating-a-post-corporate-world [Accessed on 17 August 2011].

LSX. (2011). Why sustainability, what does it mean? London Sustainability Exchange. http://www.lsx.org.uk/whysus/what%20does%20it%20mean_page2760.aspx [Accessed on 2 November 2011].

Magala, S. (2010). Ethical control and cultural change: in cultural dreams begin organizational responsibilities. *Journal of Public Affairs,* 10:139–151. [Accessed on 12 October 2011].

Maiello, A., Battaglia, M., Daddi, T. & Frey, M. (2011). Urban sustainability and knowledge: Theoretical heterogeneity and the need of a transdisciplinary framework. A tale of four towns. *Futures,* 43:1164–1174.

Martin, L., (2012). International benchmarks focusing on transdisciplinarity, Presented to the *MISTRA Transdisciplinarity Workshop*, 10 April 2012, HRSC, Pretoria.

Mbeki, M. (2011). Advocates for Change: How to Overcome Africa's Challenges. Paper presented as public lecture at Department of Politics, Faculty of Humanities, 20 September 2011, University of Johannesburg, South Africa.

McConnell, C. R. (1984). *Economics*. 9th edition. Tokyo: McGraw Hill.

McDonough, W. & Braungart, M. (2002). *Cradle to Cradle: Remaking the Way We Make Things*. NY: North Point Press.

Meadows, D. (1998). Indicators and Information Systems for Sustainable Development. Report to the Balaton Group, The Sustainability Institute, Hartland VT. http://www.biomimicryguild.com/alumni/documents/download/Indicators_and_info rmation_systems_for_sustainable_develoment.pdf Accessed on 29 August 2011.

Melleuish, G. (2004). The Clash of Civilizations, A Model of Historical Development? *Rethinking Civilizational Analysis*. Edited by S. Arjomand & E. A. Tiryakian, London: Sage, 234–244.

Merkle, R. K. W. & McKenzie, A. D. (2002). The mining and beneficiation of South African PGE ores – An Overview. In *The Geology, Geochemistry, Mineralogy and Mineral Beneficiation of Platinum-Group Elements*. Edited by L. J. Cabri. Canadian Institute of Mining, Metallurgy and Petroleum, Special Volume 54:793–809. Canada: Marc Veilleux Imprimeur.

Meyer, J. M. (2005). Ground Liberalism, Environmentalism and the Changing Boundaries of the Political: Karl Polanyi's insights. *Environmental Values in a Globalising World, Nature, justice and governance*. Edited by J. Paavola & I. Lowe. London: Routledge, 83–101.

Midgley, M. (2001). *Science and Poetry*. London: Routledge.

Minister in the Presidency. (2009). Together Doing More and Better, Medium Term Strategic Framework. (2009 to 2014). The Presidency, Republic of South Africa. http://www.thepresidency.gov.za/docs/pcsa/planning/mtsf_july09.pdf [Accessed on 5 May 2011].

Moodie, D. & Ndatshe, V. (1994). *Going for Gold: Men, Mines and Migration*. Johannesburg: Witwatersrand University Press.

Mthombeni, M. S. (2006). *The Role of Multi-national Corporations in South Africa: A Political – Economic Perspective*. Unpublished Masters dissertation. Bloemfontein: University of the Free State.

National Treasury. (2008). Introducing Public Private Partnerships in South Africa, National Treasury PPP Unit. http://www.ppp.gov.za/Documents/News/Final%20PPP%20 Introdution%20Guide%2024.01.08.pdf [Accessed on 5 September 2011].

Neumayer, E. (2003). *Weak versus Strong Sustainability: Exploring the Limits of Two Opposing Paradigms*. 2nd edition. Northampton, MA: Edward Elgar.

Nicolescu, B. (2010). Methodology of transdisciplinarity – levels of reality, logic of the included middle and complexity. *Transdisciplinary Journal of Engineering & Science*, 1(1):19–38.

Nuwarinda, H. (2007). *Air Pollution Study of a Highveld Township during a Basa ngengo Magogo Rollout*. Unpublished Masters dissertation. Johannesburg: University of Johannesburg.

Pananond, P. (2004). Thai multinationals after the crisis, trends and prospects. *ASEAN Economic Bulletin*, 21(1):106–126.

Pappu, A., Saxena, M. & Asolekar, S. R. (2007). Solid wastes generation in India and their recycling potential in building materials. *Building and Environment*, 42:2311–2320.

Philip, K. (2012). The rationale for an employment guarantee in South Africa. *Development Southern Africa*, 29(1): 177–190.

Piboolsravut, P. (2004). Sufficiency economy, Research Note. *ASEAN Economic Bulletin*, 21(1):127–134.

RBP. (2011). Corporate Profile. Royal Bafokeng Platinum. http://www.bafokengplatinum. co.za/a/profile.php [Accessed on 15 Nov 2011].

Roux, A. (2002). *Everyone's Guide to the South African Economy*. 7th edition. Cape Town: Zebra Press.

SAinfo. (2011). Education in South Africa. SouthAfrica.info. http://www.southafrica.info/ about/education/education.htm [Accessed on 27 November 2011].

Schouwstra, R. P. & Kinioch, E. D. (1999). A short geological review of the Bushveld Complex. *Platinum Metals Review*, 44(1):33–39.

Sedlacko, M. & Gjoksi, N. (2009). Sustainable development and economic growth: Overview and reflections on initiatives in Europe and beyond. European Sustainable Development Network Quarterly Reports. http://www.sd-network.eu/?k=quarterly reports&report_id=15#qr121 [Accessed on 29 August 2011].

Sen, A. (2005). Human rights and capabilities. Journal of Human Development, 6(2):151–166. http://www.unicef.org/socialpolicy/files/Human_Rights_and_ Capabilities.pdf [Accessed on 18 November 2011].

Söderbaum, P. (2008). *Understanding Sustainability Economics: Towards Pluralism in Economics*. London: Earthscan.

Söderbaum, P. (2009). Making actors, paradigms and ideologies visible in governance for sustainability. *Sustainable Development*, 17:70–81.

Sorenson, P. (2012). Sustainable development in mining companies in South Africa. *International Journal of Environmental Studies*, 69(1):21–40.

Speth, J. G. (1992). Transition to a sustainable society. *Proceedings of the National Academy of Sciences of the United States of America*, 89:870–872.

State of the Cities Report. (2011). State of the Cities Report Summary, 2011: Towards resilient cities, a reflection on the first decade of a democratic and transformed local government in South Africa 2001–2011. South African Cities Network. http://www.sacities.net/images/stories/2011/pdfs/SOCR_report_2011.pdf [Accessed on 15 October 2011].

StatsOnline. (2011). Latest key indicators. Statistics South Africa. http://www.statssa.gov.za/ keyindicators/keyindicators.asp [Accessed 02 February 2011].

StatsSA. (2012). South African language distribution maps. http://www.statssa.gov.za/ census2001/digiAtlas/index.html [Accessed 19 September 2012].

Stephens, J. (2003). *Fuelling the Empire: South Africa's Gold and the Road to War*. Hoboken, NJ: John Wiley.

Taleb, N. N. (2007). *The Black Swan: The Impact of the Highly Improbable*. London: Random House.

Taylor, J. (2002). Sustainable development: A dubious solution in search of a problem. http://www.cato.org/pub_display.php?pub_id=1308 No. 449 [Accessed 14 July 2010].

Thorn, H. (2006). The emergence of a global civil society: The case of anti-apartheid. *Journal of Civil Society*, 2(3):249–266.

UN. (2001). Co-operation between UN and all Relevant Partners, in particular the Private Sector. http://www.un.org/documents/ga/docs/56/a56323.pdf [Accessed on 5 July 2007].

UNCTAD. (1992). Transnational Corporations as Engines of Growth, Executive Summary. World Investment Report. www.unctad.org/en/docs/wir92ove.en.pdf [Accessed on 15 November 2011].

UNDP. (2002a). Capacity 2015 Local results: A global challenge (Booklet). http://www.undp.org/wssd/docs/Capacity2015-LocalResults-a-Global-Challenge.pdf [Accessed on 5 July 2007].

UNDP. (2002b). UN Type II Initiatives from the World Summit for Sustainable Development. http://www.johannesburgsummit.org/html/documents/summit_docs/2908_partnershipsummary.pdf [Accessed on 5 July 2007].

UNDP. (2011). Millennium Development Goals. http://www.undp.org/mdg/basics.shtml [Accessed on 2 May 2011].

UNEP. (2011a). Green Economy Initiative, Division of Technology, Industry and Economics, United Nations Environment Programme. Homepage. http://www.unep.org/dtie/OurWork/GreenEconomy/tabid/29699/Default.aspx [Accessed on 3 May 2011].

UNEP. (2011b). *Towards a Green Economy: Pathways to Sustainable Development and Poverty Eradication – A Synthesis for Policy Makers.* www.unep.org/greeneconomy. [Accessed on 11 May 2011].

UNEP. (2011c). Resource Efficient and Cleaner Production, United Nations Environment Programme. http://www.unep.fr/scp/cp/ [Accessed on 5 August 2011].

UNESCO. (2011). The Barberton Mountain Land, Barberton Greenstone Belt or Makhonjwa Mountains. http://whc.unesco.org/en/tentativelists/5456/ [Accessed 15 November 2011].

UNGC. (2010). United Nations Global Compact. http://www.unglobalcompact.org/docs/news_events/8.1/GC_brochure_FINAL.pdf [Accessed on 14 July 2010].

UNIDO. (2002). Business Partnerships for Industrial Development – Partnership Guide. http://business.un.org/en/documents/439 [Accessed on 5 September 2011].

Van der Bijl, H. J. (1946). Introduction. Chapter in *Post-war Prospect.* Cape Town: South African Interests Group.

Van Onselen, C. (1982). *Studies in the Economic History of the Witwatersrand, 1886–1914 (2 Vols).* London: Longman.

Van-Helten, J. J. (1977). British capital, the British state and economic investment in South Africa, 1886–1914. *Societies of Southern Africa.* 9. London: Institute of Commonwealth Studies.

Vermeulen, D. J. (1998). The remarkable Dr. Hendrik van der Bijl. *Proceedings of the IEEE*, 86(12):2445–2454.

Von Below, M. A. (1990). Nationalisation of the mines: An equitable alternative. *South African Journal of Economics*, 58(3):188–196.

WBCSD. (2010). Vision2050: The New Agenda for Business in Brief, Electronic Document, Geneva. http://www.wbcsd.org/web/vision2050.htm [Accessed on 9 May 2011].

WBCSD. (2011). World Business Council for Sustainable Development. Homepage. www.wbcsd.org [Accessed on 9 May 2011].

WEF. (2012). To Serve Society Better, Capitalism Needs a Redesign. World Economic Forum. http://www.weforum.org/news/serve-society-better-capitalism-needs-redesign [Accessed 30 January 2012].

Weiss, J. (2002). *Industrialisation and Globalisation, Theory and Evidence from Developing Countries*. London: Routledge.

Williams, R. (1958). *Culture and Society: 1780–1950*. USA: Columbia University Press.

WNA. (2011). Outline history of nuclear energy. World Nuclear Association. http://world-nuclear.org/info/inf54.html [Accessed 28 November 2011].

Worger, W. H. (1987). *South Africa's City of Diamonds: Mine Workers and Monopoly Capitalism in Kimberley, 1867–1895*. New Haven, USA: Yale University Press.

Worster, D. (1993). The Shaky Ground of Sustainable Development. In *Global Ecology: A new Arena of Political Conflict*. Edited by W. Sachs. London: Zed Books.

Zeng, S. X., Meng, X. H., Yin, H. T., Tam, C. M. & Sun, L. (2010). Impact of cleaner production on business performance. *Journal of Cleaner Production*, 18:975–983.

CHAPTER 8

BEYOND MINING: SUSTAINABILITY AND SUSTAINABLE DEVELOPMENT
SUSTAINABILITY ORIENTATED MODEL

Fátima Ferraz

SECTION A

1 INTRODUCTION

The purpose of this chapter is to address a reduction of the risk and liabilities associated with mine closure planning and retrenchments in the platinum sector and how those risks can be managed. Hence, the chapter offers a workable strategy to facilitate the emergence of an industrial sector in secondary beneficiation. Both upstream and downstream value-add processes of beneficiation can be used to develop value-add socio-economic opportunities in secondary beneficiation. However, this chapter focuses solely on secondary beneficiation of upstream mining waste residues as one vehicle for industrial development. To make it easier to plan an integrated project that is moving towards implementation, the chapter is divided into two main sections: the first looks at methodology and presents a theoretical framework and the second provides a model for secondary beneficiation of upstream mining waste residue. The chapter ends with some recommendations.

The key to assisting the platinum sector to reduce risks and liabilities when mines close, or plan to close and there are to be retrenchments, is to manage the transition; not to direct it as an act of interference, but to have structures in place to absorb workers and their co-habitants aggregated around the mine into an industry which offers them employment. That is the assumption on which this chapter is based. But the question is, how to achieve this in the context of current labour unrest, a weakening global economy and in South Africa, very high unemployment rate with few working in small firms (Magruder, 2012). Unlike the case with many 'transnational regimes' the mining industry in relation to the environmental field has emerged in South Africa with strong public regulation (Pattberg, 2008). In many countries public regulation has been absent, fragmented, or weak. Regulation is not problematic in South African mines, but conceptualising a larger idea of sustainability has been. Pattberg (2008, p. 188) is encouraging in his assessment and analysis of a number of transnational environmental regimes as he speaks of 'measurable influences of novel institutional arrangements [which] go far beyond a narrow problem-solving account. Not only have a number of unintended consequences of transnational regimes occurred, but also the existing discourse that surrounds the issue areas of…corporate environmental reporting triggered the diffusion of the organizational model and, as a consequence, the emergence of a wider organizational field of transnational governance in general'.

Pattberg's assessment is the optimistic ground for the model offered in this package. The model follows on the theoretical foundation report on sustainability, sustainable development and mining (Ferraz, 2012a), and the intention of this chapter is to bridge theory and implementation. A model is presented, the transitions for the innovation of the system are outlined, and strategies to manage those transitions are detailed. The discussion ranges around the means by which a wider involvement of stakeholders, who can conceive of sustainability as a large overarching goal, can be facilitated. Further, a mechanism to incorporate informed inputs from communities is given prominence. In conclusion, recommendations for the implementation of the model are given.

1.1　Conceptual model

Models are tricky pictures. Fortunately, there are rules of correspondence between the information given to you in a model and the object of the modelling. So, a model can be described as corresponding with, but not identical to, any idea presented (Buckland, 2000). The idea presented in this model integrates seven dimensions: ideological, institutional, economic, social, demographic, informational and technological. The model, once understood, can be unpacked to provide a toolkit for industry and government to use to help an industry emerge based on the use of mining waste residues – upstream secondary beneficiation.

The assumptions of the model are that:

1. Industry, state and specific communities can work harmoniously together towards system innovations in the mining industry as whole and in the platinum sector in particular. These innovations should serve the larger goal of sustainability.
2. Post closure obligations for mines can be managed innovatively by reducing those risks through collaboration beyond mining.
3. An industrial sector in secondary beneficiation can be created outside the ambit and governance of mining.
4. This industrial sector can result in more and different employment for former miners as well as members of the communities adjacent to mines.

1.1.1 THE MODEL

In Figure 1, the three overlapping circles in the lower half of the illustration, labelled State, Industry and Society, represent a trilateral network and/or a hybrid organisation (Ferraz, 2012b & 2013). This trilateral network and/or a hybrid organisation is conceived as functioning as a think-tank at first. The purpose of the think-tank is to come up with a hybrid business system through discussion and the use of expertise from many quarters. In this instance, *society* refers specifically to the communities aggregated around the mines; *state* and *industry* are obvious. Implementing a hybrid business system would need to happen beyond mining and it would be an *unconventional business operation* (Figure 1 at the bottom of the figure). A distinction must be made between *conventional* and *unconventional* business operations: conventional business operations function under the prevailing economic paradigm (Elkington, 1997 & 2004) of neoclassical and reformed neoclassical economics indicated on the left in Figure 1. The exact nature of the economic framework in which unconventional business operations – tri-lateral networks and hybrid organisations – are to function, however, is one that evolves, thus in the illustration there is a porous boundary between conventional and unconventional business operations.

The model, we should bear in mind, would have to be adjusted to real circumstances, because a model is a conceptual schema. The most important point to consider is that to implement the model, the networked elements – industry and government and the communities adjacent to the mines – will need a consensus understanding of sustainability.

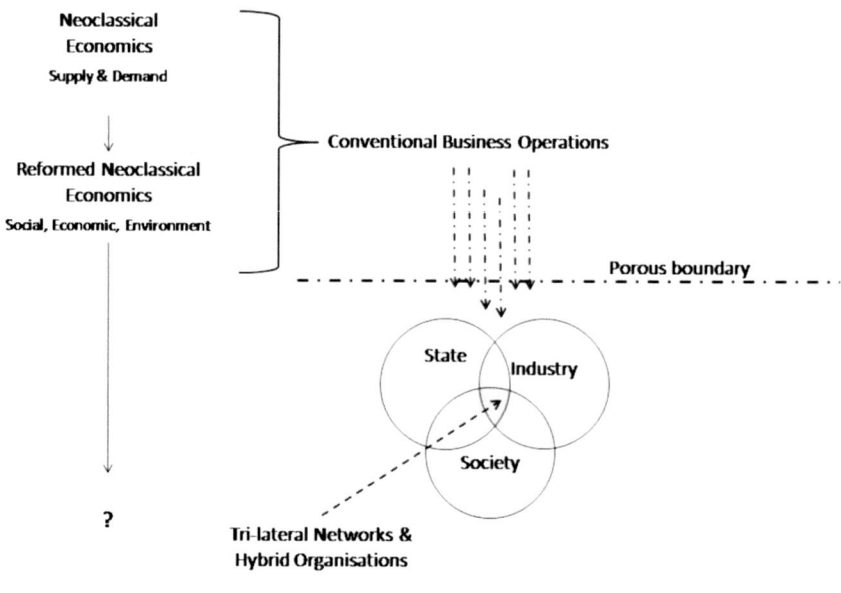

Figure 1: Illustration of model of networked elements (Ferraz & Annegarn, 2013 in prep.)

In this chapter, guidelines on a continuum of adaptations and transitions which are needed is provided. There are also extensive guidelines on the management of the processes based on successful transition management in systems as diverse as biotechnology, forestry, water and rail transport systems. Further, the knowledge gained from a small-scale research and development project which successfully piloted the use of mine waste residues in construction elements for use in building houses, is also incorporated.

Adaptations and transitions in multiple dimensions that need to be managed and directed to serve the idea of greater sustainability are outlined. These include consideration of social, environmental and ecological aspects. In transition management, the goal is delimited by a view from the future, back into the current situation. This is backcasting. The model offers a view of an ideal goal for the future which is related to the current socio-technical landscape in many ways. That is what is discussed next – the current socio-technical landscape of the platinum mining sector in South Africa in terms of management of mining waste residues.

1.2   Current socio-technical landscape

The current socio-technical landscape in South Africa is dominated by three issues: the social and economic policies of government, community activism and the relationship between government and the mining industry, especially in the platinum sector. The proposal in this chapter is that secondary beneficiation activities can be re-envisioned as the management of transitions and adaptations in a systemic overhaul. The system to be overhauled over an extended period (25 years) pertains to the relationship between the platinum mining sector, government, and the communities in the vicinity of mining in terms of the use to which mining waste residues are put. Before continuing, some background is useful.

Knowledge held back from people or knowledge freely shared can make a big difference to getting consensus, negotiating successfully and developing understanding. Describing the socio-technical landscape in this way, the role of knowledge becomes very important. The social and economic policies of government filter down inadequately to communities, including those neighbouring operational mines. Changes made to social and economic policies to drive the developmental agenda of government have been poorly communicated and understood. Detailed policy issues are not examined by the large portion of South Africa's population who are illiterate or semi-literate. The community around the mines represents a microcosm of the larger South African society: few people regularly employed, many youngsters and old people left out of the knowledge network. But family structures are retained as best as people can. This is an important fact in South African society. The human sense of security arises out of family structures, hence the effort to retain the well-being of family; but without being well-informed about what to expect, people have unrealistic expectations (Bevan, 2004).

In South Africa, when government and industry interact with communities in the vicinity of mines at times when conditions are unstable, there is antagonism between the parties. Unstable conditions usually occur at different stages of the mine life cycle. In periods of economic instability this may be reflected during life of mine, in organisational restructuring leading to increased mechanisation and retrenchments, there is a high likelihood of unrest, and post-mine closure, there is social dejection and also the possibility for unrest. The tension in the settlement areas around the mines is exacerbated by the pressure of the inward migration of people to the near-mine area in search of employment. The human settlements which arise in the near-mine vicinity inevitably have the characteristics of urban settlements with an ensuing need for infrastructure, facilities and utilities.

The mines are regulated to submit social and labour plans that would address post-closure concerns to government to obtain licences to operate. Between the plans submitted and the expectations of the communities around the mine, there is a shortfall. The particular deficiencies articulated by the communities are, in sum, that there is no employment, there is inadequate housing, and no services. The simple submission of regulated social and labour plans to government by the mines can be viewed as a first step in a process which could be further developed to ensure sustainable development. But, social resistance is evidenced by community activism which increasingly questions the effectiveness of a government mandated by the electorate to facilitate and provide opportunities as well as the sincerity of the mining corporations in addressing social and labour issues. An illustration of the displeasure at government has manifested as service delivery protests, termed by Peter Alexander as a *'rebellion of the poor'*, searching for *'the material benefits of full social inclusion'* (Alexander, 2010). The protests have also spread across different geographies including in the vicinity of mining activity. Such clashes have at times been violent as at Marikana between the platinum sector and labour (Chinguno, 2013). These disputes cut across different dimensions, including the ideological, institutional, economic, social, demographic, informational and technological dimensions.

The relationships between the actors in government and the private sector, and between the transnational mining corporations and financial institutions are much stronger than those with economically active individuals. However, collaborative partnerships between public and private entities are characterised by different goals – public entities tend to maximise a developmental goal while private entities tend to maximise profit. Consequently, the relationship between government and the mining industry is often strained. The interests of mining companies structured along a triple bottom line economic paradigm to achieve their projected financial returns contrast with that of government as a facilitator and regulator.

In the current socio-technical landscape, three key actors can be identified – they are the platinum mining sector including workers, government and communities living in the vicinity of the mines. There are levels of collaboration between these actors; nevertheless, the extent of the collaboration, it is suggested, can improve through well-managed transitions to new goals. Current problem-solving approaches have been ineffective because the underlying systems are changing rapidly. But for a move toward increased collaboration, and hence sustainability, transition management toward systemic innovation both across and within seven dimensions (put forward in a previous chapter), and at many other levels, requires commitment from all actors.

2 Methodology and Theoretical Foundation

The model in this chapter is not aimed at system improvement. It is aimed at system *innovation*. System innovation is transformation. The system which is being discussed here is the use to which mining waste residues are put. Transformations have many consequences which are unexpected. These consequences which are unexpected are the result of certain factors summarised below. By interrogating them, we hope to anticipate some of the negative consequences.

2.1 Barriers to innovation

The barriers giving rise to unexpected consequences in system innovation:
- The first is a need for a shift in perspective – the traditional idea that predefined outcomes through planning and control come before implementation for achieving particular outcomes is avoided. This is not how the model works, instead, it's focus is goal-orientated so that 'ongoing developments can be exploited strategically' (Kemp & Loorbach, 2006:109).
- The second is the reluctance to seeing transition management as open-ended and process-orientated – complexity and uncertainty in the implementation of the model are taken as a given, not as a barrier.
- The third is about narrowly conceived goals – goals as part of the management of a process aim at both a fundamental level and the wider societal aspirations for the greater good.

To sum up, implementers of innovation of a system will face three main barriers to executing change, in themselves and others: *inability to shift in perspective, reluctance to engage in a style of management which is process-orientated, and a lack of attention to wider goals.*

Barriers to system innovation occur at various levels despite a need for '[s]ome concerted action ... but public policy is highly fragmented and *orientated* towards short-term goals' (Kemp & Loorbach, 2006, p.110). The three barriers can be overcome and legitimised through a political process. System innovation is about the substance of a system and the process of bringing the innovation to fruition. But the innovation has to be seen from a holistic systems viewpoint. The process of realising greater sustainability in the mining industry is in terms of secondary beneficiation use and it is *beyond mining*.

2.2 THE PROCESS OF REALISING INNOVATION

Transition management is the process used to realise system innovation. A transition results from system innovations and other innovations and changes. It is a shift from an initial position of equilibrium or predevelopment, through a take-off, then a breakthrough phase, to a new equilibrium and stabilisation (Figure 2). Further, a transition is characterised by both fast and slow developments of interacting processes of structural change. Innovation in a part of a societal sub-system is involved in the transition (Kemp & Rotmans, 2004).

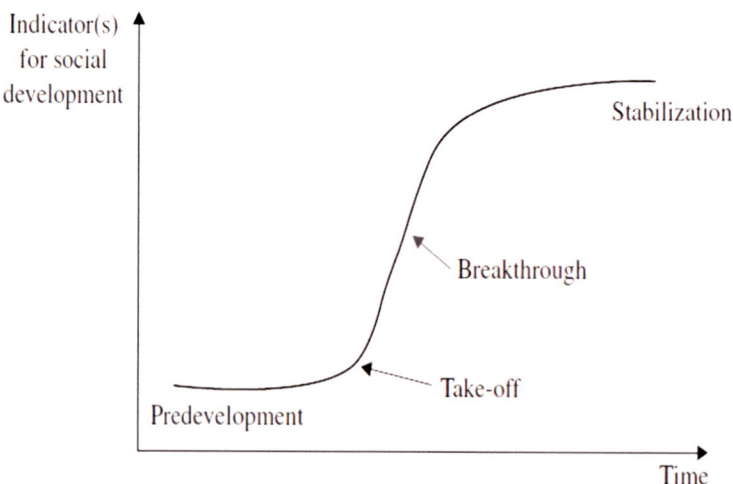

Figure 2: Four phases of transition (After Kemp & Rotmans, 2004:140).

The four phases of transition can be directed and managed. The management of transitions as *directed* change can be categorised under the headings innovation, systems thinking, collaboration, backcasting and forecasting, and finally, learning (de Bruijn, *et al.*, 2004, p. 54; Kemp & Loorbach, 2006, p. 110):

1. ***Innovation*** with a predetermined transition goal happens over a long-term period (of some 25 years) as has been discussed in a previous chapter. Long-term thinking can then act as a framework for shaping short-term policy. A prerequisite for system innovation and experimentation is a change of paradigm (to backcasting). Interactions require consensus on sustainability in the ideological and institutional dimensions. The model assumes there will be agreement among parties to partnering in hybrid networks.

2. ***Systems-thinking*** includes getting to grips with more than one dimension (multi-dimensional), with different actors (multi-actorial) and at different scale levels (multi-level); how developments in one dimension or level impact developments in other dimensions or levels is part of systems thinking; trying to change the strategic orientation of regime actors should also be considered. This last effort of changing regime actors pertains to institutional and informational structures – the set-up of research and development. In the informational dimension much has to be done to explain the process and spread knowledge.

3. ***Collaboration*** is through participation by and interaction between stakeholders. Interaction requires management and administration of the hybrid network in the institutional dimension.

4. ***Backcasting and forecasting*** involves the setting of short-term and longer term goals based on long-term sustainability visions, scenario studies, trend analyses and short-term possibilities.

5. A focus on ***learning*** and the use of a special learning philosophy – *of* ***learning-by-doing*** and ***doing-by-learning***, which includes carrying out experiments to develop required knowledge and using knowledge to do what is needed; learning about a variety of options.

Because the technological niche is often the driver for innovation, it is from here that this systemic innovation – to initiate an industry in secondary beneficiation of mining waste residues – is suggested. Although the innovation can occur at different levels it is comprehensive, takes place over a long time-frame, requires inputs from many stakeholders, and requires a change of perspective and a cultural shift among these stakeholders (de Bruijn, *et al.*, 2004).

For systemic innovation management tools and strategies, emergent factors will need to be accounted for while at the same time making existing tools emergent-proof. To give effect to change, conceptualised as transitions with specific characteristics and linkages between processes, and at multiple levels, it is useful to realise that new ideas most often grow in technological niches (Geels, 2005, p. viii).

2.2 How is system innovation different from ordinary innovation?

Ordinary innovation is understood as changes within one component of a system. In contrast, system innovation is *comprehensive*. All components are considered and not just a single component of the system. When system innovation is pursued, there is a cascade of adaptations and transitions in other components of the system. Radical revision and change of the system leads to a change in the underlying institutions as well as norms and values of those institutions (de Bruijn, *et al.*, 2004). System innovation is also associated with a restriction of the autonomy of action to those involved in the process. This means that actors must consider the bigger picture of sufficiency and greater sustainability, as it has been defined in consensus, and not simply reflect the specific viewpoint of a demographic grouping.

Backcasting is used as the methodology for the process of system innovation. The backcasting approach to future studies 'is the articulation of desired futures, and the analysis of how they might be achieved. In other words, criteria are developed for a sustainable, and desirable, future, which are then used as a guide for the design and implementation of measures that may facilitate progress towards that future' (Robinson, *et al.*, 2011, p. 756). In such scenarios, 'a target is defined for a future year without taking account of current trends' (Hoekstra & van den Bergh, 2006, p. 589). '[Backcasting] allows for a general analysis of alternative desired futures, especially those focused on sustainability, at scales ranging from the national to the local' (Robinson, *et al.*, 2011, p. 756) through either the externally defined targets or from the preferred futures as envisioned by the participants (actors).

For system innovation to take place the need for a change of paradigm is a prerequisite. A monitoring and feedback loop to track the realisation of the process is needed – the establishment of the trilateral network and hybrid organisation and subsequent adaptations – because of the need to review and adjust according to identified needs and unexpected outcomes. Monitoring has to be done – in government by an inter-ministerial committee or similar structure; in industry through the Chamber of Mines or similar structure;

and by appropriate structures in society such as the trade union movement and community organisations.

In summary, for transition management the 'substance of system innovation is that it is both comprehensive and radical. The innovation is comprehensive, because the whole system should change rather than just a component of the system. This automatically makes the change radical: the system as a whole has to be revised rather than one component, or just a few components, while underlying institutions, values and norms also have to change' (de Bruijn, *et al.*, 2004, p. 53). The model, put into practice, will provide a framework for collaboration between industry, government and society.

2.3 How to undertake transition management

Transition management is an approach and not a methodology; it has to be adapted and individualised for every specific context or problem (Kemp & Loorbach, 2006:110). The steps in transition management include:

1. **Transition arenas and multi-actor governance:** Kemp and Loorbach (2006) see transition arenas as an 'institution for interaction…considered a meta-instrument for transition management and facilitates interaction, knowledge exchange and learning between the actors. The transition arena is an open and dynamic network in which different perspectives, different expectations and different agendas are confronted, discussed and aligned where possible' (p. 111). Such a transition arena is the basis of the transition management process.

2. **Problem definition:** To define a problem we have to consider not a single, but a range of problems for which there are no ready-made solutions available. Instead, this requires trade-offs to be negotiated.

3. **Transition vision and transition goals:** A long-term vision of sustainability can function as a guide for formulating programmes and policies and the setting of short-term and long-term objectives.

4. **Transition paths and interim objectives:** Transition paths are ways to reach the final goal – there can be multiple paths.

5. **Programmes for system innovation:** The projects should be based on the vision and be designed for specific learning purposes, time-limited, and flexible.

6. **Evaluation and learning:** Monitoring and evaluation is a regular and iterative activity in transition management. This allows what has been achieved in terms of content, process dynamics, and knowledge to be evaluated.

7. **Creating public support and broadening the coalition:** Public support has to be created and retained. This helps the process continue and reduces the likelihood of a backlash should setbacks occur. 'The interactions between actors will change within the context of transition management. Transition management thus involves a change in governance, that is, the ways in which the plurality of interests is transformed into coordinated action, through deliberation, responsibilities and roles' (Kemp & Loorbach, 2006, p. 115).

Transition management is a form of reflexive governance which aims to deal with real and perceived problems or at least deal proactively with risks and negative side effects of solutions. There are five elements of transition management to be assessed at each stage of implementation (Kemp & Loorbach, 2006, p. 117).

1. **Knowledge integration:** Each actor brings different knowledge into the transition arena. Discussions between the actors integrate knowledge elements into a common understanding of the problem.
2. **Anticipation of long-term systemic effects:** In the system innovation programmes, long-term effects can be anticipated using backcasting and forecasting for scenario and trend analyses.
3. **Adaptive strategies and institutions:** Transition management is a step-by-step process which has four advantages: 1) it is not disruptive; 2) the costs of an error in a step are kept low; 3) it allows for a change in course; and 4) learnings can inform further steps. Although the process is seen as slow, it may bring faster change.
4. **Iterative participatory goal formulation.**
5. **Interactive strategy development:** This requires organising interaction and participation and simultaneously keeping governance effective. Collectively chosen long-term goals are the context in which the process unfolds with many mutual adaptations.

In summary, the first step in transition management involves establishing an arena for interaction, then elaborating the problem definition followed by the transition vision, goals, paths and objectives. These can be used to formulate programmes for system innovation. Progress can be monitored and evaluated and outputs used to grow and retain support among the actors and the public.

SECTION B

3 Toward an Industry in Secondary Beneficiation of Mining Waste Residue

Because transition management is an approach and not a methodology, it has to be customised for every specific context or problem.

The first phase of implementation of the model involves a small network which discusses and brainstorms the transition problem integrally and outlines the transition goals. The trilateral network and hybrid organisation of Figure 1 can be understood as a transition arena in which discussions occur. The network will expand at a later stage to include more actors to develop transition paths and objectives. A next step would be, after laying down goals and paths, to try short-term experiments. This is the time during which more operationally *orientated* organisations and different actors would begin to take their places in the transition.

A long-term vision of sustainability guides the approach and the setting of short-term and long-term objectives. This involves consideration of multiple components represented as dimensions within and across which transitions and adaptations are required – ideological, institutional, economic, social, demographic, informational and technological (for more on the dimensions, see previous chapter). From an implementation perspective, the transition management tasks in each dimension are interacting and integrating at many levels, and are happening recursively.

It must be noted that transition management and systemic innovation is envisaged as adaptations and changes that need to occur only in the system envisaged to facilitate secondary beneficiation of mining waste residues. The transition required is from stockpiling and dumping of mining waste residues to making these materials available as feedstock to start up an industry in secondary beneficiation of mining waste residues. The activity of secondary beneficiation of mining waste residues would occur beyond mining. The adoption of this approach would be a contribution by the platinum sector to address unemployment in the communities located in the vicinity of the mines.

In the longer term, this is an active contribution to sustainability and more meaningful corporate social responsibility. The system innovation under discussion occurs in a narrow but very complex area and while it takes the nature of platinum mining into account it does not interfere with any of the industry's existing processes. Thus, transition vision and goals, when discussed here, refer only to the specific gap of alternate uses for mining waste residues.

For each actor – platinum mining sector, government, and mining communities – the problem statement, vision and goals, will be different in each of the seven dimensions of the biosphere. Hence the transition paths and interim objectives will be different forming the basis of interventions or systemic innovation programmes. The process of evaluation and learning is iterative. This allows the system innovation programme to be monitored and adjusted particularly in terms of timelines needed for completion. For communities the transition goal is jobs being made available. For the platinum mining corporations, the transition goal is a reduction of risks and liabilities associated with regulatory obligations. For government, the goal is a stable community with employment opportunities and steady inputs into the fiscus. The common long-term goal is sustainability.

3.1 Sustainability and sustainable development

Sustainability, and its operationalisation as *sustainable development*, is a *'contested concept'* (Söderbaum, 2008, p. 1). Although there are numerous definitions of sustainable development, Söderbaum (2008, pp. 13–36) has categorised them into three ideological orientations: (i) business-as-usual; (ii) social and ecological modernisation; and (iii) a radical interpretation of sustainable development looking at alternative political economic models. For an illustrated representation of three widely used models of sustainable development, see Figure 4, Chapter 7. Each of these models reflects contemporary thinking about the significance of economics. In the business-as-usual model, the economy stands always at the heart of society, and independent of the environment, according to neoclassical thinking about economics. The proponents of the *Three-ring* model (Giddings, *et al.*, 2002) elaborated in the previous chapter take society, the environment and the economy as largely independent entities, with overlapping intersections, in which activities with a mutual impact on each other happen. Each section of the model or each entity is given the same significance – the economy, society, and the environment in which society exercises its economic interventions.

An example of a radical interpretation of sustainable development, briefly referred to in the previous chapter, advocating major shifts in paradigm, ideology and institutional frameworks is represented by the *Fuzzy boundaries* model (Figure 4, Chapter 7). This model shows a merging of society and economy and an opening up to the environment, which highlights the system interdependencies (Giddings, *et al.*, 2002). Such thinking is a result of theories moving toward a pluralistic and democratic philosophy. It is this latter definition of sustainable development, based on a system of interactions between human actors, their well-being, and the environment that is posited as ideal in this study. Sustainability and sustainable development is conceptualised as a *Fuzzy boundaries* system in which society, the economy, and the environment respond and adapt. When all seven dimensions are considered there is a better chance that all elements of the biosphere are remembered and given due attention.

3.2   Strategies for implementation

The following are a series of strategies to manage three key areas indicated in the model – public-private partnerships, role of knowledge and resistance.

3.2.1   Institutional dimension – hybrid public-private partnerships

The establishment of an effective hybrid public-private partnership – trilateral network or hybrid organisation – is of importance to the emergence of an industry in secondary beneficiation of mining waste residue. Developments such as the global economy, new communication technologies, the need for flexibility, and consumerism, have forced governments and business into collaborating. It is noted that, '*These global developments come together with an intermingling of public and private spheres, making partnerships less contested and even logical*' (Mol, 2008, p. 218). Linder (1999) has reviewed six definitions of partnering:

1. *Public–private partnership (management reform):* a tool for innovation which can change the way in which government functions through seeing how the market functions (and vice versa).
2. *Public–private partnership (the conversion of the problem):* By redefining the problems there is the opportunity to encourage investment by private profit-seeking collaborators.
3. *Public–private partnership (construed as a kind of moral regeneration):* Instead of the business as usual (meretricious) attitudes, values like self-reliance, initiative, hard work, integrity, and prudence are introduced into the profit-making enterprise.

4. **Public–private partnership (shifting risk):** There is an opportunity to move high risk from the public to the markets.
5. **Public–private partnerships (restructuring the public service):** With a move away from bureaucracy and inflexibility, and with less emphasis on union action.
6. **Public–private partnerships (operate to share power – how much? what kind?):** In changing business-government relations, the possibility arises for co-operation and trust; there can be a mutual and beneficial sharing of risk, knowledge and responsibilities; negotiation of differences becomes possible.

The public-private partnership thus envisaged in this model is a trilateral network and hybrid organisation of elements 4, 5, and 6 above – in which risk is shifted, the public and private sectors are restructured and power is shared. Considering the debate on partnerships, the major critiques can be summarised (and critically there is a general call for the environmental state to take an active role as necessary) (Mol, 2008, pp. 220–223):

1. **Conceptual criticism.** There is a tremendous variation and looseness in the definitions of partnerships and what they entail in the literature and an urgent need for analysis of partnerships in relation to their broad social context (Miraftab, 2004, Mol, 2008).
2. **Forward criticism.** In this instance, partnerships are seen and believed to rely too strongly on traditional ideas of governance. The promise of non-hierarchical multi-actor governance, especially in implementation, is often not met (Mol, 2008).
3. **Backward criticism.** Capitalism works against long-term notions of sustainability as short-term profits drive it (Hancock, 1998). Power relations jeopardise partnerships of the kind envisioned in this chapter (Mol, 2008).
4. **Political economy and democracy criticism.** Partnerships operate in the context of existing institutional structures which already have unequal balances of power. Local people might suffer (Mol, 2008).

Strategies with which to manage public-private partnerships include:

1. *One sector in the lead, the other sector follows*. There will be one sector 'in the lead' during the initial phases of the system innovation. The other sectors are subordinate. In any such hybrid system, there are many uncertainties, particularly in the early stages when the benefits are less obvious to all parties.

2. *Develop intensive interactions after some consolidation of system innovation*. The relation between leader and follower will remain unchanged during the whole period of the system innovation. Eventually, the sector that follows may play a facilitating role, thus boosting the system innovation. Government is responsible for creating favourable fiscal, physical and spatial conditions, linking these to other issues thus raising public and political support. Such a facilitating role necessitates intensive *co-operation* between government and industry, because facilitating implies being ready to act on new opportunities, and on threats to these.

   The boundary between leader and follower here is fuzzy, in the sense that national planning and definition of long-term objectives has to be initiated from the State. A hybrid system innovation is proposed: although this involves a comprehensive multi-dimensional and multi-level plan, its evolution is anticipated as an interactive process with fuzzy boundaries (bearing in mind the caveats against partnerships). The mining industry – with its triple bottom line economic paradigm and other traditional policies, such as its migrant labour, its current approach to labour unrest, its current status as a maturing mining industry over many years – has to be revisited. The plan itself has to be comprehensive and evolve from an organic process which requires regular adaptations. These adaptations lead to interaction between actors, with a constant shift in scope.

3. *Public-private partnerships must be understood as the management of competing values.* Public-private interactions can be defined in terms of a continuum of competing values; as a designed or spontaneous evolution; as demand- or supply-driven; based on fundamental or applied research; and, in terms of quality or relevance (Figure 3). These tensions remain in public-private partnerships and have to be managed.

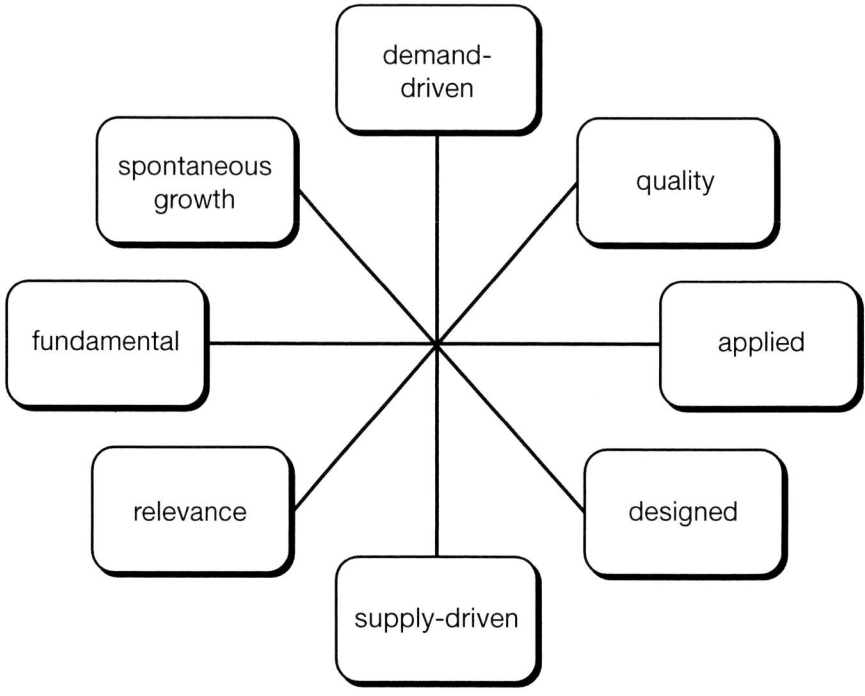

Figure 3: Competing values in designing public-private partnerships (After de Bruijn, *et al.*, 2004, p. 79)

3.2.2 Social dimension – how to manage resistance

Strategies to deal with social resistance
1. ***Use existing momentum.*** System innovation is a multidimensional exercise because many adjacent actors and factors may be found relevant in view of fuzzy system boundaries. In such a complex system, knowing which interventions will be effective beforehand is impossible so reliance on new insights to influence perceptions of those involved and, consequently, their division of interests is one way to approach the difficulty.
2. ***Identify dedicated target group.*** Make the advantages of the system innovation clear to this dedicated group. A dedicated group should first enjoy the benefits of the system innovation, while the disadvantages fall into the province of a more diffuse group. The advantage of the innovation, e.g. creating employment, involves a large group but only a

small selection from it – the unemployed – derive the most benefit. The result might be emphasis on future effects for the diffuse group. It is valuable to ensure that the proponents of the system innovation are more powerful than the opposing group.

3. ***Find out the underlying professional and financial incentives for a system innovation and try to synchronise them.*** In system innovation financial incentives for or against are important. If new commercial opportunities are clearly manifest, financial incentives will be positive and supportive of the system innovation. Professional incentives relate to the innovation from a scientific perspective. If there is scientific merit to pursue research then such incentives are also positive and can drive system innovation. Government can influence financial incentives as can other third parties.

4. ***Make advantages of system innovation visible.*** To reduce resistance the advantages of the initiative must be displayed prominently, the disadvantages not made so prominent but they will appear in the future. Perceptions are very important and need to be managed and directed, particularly as the links between the old and the new system with their concomitant failures are highly visible but not easily substantiated in terms of cause and effect. For the proposed sustainability orientated approach to succeed, the mechanism proposed for the system innovation is a designed and radical system innovation. It is important to manage the process to prevent the tendency to blame the transitions and adaptations of the system innovation for later failures. The failure of such a system would be highly visible. Previous interventions linked to system innovation might cause much resistance.

5. ***Widen system boundaries***. Porous or fuzzy system boundaries introduce more components and new issues which may be subjected to change with the result that a 'multi-issue game' emerges. Fuzzy boundaries allow exchanges between different actors. To facilitate the process of negotiating encouragement of the raising of many issues is necessary; fuzzy boundaries also incentivise collaboration as they force multi-issue decision making and coalitions have to keep adjusting to new conditions. Such an open system innovation may benefit from expanding because it allows for changes to take place. At first, the task was to undertake research on the use and displacement of strategic minerals in South Africa with a focus on PGM and fuel cell and other related technologies. These plans do not take a societal (community)

perspective into account; for example, the need to deal with unemployment after mine closure and retrenchments in mining communities. Adjusting the approach allows the interests of other stakeholders to be considered. The plan is to have actors – from the platinum mining sector, from government and from hegemonic structures in the mining communities – participate in the design for implementation and encourage restraint.

3.2.3 INFORMATIONAL DIMENSION – ROLE OF KNOWLEDGE

Without sufficient and useful knowledge no negotiations or decision-making is possible.

Strategies to deal with the role of knowledge involve three parts – analysis/instruction, variety/selection, and managing the role of knowledge. Under analysis/instruction (de Bruijn, et al., 2004, p. 63), strategies include:

1. *Framing and assembling: knowledge is framed from the stakeholders' perspectives*. Knowledge is collected, communicated in such a way as to create a sufficient number of new, attractive options for existing stakeholders. The knowledge, when it is communicated, suggests urgency, the need for change and opportunities for stakeholders.
2. *Negotiated knowledge through consensus.* Consensus and comprehensive models are not meant to underpin the need for systemic change; instead consensus serves to boost authority in pursuing the desired objective.
3. *Getting attention of decision-makers with simple and clear key messages* related to the momentum of present unstable conditions in the mining industry and platinum.
4. *Entice decision-makers to act.* Knowledge gives perspective for action. In this instance, the approach presented in earlier sections of this chapter and supported by the knowledge analysed, selected, assembled and framed in the previous chapter (Ferraz, 2012a) proffers a mechanism for implementation. However, before implementation and action, there is a need to conceptualise management strategies.
5. *Creating knowledge is an ongoing dynamic and recursive process* – it is important to have room to change in the future.
   The strategy under variety/selection (de Bruijn, *et al.*, 2004, p. 64) involves

Elements of transition management					
	Knowledge integration	Anticipation of long-term systemic effects	Adaptive strategies and institutions	Iterative participatory goal formulation	Interactive strategy development
Transition arenas & multi-actor governance					
Problem definition					
Transition vision & transition goals					
Transition paths & interim objectives					
Programmes for system innovation					
Evaluation & learning					
Creating public support					

Table 1:   Matrix of five elements to be assessed at each step of transition management (Modified after Kemp & Loorbach, 2006:117)

coping with the dynamics of variety/selection through regulating, facilitating and coupling.

The role of knowledge can be managed in two ways. First by assembling knowledge, framing it, and allowing it to take root, or second, by facilitating and coupling own interests to the dynamic described (de Bruijn, *et al.*, 2004, p. 67).

The five elements of transition also need to be taken into account in each transition management step, illustrated as a matrix of steps versus elements (Table 1).

3.3   Transition management steps

3.3.1   Conceptual work

Transition arenas and multi-actor governance
Ideological dimension – Meta-level framework: The first phase involves a small network which discusses and brainstorms the transition problem – integrally – and outlines the transition goals. The establishment and

organisation of the meta-level framework – as a tri-lateral network / hybrid organisation – is the foundation of the transition management process. The selection of participants for this activity is crucial as the complexity of the transition must be understood. The participants are selected because they have vision, like to be forerunners, and are open-minded enough to look beyond their own domains. They need to be autonomous in their own institutional capacities, but should be able to transfer their visions and initiate them in their own organisations. What this means is that they would be spending a great deal of time actively in identification, development and management of the transition arena process (Kemp & Loorbach, 2006, p. 111). The criteria for selection of participants must be specified and documented.

Institutional dimension – hybrid public-private partnerships: The construct for the business firm, a hybrid form of networked public-private partnership, would occur in the institutional dimension. This requires that organisational flexibility be one of the transitions aimed for (Fabac, 2010, p. 40). The large-scale nature of system innovations requires intensive public-private interaction. The adaptations for a hybrid system to emerge require a transition arena – a type of tri-lateral network consisting of industry and government with members of society. The hybrid system would need actors to consider the goals of its sustainability efforts being clear that the transition is not only market-economy driven. The strategies with which to handle public-private partnerships include understanding these as the management of competing values. The elements of transition also need to be taken into account in this step of problem definition (Table 1).

In the economic dimension, it is important to create employment and to consider non-monetary value. This requires that economic flexibility be one of the transitions aimed for. There needs to be consideration of non-monetary value metrics – addressing not only the market but also society. Measures of return on investment – especially for construction products from mining waste residues – need to be adjusted. The business to emerge would focus on low monetary returns but high non-monetary value (metric – number of people employed). The elements of transition also need to be taken into account in this activity (Table 1).

The mining sector addresses sustainability through what is termed 'weak sustainability', the 'three pillars', 'triple bottom line', or the 'three Es' of sustainable development (Daly 1974; Cato, 2009, 2011; Giddings, et al., 2002; Hitchcock & Willard, 2006) in business circles. Corporate social responsibility is reported on annually in terms of the triple bottom line indicators – economic, social and environmental – that quantify and qualify the sustainability initiatives of transnational mining corporations in line with the narrow consensus definition of sustainable development agreed to in the *Johannesburg Plan of Implementation* (WSSD, 2002). Additionally, mining corporations report on sustainable development using guidelines from the Global Reporting Initiative (GRI, 2011).

Managing resistance in the social dimension: This requires knowledgeable interaction to be one of the transitions aimed for in the social dimension. The adaptations for such a knowledgeable interaction to emerge require concerted efforts on the part of industry, government and members of society. Inadequate consideration of the social dimension can lead to much opposition, delays and even failure of the system innovation. Knowledge serves as a means with which to manage and mitigate resistance in the social dimension. In most cases resistance has a substantive background and is usually related to power relations. Because system innovations imply a need for change, these can affect existing interests, thus obstructing the process. Strategies to address this may evolve or need to be designed.

With a common understanding and knowledge of platinum sector employers and employees, of government and of society, information on the potential for more industrial development can be communicated. A common understanding also of mining waste residues and the possibilities for secondary beneficiation and the benefits – in terms of employment, sufficiency, and sustainability – enables more effective management and mitigation of (social) resistance. The elements of transition also need to be taken into account in this activity (Table 1).

The demographic dimension is described here as the grouping of communities living in the vicinity of mining operations. The community around the mines is representative of the larger society in South Africa – few people regularly employed, many semi-literate or illiterate, and many youngsters and old people left out of the knowledge network. Although some members of these communities work in the neighbouring mines, the majority do not. The question of unemployment is not the business of the mines only; it is also the business of government and society. The poor are represented in the proffered model which envisages a partnership with actors from more than one societal sector engaging strategically. Meadowcroft (2008, p. 195) says of partnerships of the type suggested in the model, 'the discussion does not primarily relate to partnerships that are basically commercial arrangements (for example, publicly funded but privately managed infrastructure projects), or that are essentially vehicles through which agencies distribute funding to recipients (including many of the so-called "type II" partnerships arranged in the run-up to the WSSD in Johannesburg). Instead, the concern is with "problem-solving partnerships" that take up the collective management of societal problems'.

The transition in the demographic dimension needs to move to flexibility in the roles of the actors. The elements of transition also need to be taken into account in this activity (Table 1).

The information dimension considers the role of knowledge. One of the transitions aimed for is adequate transfer of sufficient and enabling knowledge. The large-scale nature of system innovations requires intensive communication of knowledge, particularly to address the different levels of skill and knowledge and to manage expectations to ensure people have a shared point of view. Key to the facilitation of a wider involvement of stakeholders is this informational dimension. A common understanding of concepts such as sufficiency, social well-being, sustainability and, in addition the integrated nature of the biosphere, is the required first step among the actors. Hence, comprehensive tools and strategies need to be put in place to facilitate knowledgeable inputs from all stakeholder actors as well as mechanisms for distribution of knowledge and ways of checking on its uptake. The elements of transition also need to be taken into account in this activity (Table 1).

There are two patterns of knowledge production: *analysis/instruction*, in which research findings are analysed, creating knowledge which leads to instructions about the existing system and the need for a changeover to the new system; and there is *variety and selection*, in which the production of knowledge shows market-like features, with many producers and limited collaboration despite a competitive environment. Knowledge is selected from the available variety and used accordingly.

The main actors – government, the platinum sector, workers and communities in the vicinity of mines – access their knowledge in different ways. In government, the focus is on knowledge gained through analysis and instruction. In mining systems, both patterns may apply. The core business function of primary beneficiation has focused on knowledge gained through analysis and instruction while social questions are typically addressed with derived knowledge (de Bruijn, *et al.*, 2004). Communities' knowledge is derived from the variety and selected accordingly. In a network-like setting the information derived from the analyses and the instructions needs to be acceptable to different parties.

In the technological dimension, to manufacture a range of value-added products requires that different secondary beneficiation processes be considered. This requires flexibility in the use of material derived from mining waste residue stockpiles, and this must be one of the transitions aimed for. The dominant current practice of stockpiling, offers one vehicle with which to develop an alternate activity in a co-ordinated way and in alignment with government policy.

Problem definition

In the ideological dimension, a functional and operational meta-level framework is required. The starting point for transition management is the persistent problems of existing functional systems. This requires the definition of a common problem. Labour unrest and community activism point to problems that are more than just mining industry-related, and reflect the levels of marginalisation in society. In this instance, unemployment and inequality in earnings are the widely acknowledged problem for which no ready-made solution is available (Tregenna, 2011). The aim of transition management is to highlight and negotiate trade-offs. The elements of transition also need to be taken into account in this step of problem definition (Table 1).

The institutional dimension looks to a hybrid form of public-private partnerships. There is no template available currently for the evolution and establishment of a trilateral network/hybrid organisation of the nature envisaged. This will have to be developed. The elements of transition also need to be taken into account in this activity (Table 1).

In the economic dimension, creating employment and consideration of non-monetary value is important. One way to address this, the starting point for transition management, is to look at the problem of the lack of socio-economic opportunities in a particular system. This requires the definition of the common problem. More industrial and economic activity is required for new socio-economic opportunities to emerge. These may develop into employment offers for different skills levels in the vicinity of the mines. With the contraction of the current economy, the number of available jobs has reduced (Magruder, 2012). Adaptations and adjustments to bring in new employment to the mining areas under discussion require that a new template for industrialisation be developed. The elements of transition also need to be taken into account in this activity (Table 1).

How to manage resistance in the social dimension is the starting point for transition management in view of the low levels of information and knowledge amongst the actors. This requires the definition of the common problem to be provided to them. Adaptations and adjustments to bridge the knowledge gap are required. The elements of transition also need to be taken into account in this activity (Table 1).

The communities adjacent to mining operations form an important grouping in the demographic dimension. The starting point for transition management in this dimension is the low levels of information and knowledge across the actors. The common problem in this dimension is, as in the case of the social dimension, a lack of knowledge, information and understanding. Adaptations and adjustments to bridge the knowledge gap are required. The elements of transition also need to be taken into account in this activity (Table 1).

In the informational dimension, the role of knowledge is the entry point. The starting point for transition management is the low levels of information and knowledge across the actors. The common problem in this dimension is, as in the case of the social and the demographic dimensions, a lack of knowledge, information and understanding. Adaptations and adjustments to bridge the knowledge gap are required. The elements of transition also need to be taken into account in this activity (Table 1).

The starting point for transition management – in the technological dimension – is the mining waste residue stockpiles – for secondary beneficiation. This requires the definition of the common problem – the management of risk and liabilities associated with the stockpiles (current and post-closure) by the platinum miners and by government. Transitions in the management of such stockpiles to bring in new socio-economic options, require that a new template approach be developed. The elements of transition also need to be taken into account in this activity (Table 1).

TRANSITION VISION AND TRANSITION GOALS

A long-term vision of sustainability which is multi-dimensional must guide the programmes and policies and the setting of short-term and long-term objectives. The elements of transition also need to be taken into account in this step (Table 1).

Transition paths and interim objectives

Transition paths are possible routes towards the *long-term goal* and there are many ways to reach that goal. The elements of transition also need to be taken into account in this activity (Table 1).

3.3.2 Planning work
Programmes for system innovation

The projects should be based on the vision and be designed for specific learning purposes, be time-limited and flexible. The elements of transition also need to be taken into account in this step of developing programmes for system innovation (Table 1).

In the technological dimension, programmes for secondary beneficiation activities should be established. Research and development projects have identified alternate uses for platinum mining waste residues – predominantly as construction elements such as bricks and tiles. These use known process-technologies and would require simple adaptations depending on site and material composition allowing for a rapid roll-out. This would enable waste residues to be used to develop multiple income stream business models that are able to be adapted to address societal problems, as demonstrated by Gunter Pauli through his Blue Economy approach (Msimang, 2013). In contrast, research into platinum fuel cells is ongoing for downstream secondary beneficiation.

3.3.3 Reflection on work plans, paths, vision & goals, problem definition
Ongoing evaluation and learning

Monitoring and evaluation is a regular and iterative activity in transition management. This allows what has been achieved in terms of content, process dynamics and knowledge to be evaluated. The elements of transition also need to be taken into account in this evaluation and learning activity (Table 1).

3.3.4 Getting others on board
Creating public support and broadening the coalition

Public support has to be created and retained. This helps the process continue and reduces the likelihood of a backlash should setbacks occur. The elements of transition also need to be taken into account in this step of growing public support and the networks (Table 1).

4 RECOMMENDATIONS: REALISATION OF AN INDUSTRY IN SECONDARY BENEFICIATION

4.1 RECOMMENDATIONS

Systemic adaptations can move the process of system innovations toward greater sustainability in mining. But there is a great need for a series of strategies to manage three key areas – the role of knowledge, resistance, and public-private partnerships.

1. There is a need to ask questions about realising system innovation, including:
- Are the system boundaries sharp or diffuse?
- Who initiated the system innovation? (A tabulation of questions can be used to manage and monitor progress or not of the intervention (see Table 2) (de Bruijn, *et al.*, 2004)).

2. There is a need for tools and strategies in response to emergent issues and for existing tools to be made emergent-proof. Then the next step is to put strategies in place with due attention to (de Bruijn, *et al.*, 2004, p. 82):
- Porous boundaries between science and market;
- Facilitating an ongoing process of variety and selection. Strategies can target individual variety or at influencing selection. This can be achieved with stable frameworks – for example, through regulation that offers actors clear guidelines and a favourable financial climate;
- Organising the process of detecting system innovation. Links between the actors as well as between the market and science, are complex and will remain diffuse and difficult to design. Designing arrangements to facilitate innovation is easy, but identifying where they are needed and detecting the innovation is difficult (de Bruijn, *et al.*, 2004, p. 82);
- Running professional and financial incentives in parallel;
- Freezing the process of system innovation, where necessary, if the outcomes of an emergent process are unclear. However, freezing the process offers a starting point for further innovation; and
- Using process-based reasons for the possible freezing. There are limitations to how far one can manage the process. Plans trigger unpredictable backcasting, and dynamic and systemic changes which are difficult to identify.

3. When using systems management tools there is the need to reflect on:

-   A systemic overview with a perspective on timeframes to manage the entire process even though processes in implementation are not necessarily linear;
-   Changes in the relationship between industry and government (not drastically different). This requires both government and industry to do things differently in relation to the mines. Because these changes between government and industry work two ways, they can be positive or negative; and
-   From the pilot run (experiments in sustainability), setting up a hybrid organisation as a structure is the most difficult part of the process.

4. Take assumptions about systemic innovations into account:

-   One of the key issues to keep in mind is that systemic innovations rely on both the systems of mining and government responding to innovative suggestions (because innovation occurs beyond mining);
-   Another issue is choice of appropriate actors within the community — with a capacity to inform, keep up to date with the process, and accompany the process for the full duration of the process – as fully committed actors. They must be acknowledged as acting on behalf of the community and hence must understand the indigenous power structures;
-   Individual actors chosen to drive the process must share a common vision and be fully committed for the length of time required.;
-   Experiments supported a pilot process and showed that drivers in the technological dimension drive the change but this is a slow, incremental process; and
-   The geopolitical framework is also an important consideration and provides context to the drive for local value-add beneficiation initiatives. Although governments are responsible for economic upliftment, there is a view that economic interest supersedes developmental interests. The practice of transfer-pricing, for example, by the transnational corporations raises questions about accountability and governance.

4.2 TOWARDS PRACTICAL ACTION

The actors are clear – the platinum mining sector, government, and society (communities in vicinity of mining operations). The initial steps involve identifying the individuals to form the core stakeholder group, establishing a meta-level agreement between the three actors, and running three or more workshops, two of which will establish the outline adaptations for a hybrid system to emerge.

A tabulation of strategies proposed to facilitate system innovation is presented in Table 2. These questions form the basis of the guidelines and recommendations that can be used to initiate practical action. While the goal of initial interactions should be to inform, it also must guide the system through adaptive transitions to greater sustainability beyond mining. The evolution of an industry in secondary beneficiation (of mining waste residues) can be directed, managed and monitored in this way and remain the focus. The questions can also form the basis of a governance process to manage and monitor progress of the intervention or lack of it.

STRATEGIES FOR:

Realising System Innovation (SI)

Who will initiate the system innovation?

What type of system innovation is it?

What is the central approach?

Role of Knowledge

How will the knowledge for system innovation evolve?

How will this knowledge translate into action?

What will be the essence of the strategy?

When did the production of knowledge feature in the system innovation?

Dealing with resistance

Are the system boundaries sharp or diffuse?

What is the relationship between professional & financial incentives?

How are the advantages & disadvantages of the system innovation divided?

How visible are the advantages & disadvantages of the system innovation?

Role of public-private partnership (PPP)

What sector is initially in the lead?

Is the scope fixed or in motion?

What is the aim of public-private partnership?

Table 2: Strategies proposed for system innovation (Modified after de Bruijn, *et al.*, 2004:82)

Bibliography

Alexander, P. (2010). Rebellion of the poor: South Africa's service delivery protests – a preliminary analysis. *Review of African Political Economy*, 37(123):25–40.

Bevan, P. (2004). The dynamics of Africa's in/security regimes. *Insecurity and Welfare Regimes in Asia, Africa, and Latin America: Social Policy in Development Contexts*. Edited by I. Gough, G. Wood, A. Barrientos, P. Bevan, P. Davis & G. Room, Cambridge: Cambridge University Press, 202–252.

Brundtland, G. H. (Editor) (1987). *Our Common Future*. Report of the World Commission on Environment and Development of 20 March 1987. http://www.un-documents.net/wced-ocf.htm [Accessed on 18 July 2011].

Buckland, W. (2000). *Cognitive Semiotics of Film*. Cambridge: Cambridge University Press.

Cato, M. S. (2009). *Green Economics. An Introduction to Theory, Policy and Practise*. London: Earthscan.

Cato, M. S. (2011). *Environment and the Economy*. Oxon, UK: Routledge.

Chinguno, C. (2013). Unpacking the Marikana massacre. Corporate Strategy and Industrial Development, University of the Witwatersrand. http://www.polity.org.za/article/unpacking-the-marikana-massacre-february-2013-2013-02-13 [Accessed 22 February 2013].

Daly, H. E. (1974). The economics of the steady state. *The American Economic Review*, 64(2):15–21.

De Bruijn, H. Van der Voort, H., Dicke, W., de Jong, M. & Veeneman, W. (2004). Creating *System Innovation, How Large Scale Transitions Emerge*. London: A. A. Balkema Publishers.

DMR. (2011). A Beneficiation Strategy for the Minerals Industry of South Africa. Department of Mineral Resources. http://www.dmr.gov.za/publications/viewdownload/162/617.html [Accessed on 9 August 2011].

DSD. (2012). Annual Report 2011/12. Department of Social Development. *Parliamentary Monitoring Group*. http://www.pmg.org.za/report/20121017-consideration-annual-report-department-social-development-201112-fina [Accessed on 12 November 2012].

Elkington, J. (1997). *Cannibals with Forks, the Triple Bottom Line of 21st Century Business*. Oxford: Capstone.

Elkington, J. (2004). Enter the Triple Bottom Line. *The Triple Botton Line: Does it all add up? Assessing the Sustainability of Business and CSR*. Edited by A. Henriques, J. Richardson, London: Earthscan, 1–16.

Etzkowitz, H. & Leydesdorff, L. (2000). The dynamics of innovation: from national systems and 'Mode 2' to a Triple Helix of university-industry-government relations. *Research Policy*, 29(2):109–123.

Fabac, R. (2010). Complexity in organizations and environment – adaptive changes and adaptive decision-making. *Interdisciplinary Description of Complex Systems*, 8(1): 34–48.

Ferraz, M. F. F. (2012a). Sustainability, Sustainable Development and Mining: Theoretical Foundation, Work Package 01. MISTRA, Project Report. MR005: South Africa and the Global Hydrogen Economy: The Strategic Role of Platinum Group Minerals.

Ferraz, M. F. F. (2012b). Sustainability beyond Mining: Transformations in Systems for Secondary Beneficiation. Unpublished Doctoral thesis. Johannesburg: University of Johannesburg.

Ferraz, M. F. F. (2013). Beyond mining: Key transitions toward sustainability. *In prep.*

Ferraz, M. F. F. & Annegarn, H.J. (2013). The mining industry and sustainability: A different approach. *In prep.*

Fischer-Kowalski, M. & Haberl, H. (2007). Conceptualizing, observing and comparing socioecological transitions. *Socioecological Transitions and Global Change. Trajectories of Social Metabolism and Land Use.* Edited by M. Fischer-Kowalski & H. Haberl, Massachussetts: Edward Elgar, 1–31.

Geels, F. W. (2005). *Technological Transitions and System Innovations. A Co-Evolutionary and Socio-Technical Analysis.* Massachussetts: Edward Elgar.

Gell-Mann, M. (1994). *The Quark and the Jaguar, Adventures in the Simple and the Complex.* London: Little, Brown and Company.

Giddings, R., Hopwood, W. & O'Brien, G. (2002). Environment, economy and society: Fitting them together into sustainable development. *Sustainable Development,* 10:187–196. www.interscience.wiley.com [Accessed on 3 August 2011].

Glasbergen, P., Biermann, F. & Mol, A. P. J. (2008). *Partnerships, Governance and Sustainable Development Reflections on Theory and Practice.* Cheltenham, UK: Edward Elgar.

GRI. (2011). Global Reporting Initiative. Homepage. http://www.globalreporting.org/Home [Accessed 15 November 2011].

Hager, T. (2008). *The Alchemy of Air.* New York, NY: Three Rivers Press.

Hitchcock, D. & Willard, M. (2006). *The Business Guide to Sustainability: Practical Strategies and Tools for Organizations.* London: Earthscan.

Hoekstra, R. & van den Bergh, J. C. J. M. (2006). The impact of structural change on physical flows in the economy: Forecasting and backcasting using structural decomposition analysis. *Land Economics,* 82(4):582–601.

Kemp, R. & Loorbach, D. (2006). Transition management: A reflexive governance approach. *Reflexive Governance for Sustainable Development.* Edited by J. P. Voß, D. Bauknecht, R. Kemp. Cheltenham, UK: Edward Elgar, 103–130.

Kemp, R. & Rotmans, J. (2004). Managing the transition to sustainable mobility. *System Innovation and the Transition to Sustainability: Theory, Evidence and Policy.* Edited by B. Elzen, F. W. Geels, K. Green, Massachussetts: Edward Elgar, 137–167.

Linder, S. H. (1999). Coming to terms with the public–private partnership. A grammar of multiple meanings. *The American Behavioural Scientist,* 43(1): 35–51.

MacIsaac, D. (1996). An Introduction to Action Research. http://physicsed.buffalostate.edu/danowner/actionrsch.html [Accessed on 9 July 2012].

Magruder, J.R. (2012). High unemployment yet few small firms: the role of centralized bargaining in South Africa. *American Economic Journal: Applied Economics,* 4(3):138–166.

Mangena, M. (2011). *The role of old age grant in poverty alleviation within households with specific reference to Ward 31 (Lenyenye Township) under Greater Tzaneen Municipality, Limpopo Province, Republic of South Africa.* Unpublished Post graduate diploma in Social Security. University of Johannesburg.

Meadowcroft, J. (2008). Democracy and accountability: the challenge for cross-sectoral partnerships. *Partnerships, Governance and Sustainable Development Reflections on Theory and Practice*. Edited by P. Glasbergen, F. Biermann, A. P. J. Mol, Cheltenham, UK: Edward Elgar, 208–227.

Mhone, G. C. Z. (2003). The socio-economic context and social policy needs in South Africa. Social Security: A Legal Analysis. Edited by M. P. Olivier, N. Smit, E. R. Kalula, Durban: LexisNexis Butterworths. 1–47.

Miraftab, F. (2004). Public–private partnerships. The Trojan horse of neoliberal development? *Journal of Planning Education and Research*, 24(1): 89–101.

Mol, A. P. J. (2008). Bringing the environmental state back in: partnerships in perspective. *Partnerships, Governance and Sustainable Development Reflections on Theory and Practice*. Edited by P. Glasbergen, F. Biermann, A. P. J. Mol Cheltenham, UK: Edward Elgar, 214–236.

O'Brien, R. (2001). Um exame da abordagem metodológica da pesquisa ação [An Overview of the Methodological Approach of Action Research]. *Teoria e Prática da Pesquisa Ação [Theory and Practice of Action Research]*. Edited by R. Richardson, João Pessoa, Brazil: Universidade Federal da Paraíba. http://www.web.ca/~robrien/papers/arfinal.html [Accessed on 9 July 2012].

Pattberg, P. (2008). Partnerships for sustainability: An analysis of transnational environmental regimes. *Partnerships, Governance and Sustainable Development Reflections on Theory and Practice*. Edited by P. Glasbergen, F. Biermann, A. P. J. Mol, Cheltenham, UK: Edward Elgar, 173–213.

Robinson, J., Burch, S., Talwar, S., O'Shea, M. & Walsh, M. (2011). Envisioning sustainability: Recent progress in the use of participatory backcasting approaches for sustainability research. *Technological Forecasting and Social Change*, 78(5):756–768.

Söderbaum, P. (2008). *Understanding Sustainability Economics: Towards Pluralism in Economics*. London: Earthscan.

Speth, J. G. (1992). Transition to a sustainable society. *Proceedings of the National Academy of Sciences of the United States of America*, 89:870–872.

Tregenna, F. (2011). Earnings inequality and unemployment in South Africa. *International Review of Applied Economics*, 25(5):585–598.

Van der Berg, S., Siebrits, K. & Lekezwe, B. (2010). *Quantifying efficiency and equity effects of social grants in South Africa*. Report prepared for the Financial and Fiscal Commission, South Africa.

Von Werlhof, C. (2000). Globalization and the permanent process of primitive accumulation: The example of the MAI, the Multilateral Agreement on Investment. Translated from the German by U. M. Ernst. *Journal of World-Systems Research*, 6(3):728–747.

Williams, M. J. (2007). The Social and Economic Impacts of South Africa's Child Support Grant. *Economic Policy Research Institute Working Paper* #40. http://epri.org.za/wp-content/uploads/2011/03/rp40.pdf [Accessed 20 December 2012].

WSSD. (2002). Plan of Implementation, Johannesburg, World Summit for Sustainable Development, Johannesburg, September 2002. www.johannesburgsummit.org [Accessed on 13 May 2011].

CHAPTER 9

POLICY IMPLICATIONS

Velaphi Msimang

1 INTRODUCTION

This study is informed by the strategic intent of stakeholders in the South African government, the mining industry, and society generally, to utilise the country's mineral resource endowments towards achieving the country's developmental objectives. As outlined in the various chapters, South Africa has extensive Platinum Group Metals (PGM) endowments; and their strategic significance in the development of green technologies cannot be overemphasised.

The various chapters have laid out local and global aspects of the PGM industry value chain, teased out the barriers to be overcome, identified some of the key drivers of market development, and outlined attendant policy implications.

The predominant focus of the report is on PGM and their application in fuel cells and the variety of implications this has for PGM mining and beneficiation in South Africa, as well as the global networks and partnerships required for the country to attain its objectives.

The purpose of this section is to draw highlights from the contributions detailed in the preceding chapters, and to synthesise a few key messages that emerge from these.

ON THE GLOBAL SIGNIFICANCE OF PLATINUM GROUP METALS

The special physical and chemical properties of PGM – the largest terrestrial reserves of which are within the borders of South Africa – accords them global utility for numerous applications. To the extent that they are inimitable, and assuming markets for them are not disrupted, demand for these minerals can only trend upwards in a world with a growing population, growing economies, and growing concerns about the health and environmental impacts of economic activity. This demand can be met from direct mining as well as recycling of the PGM which are largely non-perishable.

Recognition of these realities and prospects is reflected in the numerous interventions – actual and proposed – by critical entities around the world and in South Africa, including governments, research institutions and industry stakeholders. The interventions assume the form of explicit policy statements and implied measures, including the categorisation of PGM as 'critical resource minerals' (CRM) and the mobilisation of related policy responses which assume variations of legal, technological, economic, trade, environmental, and military dimensions.

Especially in view of the potential role of PGM in the transition towards a hydrogen economy, prospects for appreciable growth in demand for these minerals only enhance their significance.

Given South Africa's strategic positioning in the PGM value chain, the importance of continually investigating and investing in options for leveraging this endowment for the purpose of achieving national policy objectives, cannot be overstated. Related to this is the country's responsibility – along with other PGM producers – to ensure global security of supply.

Considering the large range of possible PGM applications involved, including in the transportation, energy, manufacturing and jewellery sectors, well-co-ordinated and cross-sectorial activities, within and across the public and private sectors, are prerequisites for successful outcomes. For this purpose, an appropriate vehicle is well worth investigating.

On hydrogen and fuel cells in South Africa and globally

Despite its notably 'glacial pace' of development (Nail, *et al.*, 2005), the battle for entry into mainstream energy markets is not unique to hydrogen and fuel cells, and lessons from the history of similar experiences by technologies that are part of the current dominant regime are worth heeding. Described and explained elsewhere (Unger, 2010), these include the nurturing of niche markets whereby the technologies can be incubated until entry is realised.

The chapter on knowledge networks makes it clear that the South African effort to build PGM-based hydrogen and fuel cell capabilities need to be ramped up and strategic international networks leveraged towards this purpose. Simultaneously, as a complementary measure to strengthen its capabilities in hydrogen and fuel cell value chains, markets need to be created within South Africa in order to stimulate bottom-up efforts to commercialise PGM-based fuel cell products and services.

Emergent South African enterprises should be supported so as not to be crowded out by their more established international peers who have enjoyed support either from their national governments or space/defence industries. Support should include assisting enterprises to play an increasingly prominent role in global supply chains. The translatability of skills from the catalytic converter industry could, in the medium-term, be leveraged towards this purpose.

On industry and knowledge factors in the emergence of new industries

Notwithstanding the uncertainty that characterises the development of new industries, and the fierce rivalry from alternative options, patent and bibliometric data analyses have uncovered the factors shaping the emergence of the global fuel cell industry.

By reducing risks of the different stakeholders, the combination of supportive government policy and networks made up of entities with complementary capabilities – within and across national borders – provides the recipe necessary to seed the growth of a fuel cell industry. Indeed, to the extent that it involves both existing and new fields of knowledge, the fuel cell knowledge base is neither continuous nor entirely disruptive.

For South Africa to have a chance to supply 25 per cent of global demand for platinum-based catalysts to the hydrogen and fuel cell industry, it needs to set up and nurture the industry that will enhance understanding of the knowledge factors crucial for the industrialisation of fuel cell technologies.

This involves both the development of strategically networked R&D capacity, creation and sustenance of the markets that will mobilise entrepreneurial activity, and fostering linkages between these.

On fuel cell manufacturing

South Africa supplies more than 14 per cent of global markets for auto catalysts, and is home to the complete value chain of catalytic converter manufacturing. This multi-billion rand industry accounts for more than 50 per cent of foreign currency earned from the export of automotive components (Conceivious, 2010).

Globally, the fuel cell manufacturing industry is mainly about research, development and the manufacture of prototypes. While companies like Altergy Systems have automated fuel cell manufacturing facilities, large-scale manufacturing that is comparable to that of catalytic converters does not exist. This stems from lack of demand (for fuel cells) rather than inability on the part of the manufacturers.

In order to facilitate markets for fuel cells, governments in some countries are setting policy incentives, including (in Japan) a declining subsidy scheme, and (in South Korea and Germany) renewable portfolio standard (RPS) and a declining feed-in-tariff (FIT).

Among commercial agreements announced in South Africa in 2010 is a joint venture (JV) between the Department of Science and Technology (DST), Anglo Platinum's Platinum Growth Metals Development Fund, and California-based Altergy Systems. The JV – in which each of the parties holds equity – established Clean Energy Investments, 'whose principal objective is to manufacture and market Altergy fuel cell systems in the Republic of South Africa and other Sub-Saharan countries' (Altergy Systems, 2010). The collaboration was reported to 'mark the launch of the South African government's Hydrogen South Africa (HYSA) strategy to develop a manufacturing-based "hydrogen economy" and transform and expand uses for the country's national resources, including platinum' (Altergy Systems, 2010).

In 2012, a related announcement was made by the then DST Minister, Naledi Pandor, who confirmed the sale of 18 platinum-driven fuel cells to the

South African mobile phone sector (Creamer Media Reporter, 2012). A further MISTRA interview with a relevant official indicated that 1,000 fuel cell systems would have to be sold before the establishment of a manufacturing plant could be justified.

ON POSSIBLE PATHWAYS TO A HYDROGEN FUEL CELL TRANSITION IN THE SOUTH AFRICAN ROAD TRANSPORT SECTOR

The modelling employed assumed centralised generation of hydrogen in considering the potential for fuel cell adoption in the South African road transport sector. It also catered for the architecture of the South African energy system, and used relevant current data on its transport and economic sectors. Hydrogen generation options considered were limited to coal gasification, steam methane reforming, and water electrolysis.

The attraction to water electrolysis is informed by the potential complementarity of electricity and hydrogen. Particularly with the advent of intermittent renewable energy sources, excess power could be used to electrolyse water and generate hydrogen, which could be used to mitigate for their intermittence.

CONSIDERING FUEL, INFRASTRUCTURE, TECHNOLOGY PRICES, AND OTHER CRITICAL PARAMETERS IN MAINSTREAMING HYDROGEN AND FUEL CELLS

Characterised by Chu as the 'four miracles' that need to be realised for hydrogen and fuel cells to make commercial sense, the hurdles to be overcome before the transport sector can adopt fuel cells include: price parity of fuel cell cars with the internal combustion engine (ICE) option; price parity of hydrogen fuel with gasoline; hydrogen storage; and the hydrogen distribution infrastructure.

Even though it requires high energy prices to become commercially viable, of the three options, coal gasification seems currently to be the cheapest hydrogen generation option for South Africa.

The network of hydrogen supply infrastructure, popularly branded as a 'hydrogen highway', is an important factor affecting the rate and extent of uptake for fuel cell-powered mobility options. Lack of it partly explains the inertia that characterises the energy system, with car manufacturers citing it as a reason for putting off investment in the mass production of fuel cell cars. While the two options can complement each other in the hybrid configuration – which is an important consideration as it may enhance the extent of uptake – the fact that the battery-powered electric car is ahead on this front therefore may not bode well for the rapid uptake of the hydrogen fuel cell option.

Largely due to their logistical convenience and longer lifetimes, bus fleets offer the best opportunity for exploring and leveraging options to transition the transport sector towards a hydrogen economy. Capping carbon emissions also offers the best policy option for ensuring rapid uptake of this technology.

CONSIDERING THE GREENHOUSE GAS MITIGATION POTENTIAL OF A SIGNIFICANT SHIFT TO HYDROGEN FUEL CELL CARS IN SOUTH AFRICA

Clearly, the fact that hydrogen is not a fuel source suggests that the mitigation potential of hydrogen and fuel cells is largely determined by the rate and extent of decarbonisation of the South African energy supply mix. The significant size and high carbon intensity of the current system suggests that even a costly and highly-aggressive rate of penetration of low carbon fuels is not likely to change the mix appreciably by 2050. This has implications on the viability of achieving long-term mitigation scenarios (LTMS) objectives via the hydrogen economy. The related cost premium also raises affordability issues.

ON THE GEOPOLITICAL IMPLICATIONS OF A HYDROGEN ECONOMY

As is the case with energy carriers, hydrogen is only one component of the energy mix. When considering a hydrogen economy from the perspective of minerals beneficiation, the demand for PGM is a critical factor.

Geopolitical concerns impact on any potential uses of the resources of a country. Globally, it is the potential demand for PGM as a resource which drives the strategies with which to inform policy. Policies are developed to facilitate access to a country's PGM resources in return for benefits to the country. Threats to access, however, are multiple, such as market and trade barriers, conflicts and so on. In terms of the hydrogen economy, products containing PGM are not the only ones, though they are indispensable for specific critical fuel cell applications. Opportunities from the perspective of South Africa's position as a leading supplier of PGM are plentiful, even without partnering with the other PGM suppliers. However, the leading supplier position has its own limitations as demand and supply are governed by private contracts. Demand will increasingly be met by other suppliers, from recycling, and be offset by substitution and, as such, this will ensure that global security of supply is in South Africa's self-interest. In this way, the country can leverage opportunities in beneficiation, provided that these are underpinned by domestic social compacts and some form of global partnership.

Post-Cold War geopolitical discourse is in part about the 'New World Order'. This is characterised by globalisation, an element of which is the role of transnational corporations that transcend the nation state. This in part implies a shift in the locus of power beyond the geographic location of mineral endowments.

As a leading supplier, South Africa could leverage opportunities in beneficiation and recycling. This would require a systemic approach, which is different to current practice. In order to exploit PGM resources for use in a hydrogen economy to the full, the country will need to mitigate the risks. Part of the country strategy should include sustainable development principles throughout the metal life cycle. As (Wager, *et al.*, 2012) states:

> *If scarce metals are applied in technologies supposed to support a transition towards a sustainable society, then their supply likewise has to comply with principles of sustainable development: the preservation of natural resources, the minimisation of adverse environmental and social impacts along the metals life cycle, as well as the robustness/resilience of the supply system against changing socio-technical boundary conditions. Transforming existing scarce metals life cycles into more sustainable ones requires interventions into socio-technical systems ranging from specific technological optimisations to the adaptation of behaviour and lifestyles (p. 301).*

However, the use of scarce metals in a global energy transition of the kind described in this study would mean, in geopolitical terms, a migration from 'New World Order' geopolitics through 'environmental geopolitics' to 'sustainability geopolitics'. Sustainability geopolitics, understood in this instance as the transformation of metals life cycles from simple exploitation to the complexity of sustainable cycles, is outlined elsewhere (Wager, *et al.*, 2012).

Some of the implications for the PGM sector are summarised below.

- There has to be a suite of policy measures to guide decisions about the exploitation of resources in the country to make certain that:
  - mining of scarce metals benefits the country and its citizens;
  - mineral extraction and processing are technically efficient;
  - the environment is not degraded; and
  - governance and transparency in reporting and accountability are ensured.
- There has to be security of electricity supply and diversification of the power mix supporting mineral extraction and processing. For instance, power supply disruptions may impact global supply, and the continued reliance on coal would be detrimental to the economy if carbon penalties were to be exacted on exported products.
- There has to be enough water. In a water scarce country, measures to enhance the utilisation of resources, especially water supply, have to be developed.
- To mitigate volatility of prices and ensure management of steady global supplies, a metals exchange mechanism is suggested. This should be developed in partnership with other suppliers, and it should operate transparently, taking sensitivities of consumers into account.
- There have to be benefits for the citizens aligned with the developmental goals of the country. Partnerships among the PGM industry stakeholders are required. An important consideration in this regard should be mechanisms in the execution of policies on Broad Based Black Economic Empowerment (BBBEE) that ensure that workers in the PGM (and other) enterprises become and feel a full part of the stakeholders through, for instance, employee share ownership programmes (ESOPs).
- There has to be recognition of the fact that countries and enterprises outside South Africa have understood the critical role of PGM in their economies and are developing responses to mitigate price volatility and risks to security of supplies. These include mechanisms to lock in supplies from the source, and investment in recycling and substitution, as exemplified by recent related European Commission legislation.

There has to be acknowledgement that the most important strength – the spatial location of the South African PGM endowments – is balanced out by initiatives in recycling and substitution.

On Sustainability and Sustainable Development

The need to address the question, 'When PGM mines close or effect economically-driven retrenchments, what will mine-workers and neighbouring mining communities do?' drives the chapters on sustainability and sustainable development. There is a need for alternate socio-economic opportunities.

The sustainability-orientated approach offers a strategy to facilitate the emergence of an industrial sector in secondary beneficiation of mining waste residues as one vehicle for industrial development occurring beyond mining. The strategy considers mining, sustainable development and sustainability and envisages transitions for the innovation of the system and strategies with which to manage those transitions. The approach comprises networked elements – platinum mining sector, government and society (communities aggregated around the mines) functioning as a think-tank first, and then developing a hybrid business system. This hybrid system would be in the form of an unconventional business operation, the economic framework of which needs to evolve. The nature of the hybrid system readily allows for BBBEE involvement. The approach proposed is based on a broader understanding of 'sustainability' to move to a different point in the continuum that truly represents sustainability.

Sustainable development is the mechanism by which sustainability (as the larger concept) is believed to be achieved in the business world, including in the mining industry. This is aligned with the definition accepted by the World Business Council for Sustainable Development (2010) as '…development that meets the needs of the present without compromising the ability of future generations to meet their own needs' (Brundtland, 1987).

However, sustainable development remains 'a contested concept' (Soderbaum, 2008) as it lacks a normative definition as well as a temporal component. There are four main approaches to sustainable development: needs approach; limits-to-growth approach; capital-based approach and human development approach. The approaches can be considered as strong and weak sustainability depending on the extent of the exploitation of natural resources. Strong sustainability is where natural capital cannot be substituted for other types of capital. The different approaches are connected and overlap, emphasising the economic paradigm in which they function.

Organisations define sustainable development in terms of an environmental, an economic and a social dimension to measure success. This is referred to as the three pillars, the triple bottom line or as the three Es (where social is translated to equity/equality). Because there are many definitions of sustainable development, approaches can be categorised in three ways depending on the economic paradigm: business-as-usual approach based on neoclassical economics; social and ecological modernisation or reformist neoclassical economics approach; and a radical approach which considers alternative political economic models.

Implementation of the sustainability-orientated approach is through a series of systemic transitional steps in multiple dimensions to improved levels of social well-being and sufficiency. However, key assumptions include:

1. The willingness of the platinum sector – the state and specific communities work together toward system innovations which serve the larger goal of sustainability.
2. The risk associated with post-closure obligations for mines can be reduced through collaboration beyond mining.
3. An industrial sector in secondary beneficiation – of mining waste residue – can be created outside mining.
4. The industrial sector can result in more and different employment for former miners and members of communities.

Of note is that the system innovation under discussion occurs in a narrow, but complex area of alternate uses for mining waste residues. It also takes the nature of PGM mining into account but does not interfere with industry's existing processes. If the sustainability-orientated approach is implemented, it can be considered as the PGM sector facilitating secondary beneficiation of mining waste residues beyond mining as a contribution to sustainability, not merely sustainable development.

CONCLUSIONS AND POLICY RECOMMENDATIONS

The government's HySA initiatives are laudable both in the goals they seek to achieve and the execution thereof. Nevertheless, more needs to be done to enhance and further leverage the emergent capabilities as the opportunity space (for South Africa to carve a significant niche in the prospective global hydrogen economy) is quite large. Specifically, more aggressive development and acquisition of the know-how necessary to ensure successful local development of a hydrogen economy and penetration of global markets therefore is necessary.

Especially in view of the opportunity provided by rising costs of energy, and the need to enhance the capacity of the country's energy system, experimenting with the option of platinum-based fuel cells could lead to surprising outcomes on their potential to meet some of the country's energy policy objectives and support economic growth while reducing environmental emissions.

The nascence of the fuel cell industry suggests the dominant design option for the post-ICE paradigm has not yet emerged. The window of opportunity for South Africa to influence the trajectory towards a platinum-based fuel cell option – and thus to expanded markets for its PGM endowments – remains open. However, investment (from both the private and public sectors) towards deepening and broadening the capabilities associated with this technology, and further sensitisation of the public in order to enhance market development, are necessary to make better informed policy decisions. Towards this end, the implementation of visible demonstration projects can be helpful. As stated elsewhere, but worth repeating,

Demonstration products and projects are particularly valuable for a radical new technology, such as fuel cells, in reducing uncertainty. This uncertainty encompasses a range of issues – technology performance, product standards that may be imposed, uptake by potential markets, and the source of funds to develop and commercialise the technology. Demonstration products allow developers to reduce some of this uncertainty by testing the process and learning about the drivers and barriers that the new technology will face. The opportunity to 'observe' and 'trial' aids innovation adoption by allowing potential customers to experience the innovation without purchasing – judging key factors such as complexity, compatibility with lifestyle or work practices, and whether it offers advantages over existing products.

Potential financiers gain a better understanding of the new technology and its applications, thus removing one of the 'roadblocks' that impede the funding of radical innovation (Harborne, et al., 2011, p. 168).

Bibliography

Altergy Systems. (2010). *PR Newswire*. [Online] Available at: http://www.prnewswire.com/news-releases/altergys-freedom-power-fuel-cell-systems-to-power-africa-101632243.html [Accessed 28 February 2013].

Anglo American. (n.d.). *Approach and policies – Extractive Industries Transparency Initiative.* [Online]

Available at: http://www.angloamerican.com/development/approach-and-policies/partnerships-and-collaboration/collaboration/eiti [Accessed 19 July 2013].

Bakker, S. (2010). The car industry and the blowout of the hydrogen hype. *Energy Policy*, July-September, 38(11), pp. 6540–6544.

Bakker, S., van Lente, H. & Engels, R. (2012). Competition in a technological niche: the cars of the future. *Technology Analysis & Strategic Management*, 24(5), pp. 421–434.

Ballard. (2012). *Fuel cell seminar & exposition*. [Online] Available at: http://www.fuelcellseminar.com/media/51071/b2b23-1.pdf [Accessed 27 February 2013].

Ballard. (2013). *EX-99.2 3 exhibit99-2.htm MANAGEMENT'S DISCUSSION AND ANALYSIS*. [Online] Available at: http://www.sec.gov/Archives/edgar/data/1453015/000120677413000708/exhibit99-2.htm [Accessed 27 February 2013].

Baxter, R. (2011). [Online] Available at: http://www.saimm.co.za/Conferences/JhbBranch/RogerBaxter-17Feb2011.pdf [Accessed 26 February 2013].

Behling, N. H. (2013). *Fuel Cells: Current Technology Challenges and Future Research Needs*. s.l.: Elsevier.

Bush, V. (1945). *Science – the Endless Frontier*, Washington, D.C.: U.S. National Science Foundation.

Butler, J. (2011). *Platinum 2011*, s.l.: Johnson Matthey Plc.

Butler, J. (2012). *Platinum 2012* – Interim Review, s.l.: Johnson Matthey Plc.

Christensen, C. (2006). *The Innovator's Dilemma*. New York: Collins Business Essentials.

Collantes, G. & Sperling, D. (2008). The origin of California's zero emission vehicle mandate. *Transportation Research Part A*, Volume 42, pp. 1302–1312.

Conceivious, H. I. (2010). The impact of customer-specific requirements on supply chain management. *Journal of Transport and Supply Chain Management*, pp. 57–68.

Crabtree, G. W., Dresselhaus, M. S. & Buchanan, M. V. (2004). The Hydrogen Economy. *Physics Today*, pp. 39–44.

Cramer, D.(2011). *mybroadband*. [Online] Available at: http://mybroadband.co.za/news/cellular/39627-vodacom-tests-fuel-cell-powered-base-station-at-cop17.html [Accessed 27 February 2013].

Creamer Media Reporter.(2012). *Mining Weekly.com*. [Online] Available at: http://m.miningweekly.com/article/pilot-plants-on-way-for-platinum-titanium-value-add-science-minister-2012-09-25 [Accessed 28 February 2013].

Creamer, M. (2012). *Mining weekly.com*. [Online] Available at: http://www.miningweekly.com/article/form-platinum-exchange-to-save-price-stricken-platinum-mining-iraj-abedian-2012-06-19 [Accessed 08 March 2013].

Crouch, M. (2012). *Fuel cell systems for base stations: Deep dive study*, London: GSMA Development Fund.

Department of Science and Technology. (2007). *South African National Hydrogen and Fuel Cells Research, Development and Innovation Strategy*, s.l.: s.n.

Domfeh, K. A. (2011). *Transfer pricing transgression erodes Africa's tax revenue*. [Online] Available at: http://business.myjoyonline.com/pages/news/201111/77461.php [Accessed 19 July 2013].

Duhigg, C. & Bradsher, K. (2012). [Online] Available at: http://www.nytimes.com/2012/01/22/business/apple-america-and-a-squeezed-middle-class.html?emc=eta1 [Accessed 25 February 2013].

EERE. (n.d.). *Types of Fuel Cells*. [Online] Available at: http://www1.eere.energy.gov/hydrogenandfuelcells/fuelcells/fc_types.html [Accessed 21 February 2013].

Elvis, M. (2012). Let's mine asteroids – for science and profit. *Nature*, 31 May, Volume 485, p. 549.

European Commission. (2003). *Hydrogen energy and fuel cells, a vision for our future*, Brussels: European Commission.

Farley, H. R. & Mesiti, P. (2012). [Online] Available at: http://cepgi.typepad.com/files/cepgi-3d-quarter-2012-1.pdf [Accessed 22 February 2013].

Friedman, D., Masciangioli, T. & Olson, S. (2012). *The role of the chemical sciences in finding alternatives to critical resources – a workshop summary*, s.l.: National Academies Press.

FuelCellWorks. (2013). [Online] Available at: http://fuelcellsworks.com/news/2013/02/19/oorja-protonics-and-hysacatalysis-forge-strategic-partnership-for-commercialization-of-methanol-fuel-cells-in-south-africa [Accessed 21 February 2013].

Grubler, A. (2012). Grand Designs: Historical Patterns and Future Scenarios of Energy Technological Change. In: The Global Energy Assessment. Cambridge, UK: Cambridge University Press.

Grubler, A., *et al.* (2012). Policies for the energy technology innovation system (ETIS). In: Global Energy Assessment. s.l.: Cambridge University Press, pp. 1551–1743.

Halper, M. (2013). [Online] Available at: http://www.smartplanet.com/blog/bulletin/want-a-cheap-hydrogen-fuel-cell-wash-out-the-platinum/13250 [Accessed 21 February 2013].

Hamilton, T. (2012). Is the fuel-cell industry really near a tipping point?. [Online] Available at: http://www.thestar.com/business/2012/06/08/is_the_fuelcell_industry_really_near_a_tipping_point.html [Accessed 27 February 2013].

Harborne, P., Hendry, C. & Brown, J. (2009). Fuel cells as disruptive innovation. The power to change markets. In: *Innovation, Markets and Sustainable Energy – The Challenge of Hydrogen and Fuel Cells*. Northampton, Massachusetts: Edward Elgar Publishing, Inc, pp. 34–51.

Harborne, P., Hendry, C., Brown, J. & pg. 168. (2011). The development and diffusion of radical technological innovation: the role of bus demonstration projects in commercializing fuel cell technology. *Technology Analysis & Strategic Management*, 19(2), pp. 167–188.

Heistein, P. (2012). *Business Report*. [Online] Available at: http://www.iol.co.za/business/business-news/sa-will-miss-benefits-by-pulling-plug-on-joule-car-1.1334575#.USsxjDLPPo [Accessed 28 February 2013].

IEA. (2012). *Energy Technology Perspectives 2012 – Pathways to a Clean Energy System*, Paris: OECD/IEA.

Keane, A. G. (2012). *Bloomberg.* [Online] Available at: http://www.bloomberg.com/news/2012-10-16/obama-s-5-billion-slow-to-charge-electric-car-purchases.html [Accessed 21 February 2013].

Ko, V. (2012). [Online] Available at: http://edition.cnn.com/2012/11/25/business/eco-hydrogen-fuel-cell-cars [Accessed 21 February 2013].

Kosich, D. (2012). *Mineweb.* [Online] Available at: http://www.mineweb.co.za/mineweb/content/en/mineweb-exploration?oid= 150034&sn=Detail [Accessed 20 March 2013].

Kosich, D. (2013). *Canada to mandate extractive industries transparency reporting.* [Online] Available at: http://www.mineweb.com/mineweb/content/en/mineweb-political-economy?oid=194036&sn=Detail [Accessed 19 July 2013].

Kraemer, K. L., Linden, G. & Dedrick, J. (2011). *Capturing value in global networks: Apple's iPad and iPhone,* s.l.: CISE/IIS.

Letourneau, A. (2013). Asteroid mining becoming more of a reality. *Forbes,* 25 January.

Marchetti, C. (1973). Hydrogen and Energy. *Chemical Economy & Engineering Review,* 5(1), pp. 7–15.

Marwala, T. (n.d.). *Academia.edu.* [Online] Available at: http://www.academia.edu/1481269/The_Platinum_Group_Metals_and_the_National_Democratic_Society [Accessed 10 March 2013].

Miningreview.com (2010). [Online] Available at: http://www.miningreview.com/node/17614 [Accessed 26 February 2013].

Ministerial Review Committee. (2012). *Ministerial Review Report on the Science, Technology and Innovation Landscape in South Africa,* Pretoria: Department of Science and Technology.

Mock, P. & Schmid, S. A. (2009). Fuel cells for automotive powertrains – a techno economic assessment. *Journal of Power Sources,* Volume 190, pp. 133–140.

Morgan, T. (2006). *The Hydrogen Economy – A non-technical review,* s.l.: United Nations Environment Programme.

Mytelka, L. (2008). Hydrogen fuel cells and alternatives in the transport sector: A framework for analysis. In: *Making choices about hydrogen.* s.l.: United Nations University, pp. 35–63.

Nail, J. M., Anderson, G., Ceaser, G. & Hansen, C. J. (2005). *The Role of the U.S. National Innovation System in the Development of the PEM Stationary Fuel Cell,* Washington: US Department of Commerce.

OECD. (2013). Economic growth in South Africa: Getting to the right shade of green. In: *OECD Economic Surveys: South Africa 2013.* s.l.: OECD Publishing, pp. 91–120.

Ohadi, M. M. & Jianwei, Q. (2008). Alternative Energy Technologies: Price Effects. In: *Encyclopedia of Energy Engineering and Technology.* s.l.: s.n., p. 36.

Pak, S. J. (2012). *Lost Billions – Transfer Pricing in the Extractive Industries,* Oslo, Norway: Publish What You Pay Norway.

PMG. (2012). [Online] Available at: http://www.pmg.org.za/report/20120912-hydrogen-fuel-cell-technology-research-development-programme-briefing [Accessed 21 February 2013].

Reuters. (2013). *Fuel cells are driving electric cars into dead end.* [Online] Available at: http://www.iol.co.za/business/business-news/fuel-cells-are-driving-electric-cars-into-dead-end-1.1464110?goback=%2Egde_3284455_member_212041266#.UXo4h7VR_rx [Accessed 27 April 2013].

Rifkin, J. (2011). *The Third Industrial Revolution – How Lateral Power is Transforming Energy, the Economy, and the World.* New York: Palgrave Macmillan.

Satyapal, S. (2012). *Energy.gov.* [Online] Available at: http://energy.gov/articles/calling-all-fuel-cells [Accessed 27 February 2013].

Saurat, M. & Bringezu, S. (2008). Platinum Group Metal Flows of Europe, Part I. *Journal of Industrial Ecology*, 12(5/6), pp. 754–767.

ScotiaMocatta. (2012). *Scotiabank.* [Online] Available at: http://www.scotiamocatta.com/scpt/scotiamocatta/prec/PGMForecast2013.pdf [Accessed 27 February 2013].

Seccombe, A. (2013). *Business Day BDlive.* [Online] Available at: http://www.bdlive.co.za/business/energy/2013/02/28/nersa-grants-eskom-8-annual-increases-over-next-five-years [Accessed 28 February 2013].

Shinnar, R. (2004). Demystifying the hydrogen myth. *Chemical Engineering Progress*, November, pp. 5–6.

Smil, V. (2012). Far from Electrifying: Electric car hopes never die – but electric realities keep intervening. *The American*, 26 November.

Spath, P. L. & Mann, M. K. (2001). *Life cycle assessment of hydrogen production via natural gas steam reforming*, s.l.: NREL.

Steinemann, P. (1999). *R&D Strategies for New Automotive Technologies: Insights From Fuel Cells*, s.l.: Massachussetts Institute of Technology.

Stillwater Mining Company. (2012). *Palladium Fundamentals.* [Online] Available at: http://www.corporatereport.com/stillwater/Palladiam_Fundamentals_9-26-2012.pdf [Accessed 27 February 2013].

Stroeben, J. & Sterman, J. (2008). Transition challenges for alternative fuel vehicle and transportation systems. *Environment and Planning B: Planning and Design*, 35(6), pp. 1070–1097.

Tuttle, B. (2012). *Time.* [Online] Available at: http://business.time.com/2012/09/07/is-it-time-to-declare-the-nissan-leaf-a-flop/ [Accessed 28 February 2013].

UN ECA. (2012). *Report blames multinationals for illicitly transferring most of the $1.5 trillion made in Africa each year.* [Online] Available at: http://www1.uneca.org/TabId/3018/Default.aspx?ArticleId=1671 [Accessed 19 July 2013].

Unger, D. (2010). Innovation and market entry in the energy industry: Lessons for fuel cells and new technologies. *Journal of Business & Economic Research*, 8(10), pp. 63–72.

Utterback, J. (1994). *Mastering the Dynamics of Innovation.* s.l.: Harvard Business School Press.

van den Hoed, R. (2005). Commitment to fuel cell technology? How to interpret carmakers' efforts in this radical technology. *Journal of Power Sources*, Volume 141, pp. 265–271.

van Ravenswaay, J., et al. (2009). *South Africa's Nuclear Hydrogen Production Development Programme.* Oakbrook, Illinois, Organisation for Economic Cooperation and Development, pp. 205–212.

Verne, J. (1874). *The Mysterious Island.* s.l.: s.n.

Wager, P. A., et al. (2012). Towards a More Sustainable Use of Scarce Metals. *GAIA*, 21(4), pp. 300–309.

Wesoff, E. (2012). *greentechmedia.* [Online] Available at: http://www.greentechmedia.com/articles/read/bloom-energy-fuel-cell-financials-revealed [Accessed 26 February 2013].

Wild, S. (2013). SA set to roll out prototype hydrogen golf car this year. *Business Day*, 18 February , p. 4.

Xing, Y. & Detert, N. (2010). How the iPhone widens the United States trade deficit with the People's Republic of China. *ADBI Working Paper Series*, Issue 257.

APPENDICES

CHAPTER 3
APPENDIX A

METHODOLOGY FOR THE BIBLIOMETRIC ANALYSIS

The citation databases that were used were derived from the Web of Science (WoS), which includes the Science Citation Index Expanded, the Social Sciences Citation Index and the Arts & Humanities Citation Index, all from 1988 to 2012 (December). The Web of Science is the online version of the citation indexes of the Arts and Humanities Citation Index (A&HCI), Social Sciences Citation Index (SSCI) and Science Citation Index Expanded (SCIE). The Science Citation Index Expanded is a multidisciplinary index to the journal literature of the sciences and indexes 5,900 major journals across 150 scientific disciplines. The Social Sciences Citation Index is a multidisciplinary index to the journal literature of the social sciences, which indexes more than 1,725 journals across 50 social sciences disciplines. Arts and Humanities Citation Index is a multidisciplinary index covering the journal literature of the arts and humanities, and fully covers 1,144 of the world's leading arts and humanities journals.

PATENT DATABASE

As for the patent research the EPO Worldwide Patent Statistical Database, October 2011 was used. The PATSTAT, also known as the EPO Worldwide Patent Statistical Database, is a snapshot of the EPO master documentation database (DOCDB) with worldwide coverage. It contains more than 20 tables with bibliographic data, citations and family links of about 70 million applications from more than 80 countries.

Choice of Keywords

Frequently used keywords are selected according to seven technological sub-fields corresponding to the most common types of fuel cells (Klitkou, *et al.*, 2007): alkaline fuel cells; direct methanol fuel cells; molten carbonate fuel cells; phosphoric acid fuel cells; proton exchange membrane fuel cells; solid oxide fuel cells and regenerative (reversible) fuel cells. The study also used these keywords, but additionally trailed a wider method following the approach from the Australian Academy of Science (2008). Keywords can be consulted on page 362.

For the patent analysis, main patent classes (appendix section) were focused on.

Choice of Publication Types

A wide-range approach was undertaken, using all types of references inside WoS. Nevertheless, most results returned were journal articles.

Time-period

The time period covered in this bibliometric research is 1988 to 2012 and includes all available records.

Data Retrieval and Standardisation

A keyword approach detailed in the paper (page 362) was followed. Using this methodology one was able to retrieve 41 274 observations (articles) for fuel cells and 13 958 observations (articles) for hydrogen storage. Data was standardised and cleaned and was categorised using several identifiers, such as country, authors, key words, WoS category, etc.

Using patent data from PATSTAT it was decided to create a network that describes the knowledge flows between countries. The nodes (vertices) in the network will be inventor countries, whereas the edges are either based on backward or forward citations between patent families[1].

Backward citations were focused on as citations exist between patent publications. Publications have a many-to-zero relationship with applications. In other words, a patent application can have any number of publications, even zero, because not all applications are published. In their turn, applications have a many-to-one relationship with patent families. Or, a 'family' consists of one or more applications. In our case we used the

---

1. A patent family is 'a set of patents taken in various countries to protect a single invention (when a first application in a country – the priority – is then extended to other offices' (OECD, 2001, p. 60).

DOCDB family, which comes predefined with PATSTAT. In this family all patents that protect the same technical content are grouped together. There are two problems with the inventor country:

1. Within the 'family', the inventors may change. It is probably the effect of continuations, and this might occur less when opting for equivalents. To account for this, we have chosen to look at the first filing within the closer 'family' and take its inventors as the inventors for the entire family.[2]

2. Not all patents have country information for the inventors. While one could tag the country as 'unknown', two extra searches to find country information elsewhere were implemented. The EPO standardised the names of inventors and applicants, and if there were any inventors with the same standardised name that have a country affiliation, these were checked.

APPENDIX B

DESCRIPTIVE STATISTICS

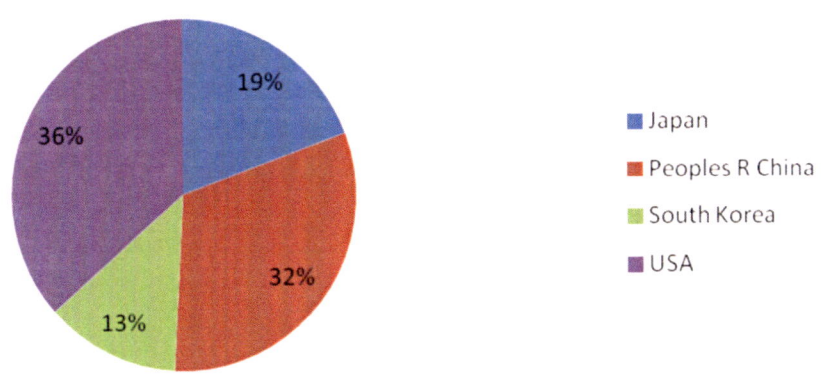

Figure 5: Publications per country (above 5,000)
Source: Web of Science (WoS) between 1988–2011

---

2. It is worth mentioning that a patent can have multiple inventors, and for this we used fractional counts.

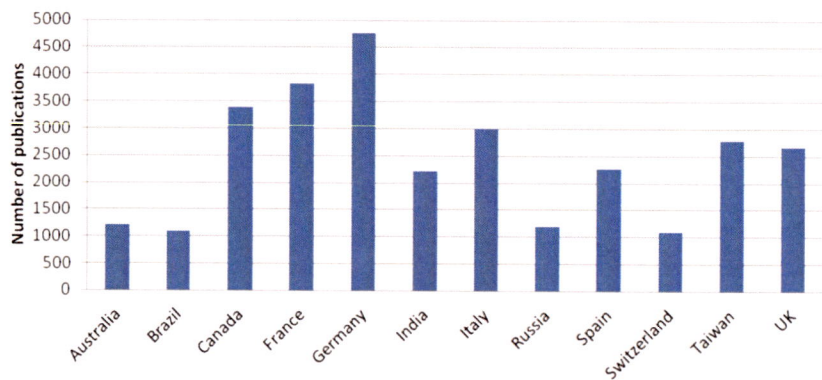

Figure 6: Publications per country (between 1,000–5,000)
Source: Web of Science (WoS) between 1988–2011

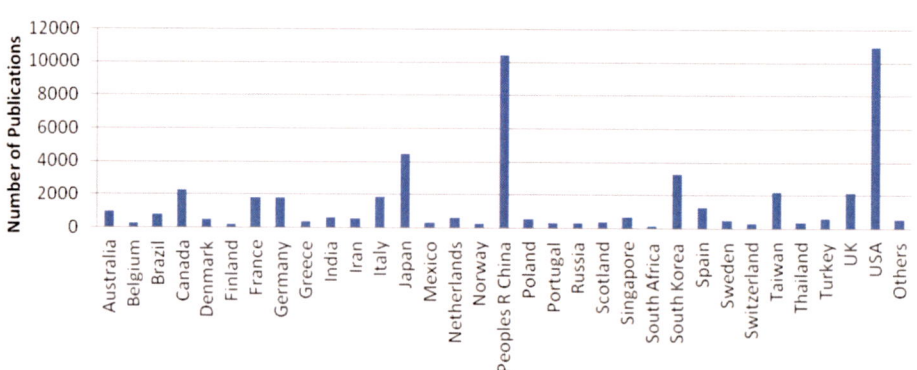

[3]Figure 7: WoS selected University publications per country 1988–2011 (0–12 000)
Source: Web of Science (WoS): 1988–2011

---

3. Others in figure 7 include Malaysia, Jordan, Austria and Argentina as their publications were relatively low during the period 1988–2011.

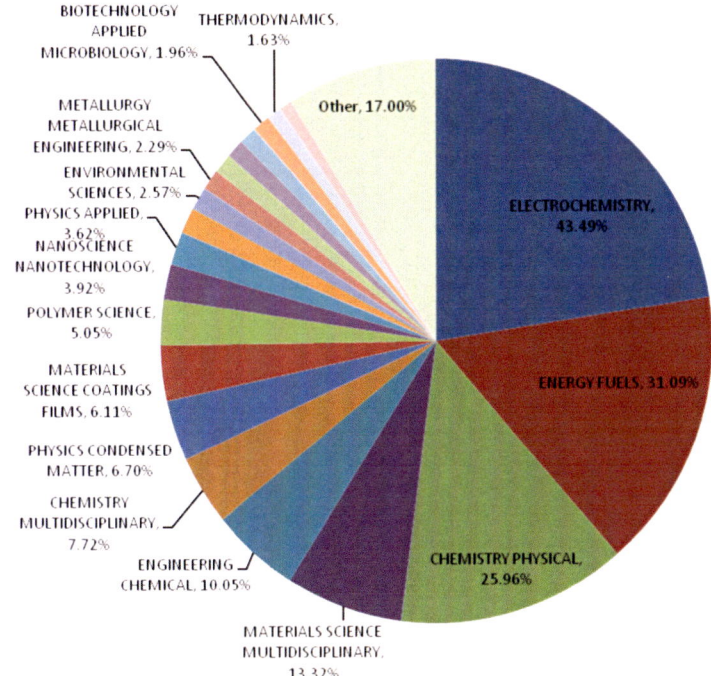

Figure 8: Distribution of articles by field – fuel cells (1988–2011)
Source: Web of Science (WoS): 1988–2011

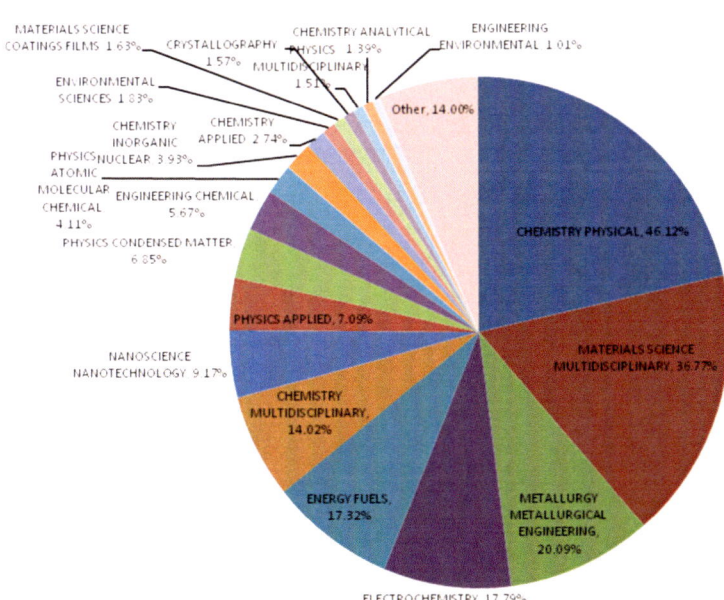

Figure 9: Distribution of articles by field – hydrogen storage (1988–2011)
Source: Web of Science (WoS): 1988–2011

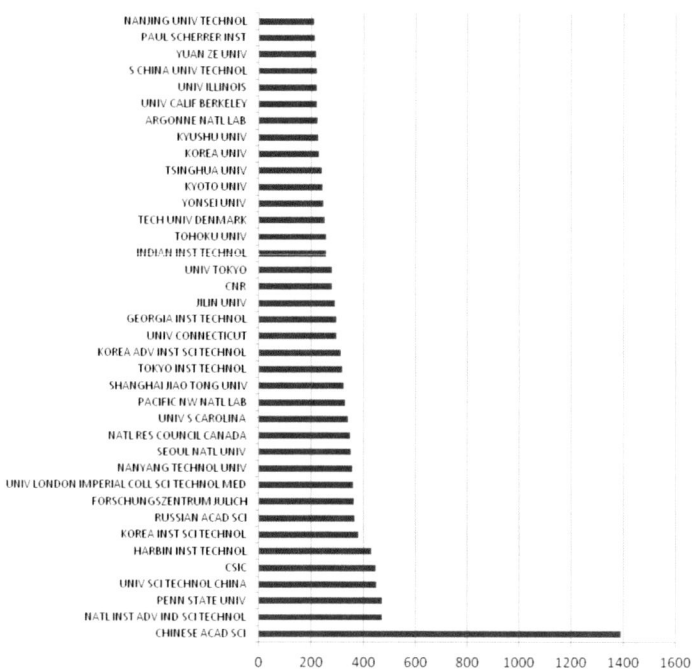

Figure 10: Main research organisations for fuel cells (1988–2011)
Source: Authors' Own

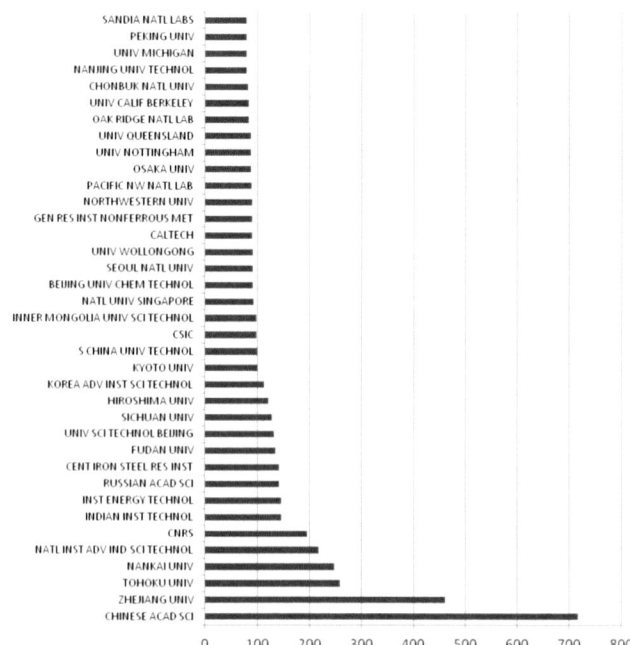

Figure 11: Main research organisations for hydrogen storage (1988–2011)
Source: Web of Science (WoS): 1988–2011

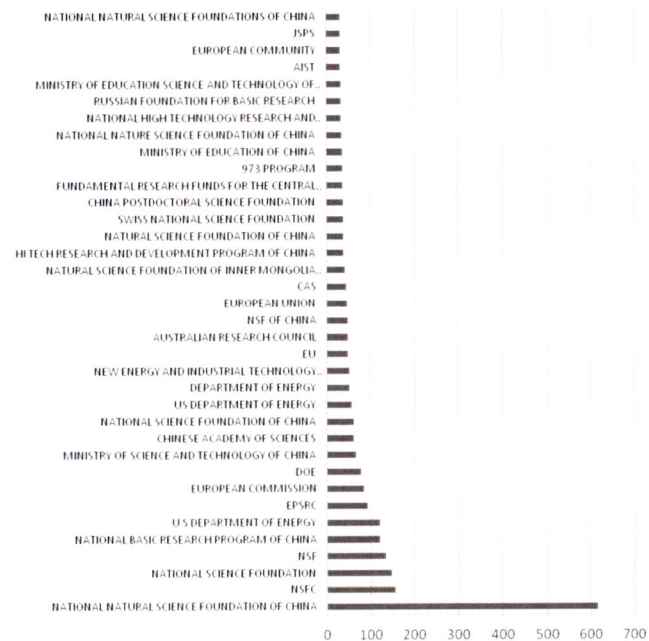

Figure 12: Main funding organisations by total number of publications
Source: Web of Science (WoS): 1988–2011

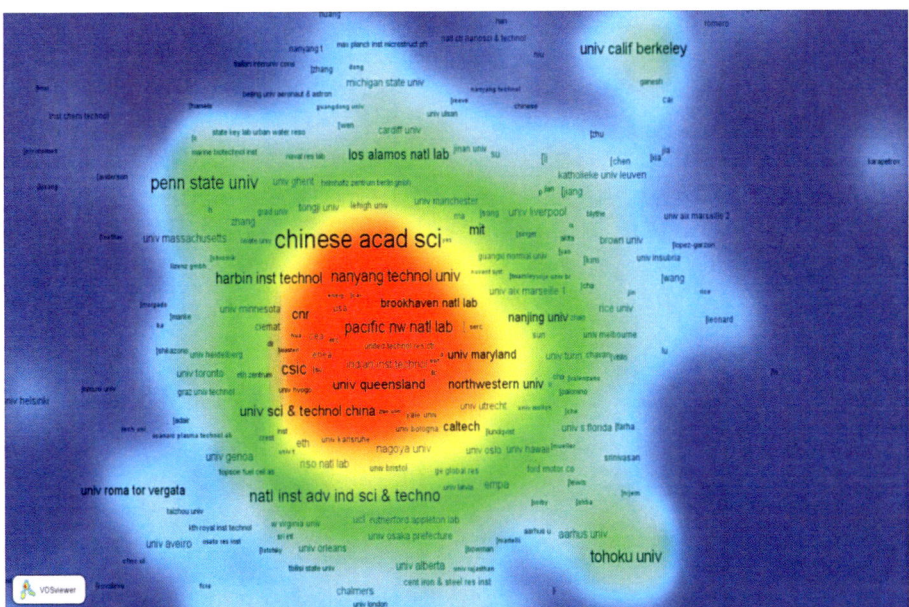

Figure 13: Bibliographic coupling of organisations – density view

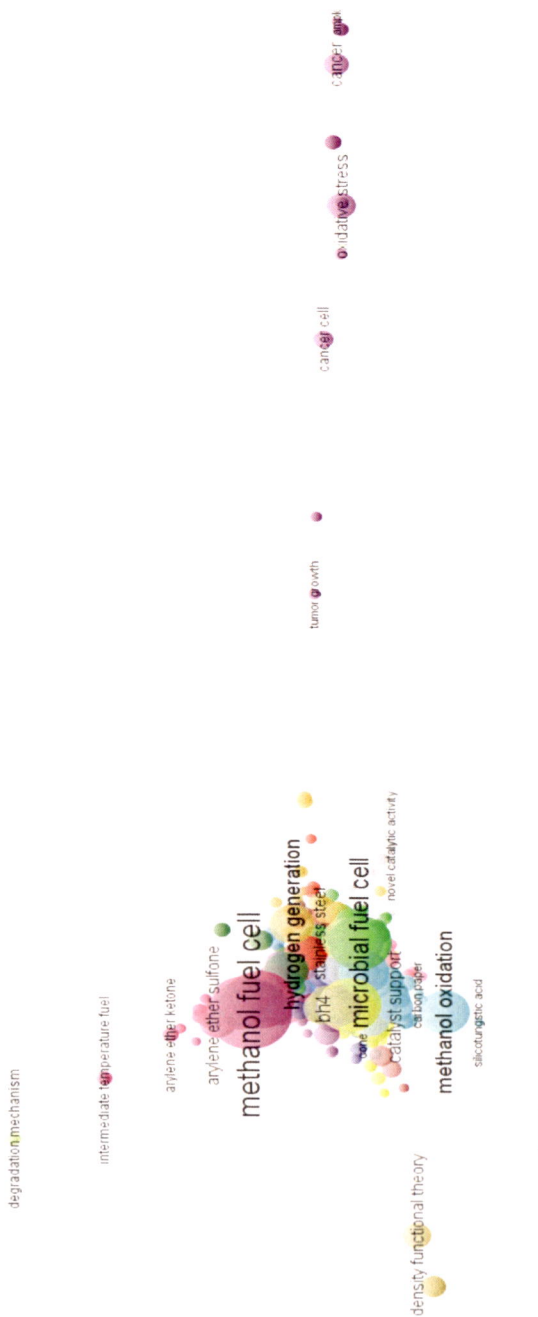

Figure 14: Main research areas – word analysis – label view

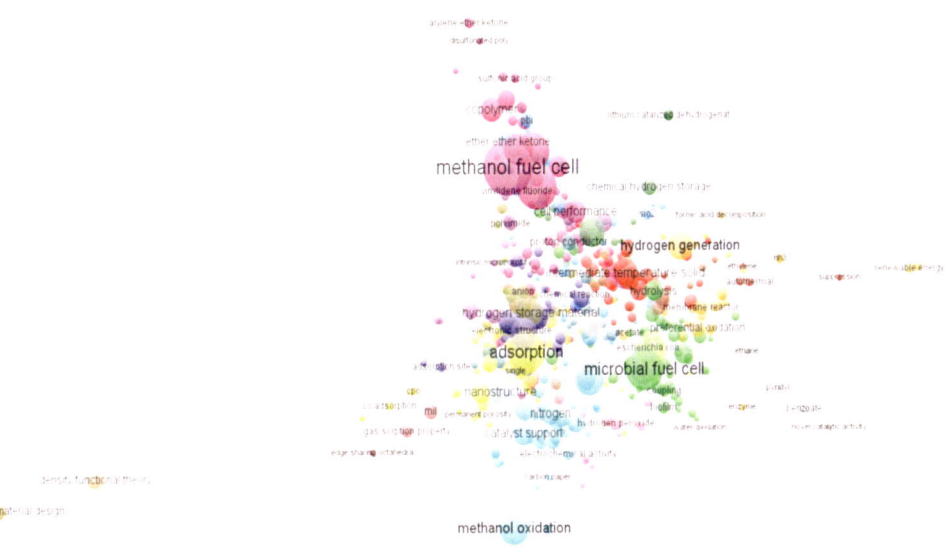

Figure 15: Main research areas – word analysis – label view amplified

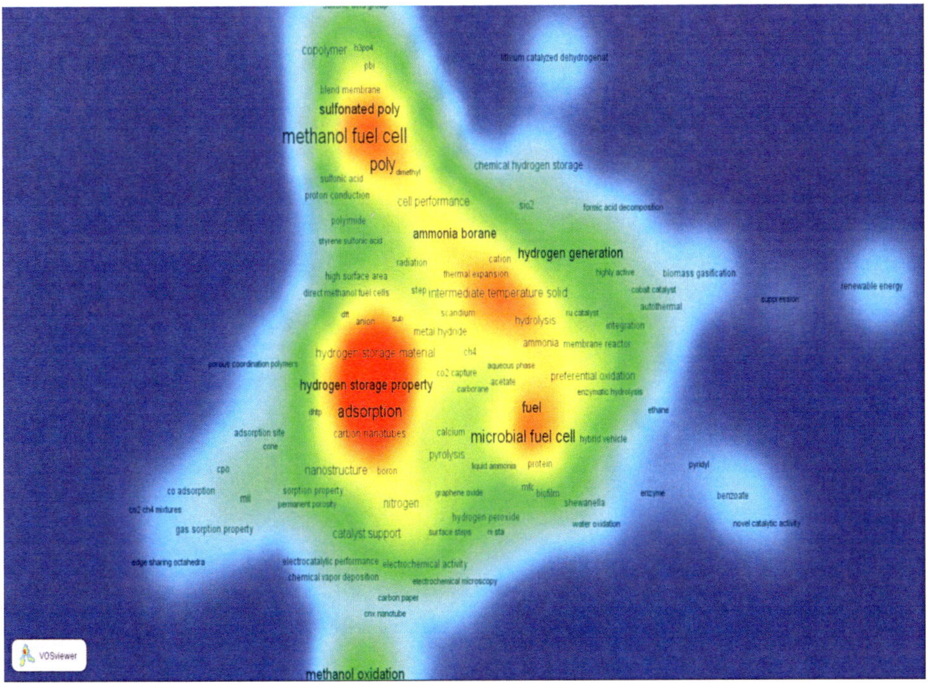

Figure 16: Main research areas – word analysis – density view

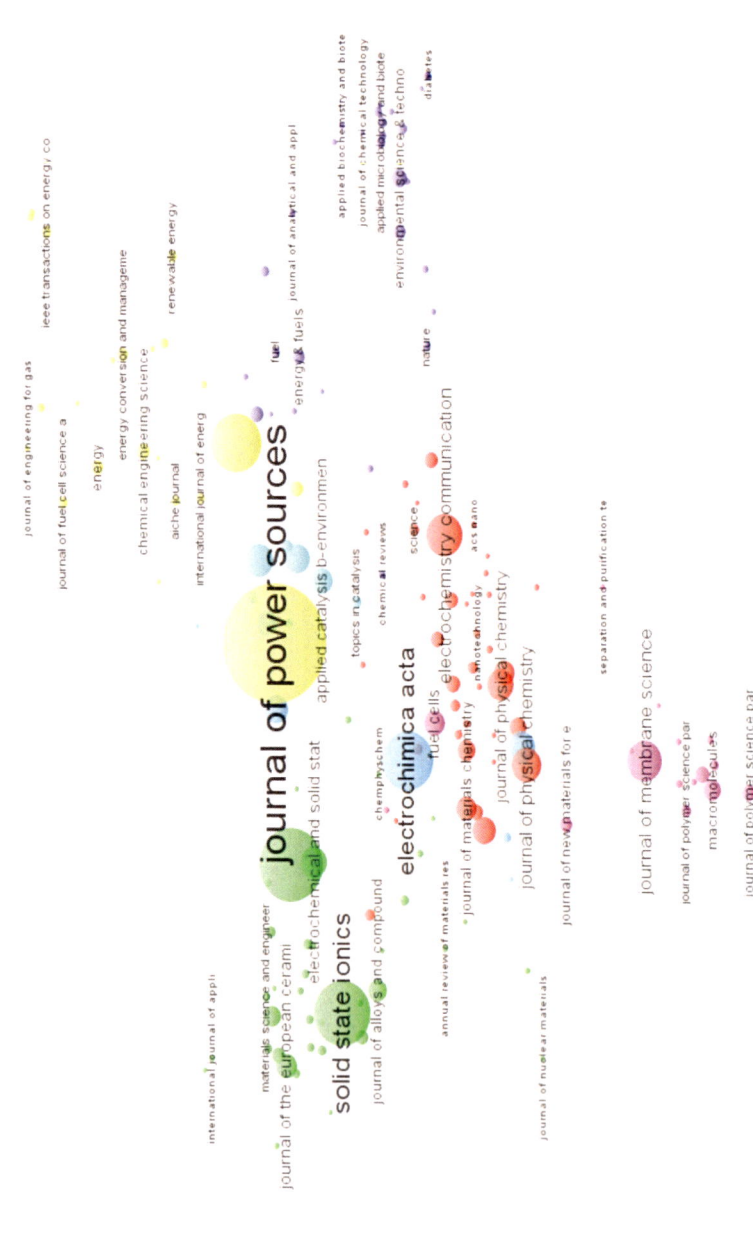

Figure 17: Co-citation of sources – fuel cells – label view

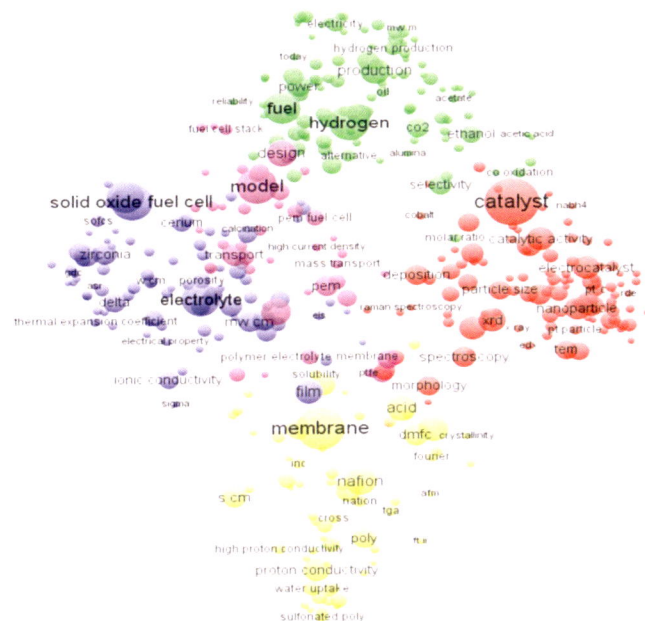

Figure 18: Main research areas fuel cell – word analysis – label view amplified

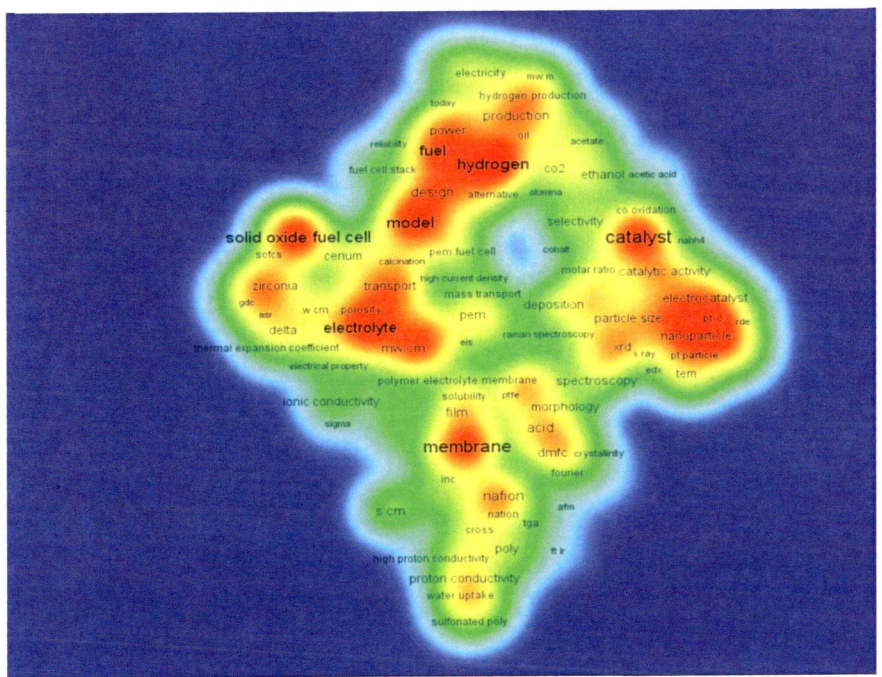

Figure 19: Main research areas fuel cell – word analysis – density view

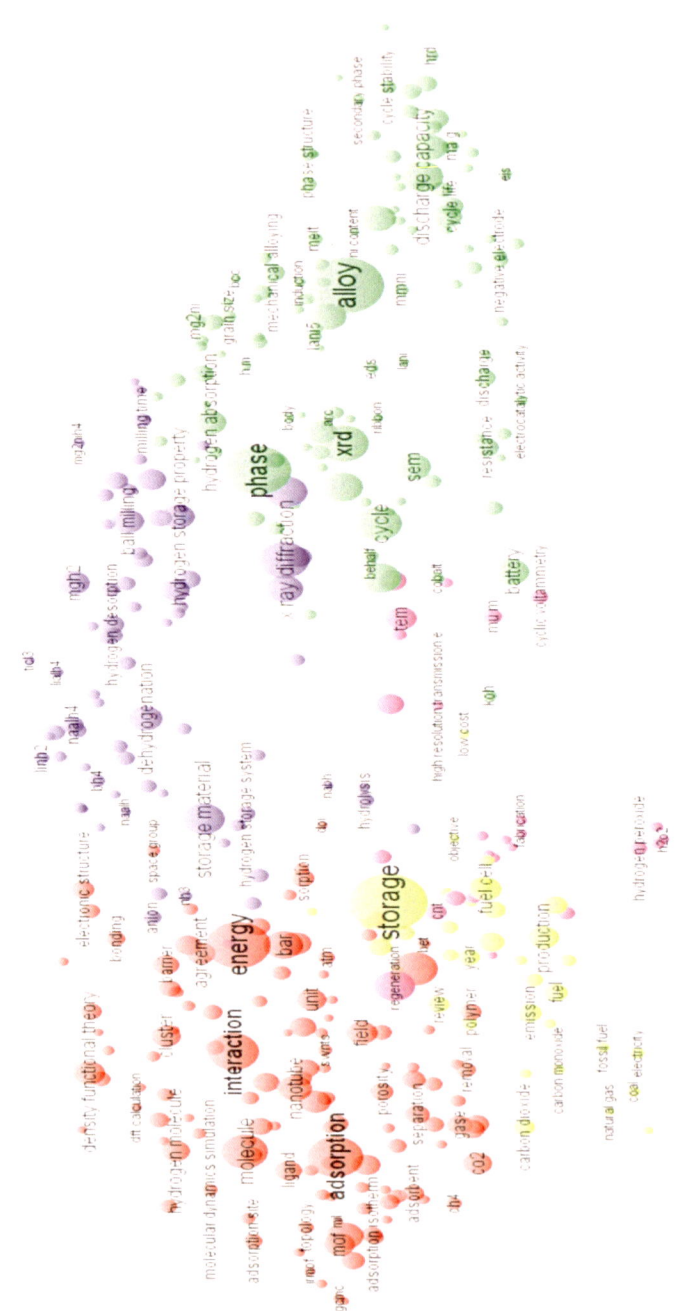

Figure 20: Main research areas storage – word analysis – label view

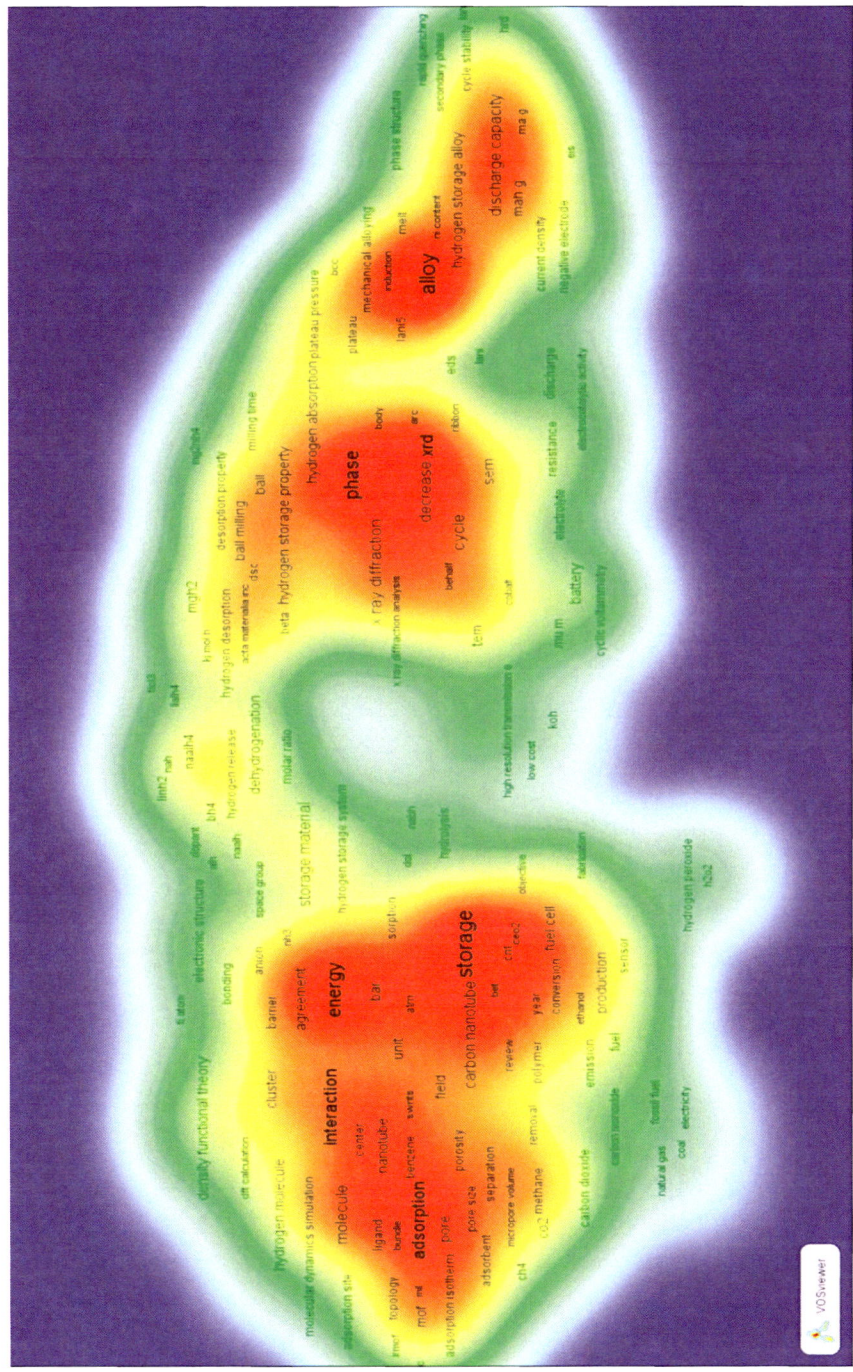

Figure 21: Main research areas storage– word analysis – density view

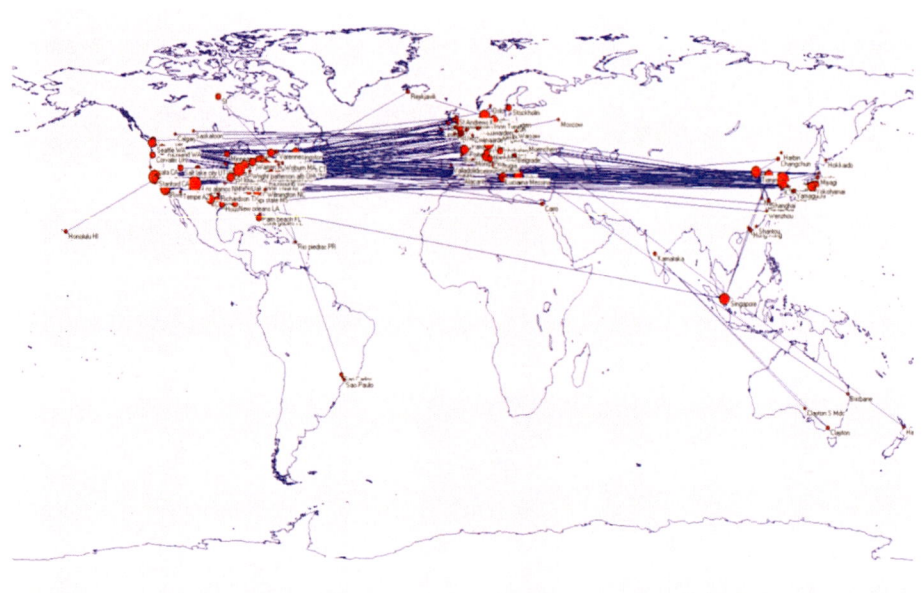

Figure 22: Collaboration network – top 500 papers (by citation) on hydrogen fuel cells

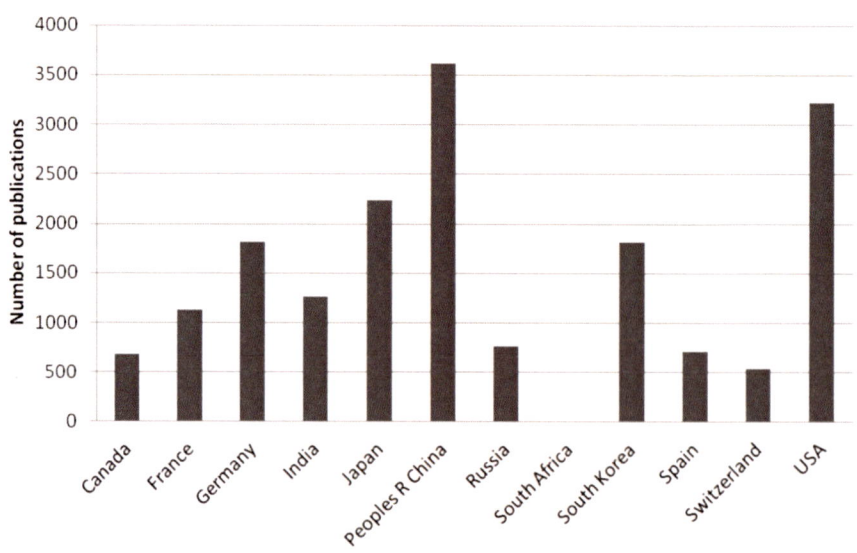

Figure 23: Government publications per country (above 500 with the exception of South Africa with one publication). Source: Web of Science (WoS): 1988–2011

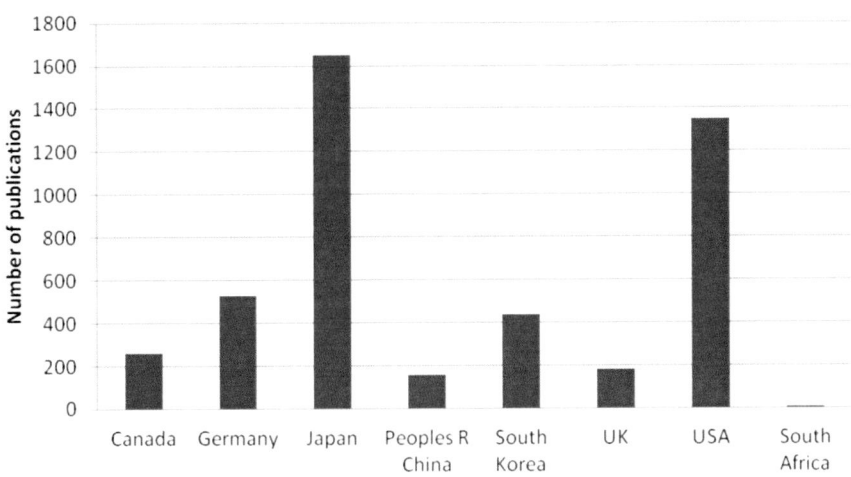

Figure 24: Industry publications per country (above 150 with the exception of South Africa with two publications). Source: Web of Science (WoS): 1988–2011

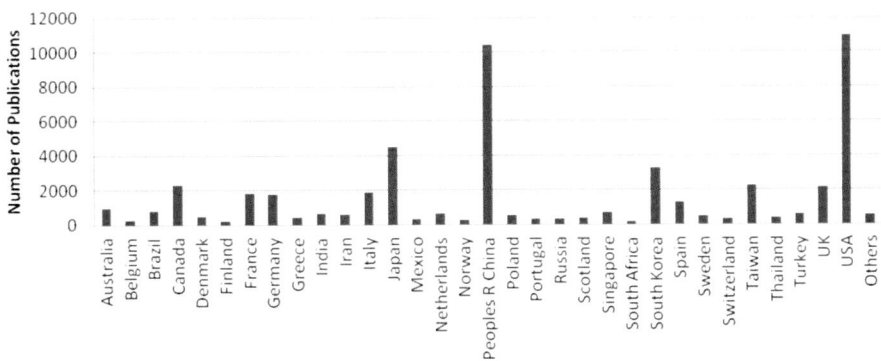

Figure 25: Total publications by universities in each country
Source: Web of Science (WoS): 1988–2011

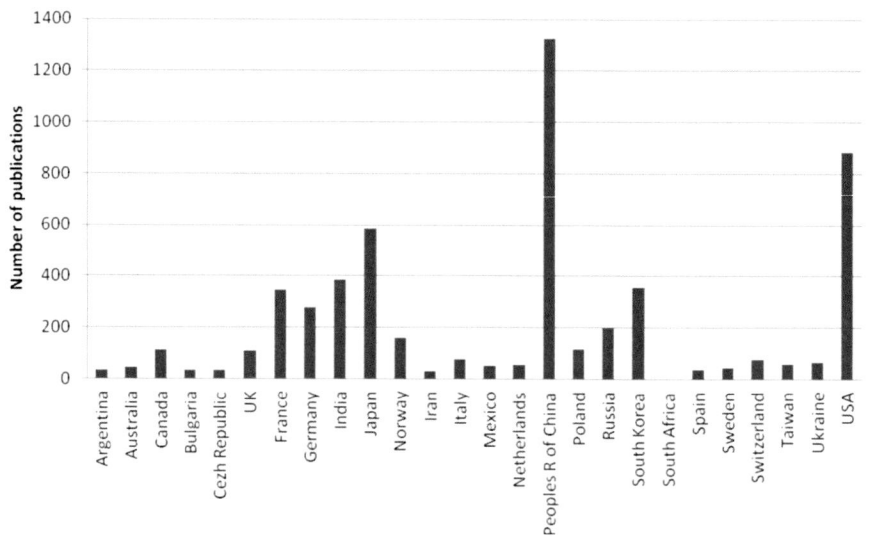

Figure 26: Total publications by government on hydrogen storage
[6]Source: Web of Science (WoS): 1988–2011

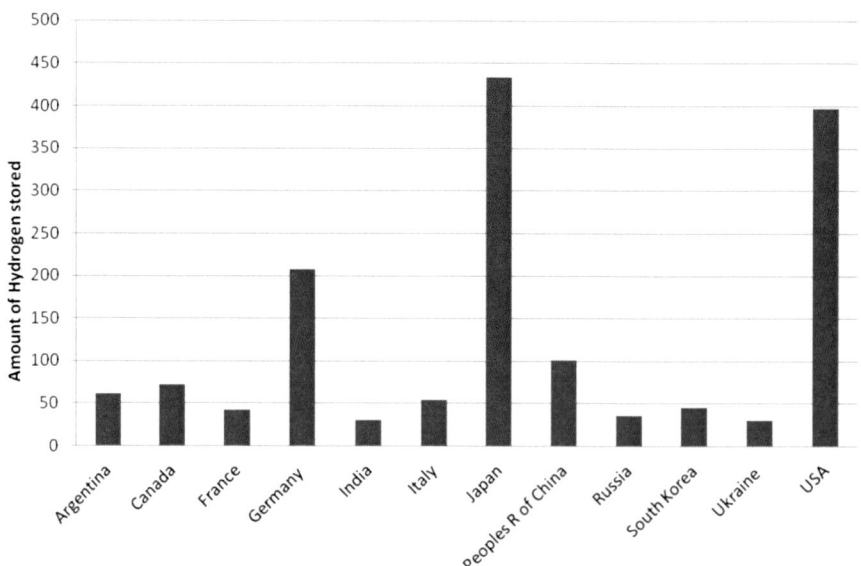

Figure 27: Publications by industry for each country on hydrogen storage
Source: Web of Science (WoS): 1988–2011

6. Others in Figure 29 include Malaysia, Jordan, Austria and Argentina. Their counts are relatively low.

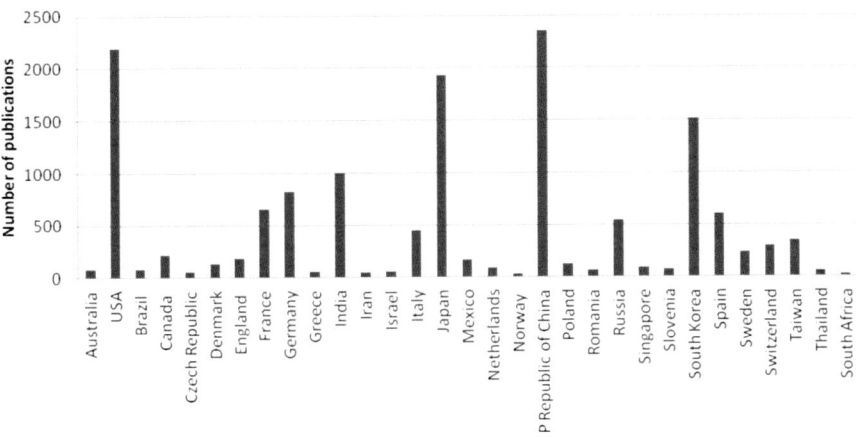

Figure 28: Publications by university of each country on hydrogen storage
Source: Web of Science (WoS): 1988–2011

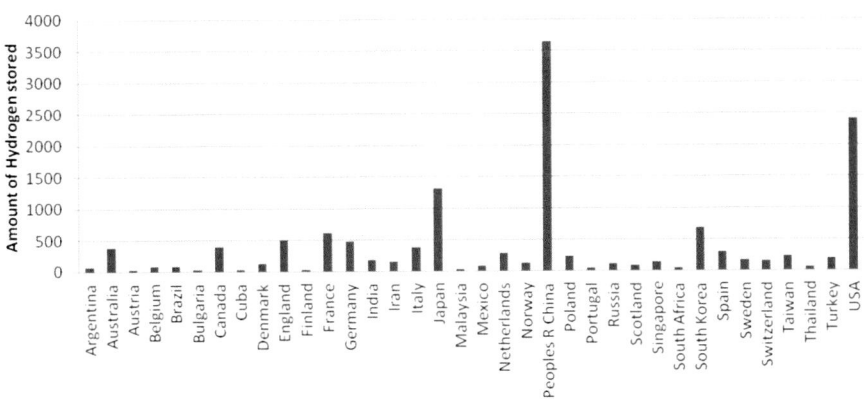

Figure 29: Publications by government on hydrogen fuel cells
Source: Web of Science (WoS): 1988–2011

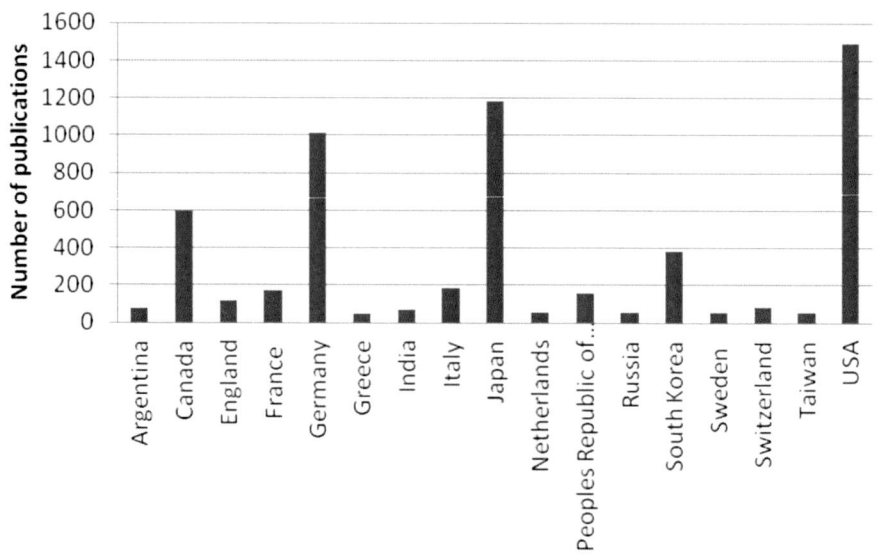

Figure 30: Publications by industry in each country on hydrogen fuel cells
Source: Web of Science (WoS): 1988–2011

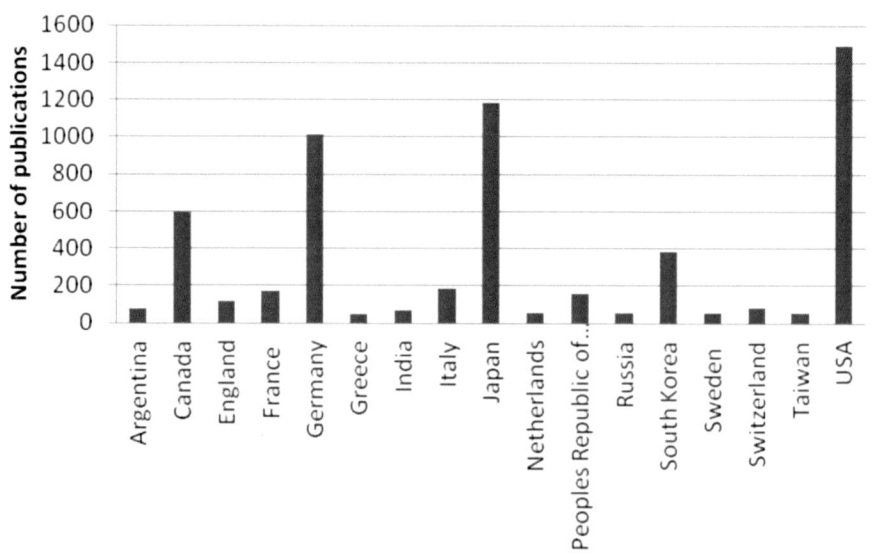

Figure 31: Publications by university in each country on hydrogen fuel cells
Source: Web of Science (WoS): 1988–2011

Descriptive data Patents

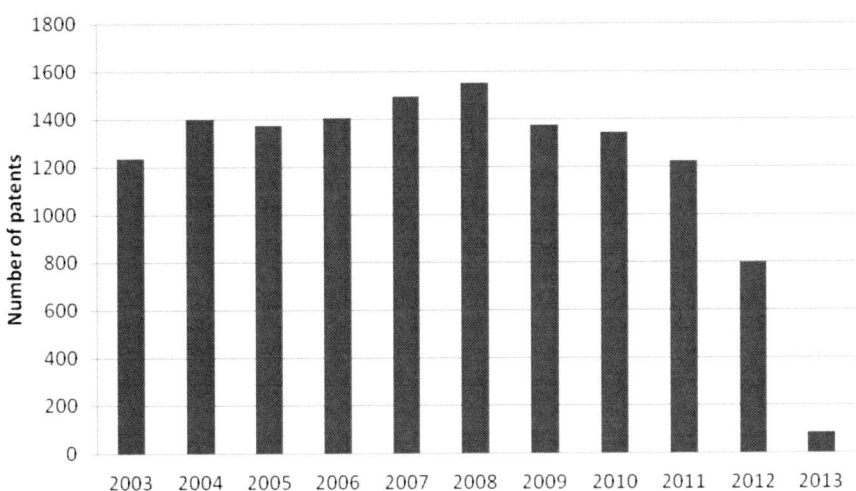

Figure 32: Patents per year for hydrogen storage patents (number of patents)
Source: WIPO (2013)

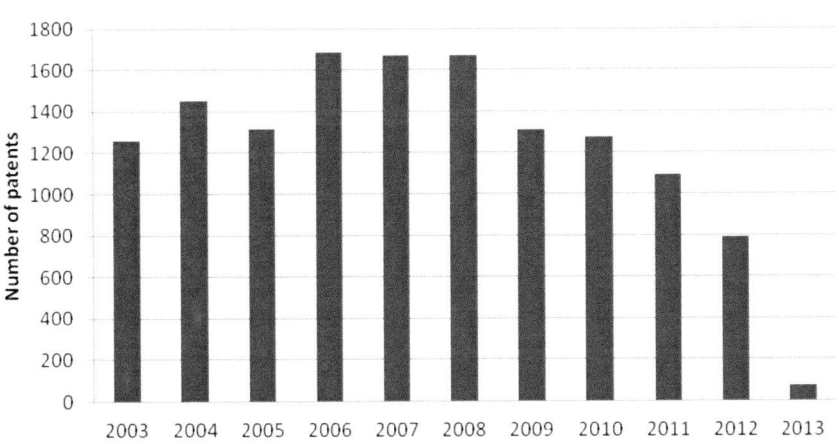

Figure 33: Patents per year for hydrogen production patents (number of patents)
Source: WIPO (2013)

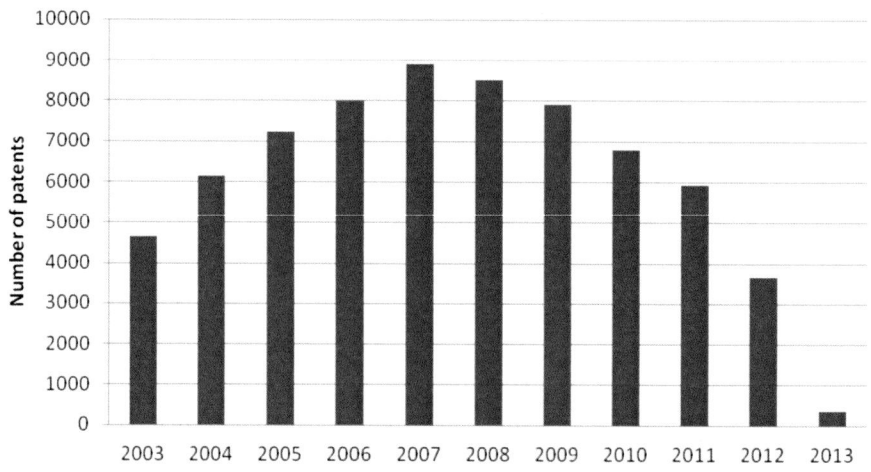

Figure 34: Patents per year for fuel cells patents (number of patents)
[7]Source: WIPO (2013)

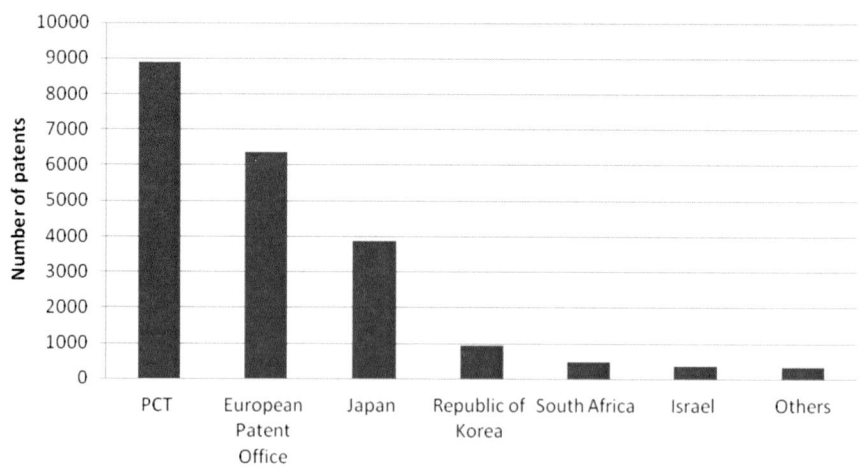

Figure 35: Main applicants for hydrogen storage patents (number of patents)
[8]Source: WIPO (2013)

7. Others: Russian Federation, Mexico, Russian Federation (USSR data), Singapore, ARIPO and Spain.
PCT: Patent Cooperation Treaty
8. Others: Singapore, Russian Federation (USSR Data), Spain, Jordan, Argentina, Kenya, ARIPO.

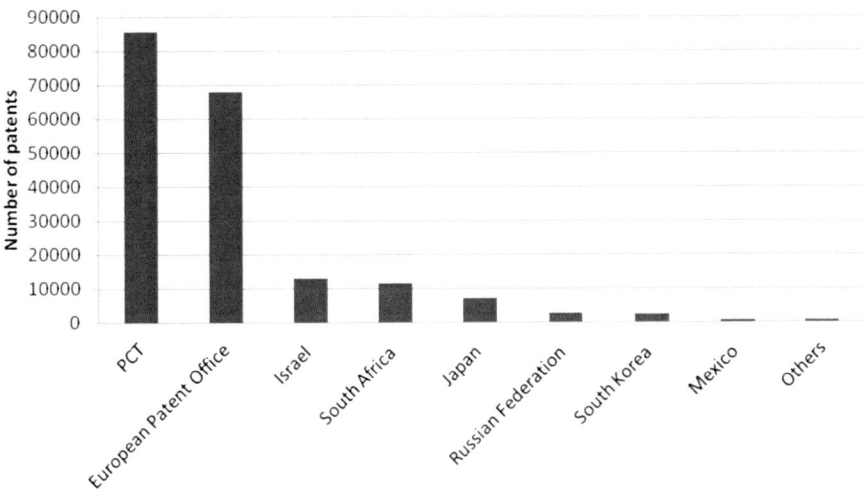

Figure 36: Main applicants for hydrogen production patents (number of patents)
Source: WIPO (2013)

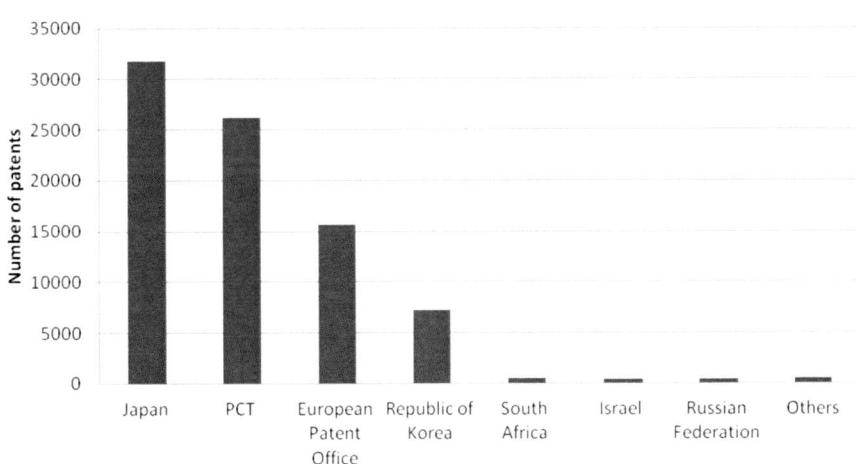

Figure 37: Main Applicants for fuel cells patents (number of patents)
Source: WIPO (2013)

ARIPO: African Regional Intellectual Property Rights Organisation
PCT: Patent Cooperation Treaty

Key Words Used in Bibliometric Data Search

Hydrogen Fuel Cells

Hydrogen; membrane*; ceramic; polymer electrolyte membrane fuel cell; PEMFC; solid oxide; SOFC; phosphoric acid fuel cell; POFC; alkaline fuel cell; AFC; molten carbonate fuel cell; MCFC; direct methanol polymer electrolyte membrane; DMPEM; proton electrolyte membrane fuel cell; solid polymer electrolyte; Nafion; silica; zirconia; micro fuel cells; micro-react; micro-PIV; titanium phosphate; ruthenium; CMK-3; Tio2; platinum; palladium; Pd; V; magnesium; Ni; MG(0001); LIFEPO4; CeO2; zirconium; ceria catalysts; noble-metal catalysts; cermet membrane*; gas-diffusion; Laves phase alloy; uranium; (yttri*)

Hydrogen Storage

Ads;ption; ads;b*; metal hydride; complex hydrides;
hydride*; nanomaterials; nanoparticles; carbon; carbon nanofibres; carbon nanotubes; batteries; lead-acid battery; b;on; La; Mg; Ni; alloy)

IPC codes on hydrogen fuel cells, production and storage

Hydrogen production C01B3, C25B1/04, C07C4/20

Hydrogen, produc*, generat*, obtain*, reform*, preparat*, manufacture*

Hydrogen storage B01D53/02, C01B3/0*, C01B3/1*, C01B3/2*, C01B3/3*, C01B3/4*, C01B3/5*, C22C22/00, C22C33/00, F25B17/12, H01M4/38, H01M8/06

Hydrogen, storage*, reservoir*, alloy*, absorb*

Proton-exchange membrane fuel cells: H01M4/00, H01M4/86, H01M4/88, H01M4/90, H01M8/0*, H01M8/1*, H01M8/20, H01M8/22, H01M4/24

Fuel-cell*, fuel-batter*, fuel cell*, fuel batter*, PEM, PEMFC, polymer*, proton, ion, exchange*, membrane

Solid oxide fuel cells: H01M4/00, H01M4/86, H01M4/88, H01M4/90, H01M8/0*, H01M8/1*, H01M8/20, H01M8/22, H01M4/24

Fuel-cell*, fuel-batter*, fuel cell*, fuel batter*, SOFC*, solid oxide*, solid oxid*, zirconium, Zr0

Molten carbonate fuel cells: H01M4/00, H01M4/86, H01M4/88, H01M4/90, H01M8/0*, H01M8/1*, H01M8/20, H01M8/22, H01M4/24

Fuel-cell*, fuel-batter*, fuel cell*, fuel batter*, MCFC, molten, melt*, carbonat*

Other types of fuel cells (H01M4/00, H01M4/86, H01M4/88, H01M4/90, H01M8/0*, H01M8/1*, H01M8/20, H01M8/22, H01M4/24)

Main Patent Classes Used[4]

H01M 4/86, H01M 4/88, H01M 4/90, H01M 4/92, H01M 4/94, H01M 4/96, H01M 4/98,

---

4. http://www.wipo.int/ipcpub/#refresh=page&notion=scheme&version=20060101&symbol=H01M

H01M 8/*, H01M 12/*. There are 157 166 hydrogen patents, with time coverage 1902–2011.

Hydrogen production: C01B3, C25B1/04, C07C4/20

Hydrogen storage: B01D53/02, C01B3/0*, C01B3/1*, C01B3/2*, C01B3/3*, C01B3/4*, C01B3/5*, C22C22/00, C22C33/00, F25B17/12, H01M4/38, H01M8/06

Proton-exchange membrane fuel cells: H01M4/00, H01M4/86, H01M4/88, H01M4/90, H01M8/0*, H01M8/1*, H01M8/20, H01M8/22, H01M4/24

Solid oxide fuel cells: H01M4/00, H01M4/86, H01M4/88, H01M4/90, H01M8/0*, H01M8/1*, H01M8/20, H01M8/22, H01M4/24

Molten carbonate fuel cells: H01M4/00, H01M4/86, H01M4/88, H01M4/90, H01M8/0*, H01M8/1*, H01M8/20, H01M8/22, H01M4/24

Other types of fuel cells: H01M4/00, H01M4/86, H01M4/88, H01M4/90, H01M8/0*, H01M8/1*, H01M8/20, H01M8/22, H01M4/24

Data categorisation for bibliometric data

PT = Publication Type (J=Journal; B=Book; S=Series)
AU = Authors
BA = Book Authors
BE = Editors
GP = Book Group Authors
AF = Author Full Name
CA = Group Authors
TI = Document Title
SO = Publication Name
SE = Book Series Title
LA = Language
DT = Document Type
CT = Conference Title
CY = Conference Date
CL = Conference Location
SP = Conference Sponsors
HO = Conference Host
DE = Author Keywords
IDX = Keywords Plus®
AB = Abstract
C1 = Author Address
RP = Reprint Address
EM = E-mail Address
FU = Funding Agency and Grant Number
FX = Funding Text

CR = Cited References
NR = Cited Reference Count
TC = Times Cited
PU = Publisher
PI = Publisher City
PA = Publisher Address
SN = ISSN
BN = ISBN
J9 = 29-Character Source Abbreviation
JI = ISO Source Abbreviation
PD = Publication Date
PY = Year Published
VL = Volume
IS = Issue
PN = Part Number
SU = Supplement
SI = Special Issue
BP = Beginning Page
EP = Ending Page
AR = Article Number
DI = Digital Object Identifier (DOI)
PG = Page Count
WC = Web of Science Category
SC = Subject Category
GA = Document Delivery Number
UT = Unique Article Identifier

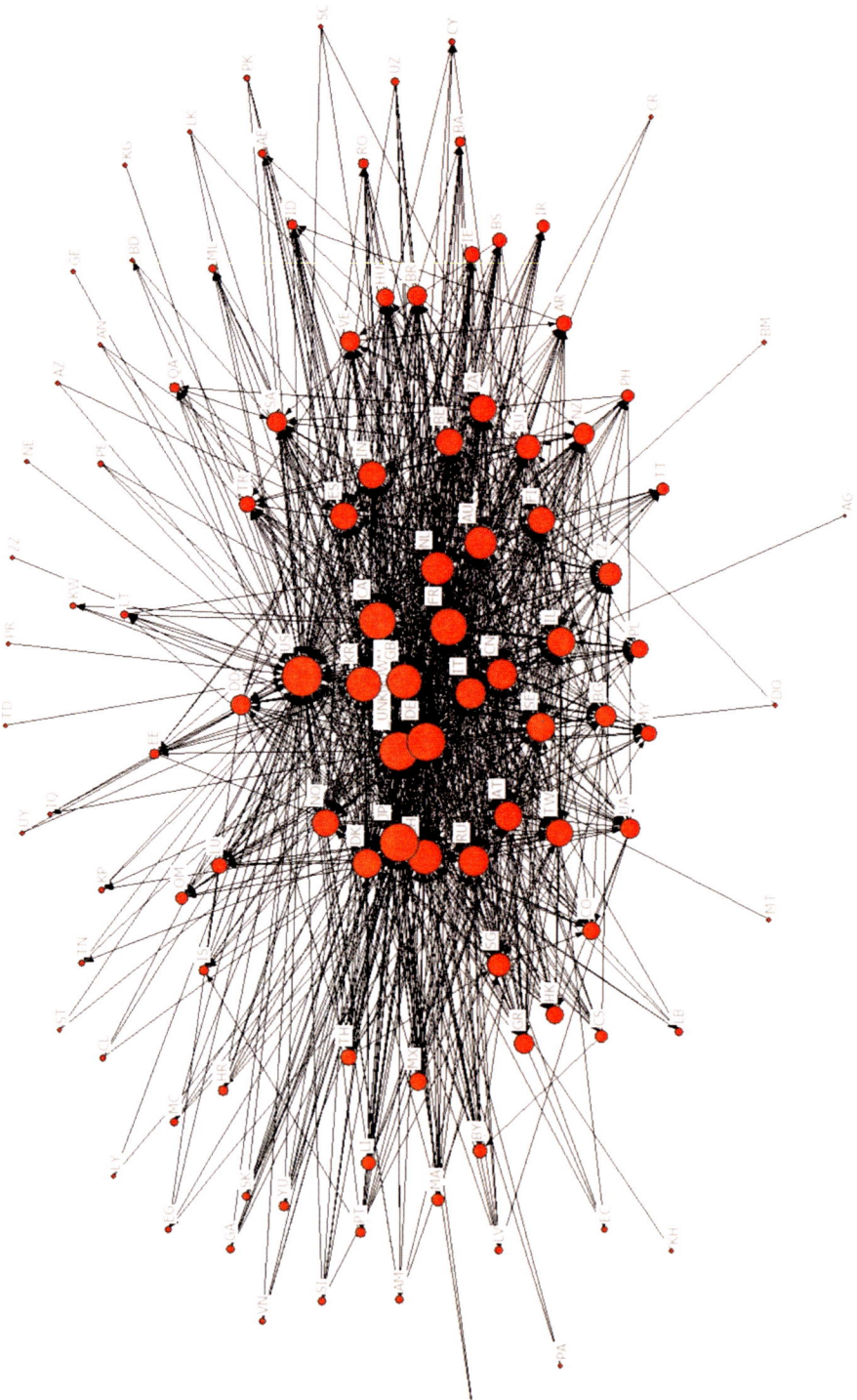

Figure 38: Country citation network (weighted by Eigenvector Centrality)

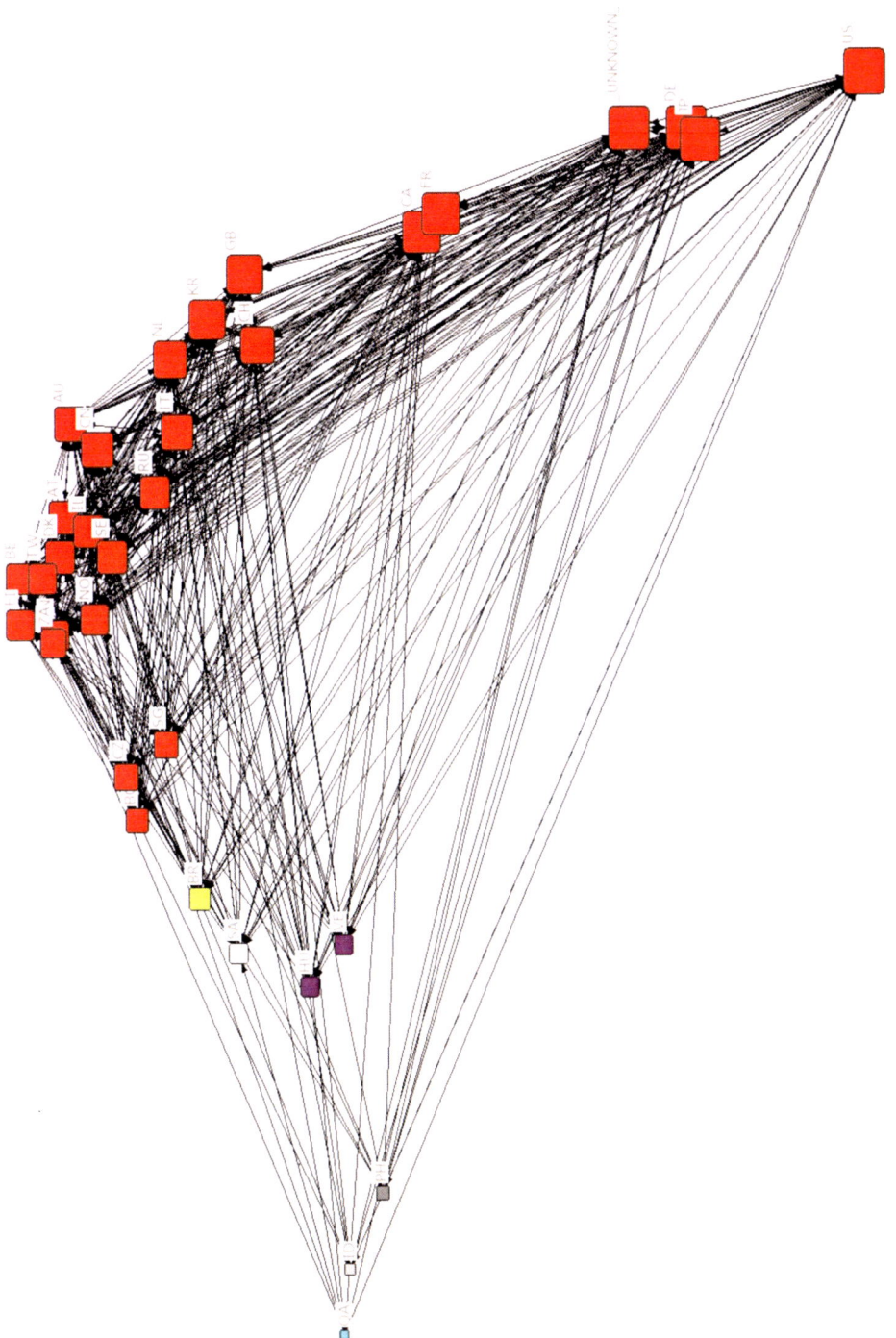

Figure 39: South African citation 'ego' network (weighted by Eigenvector Centrality)

CHAPTER 5
APPENDIX A

MODELLING METHODOLOGY

The penetration and impacts of new energy technologies have been studied with many types of models including accounting models, agent-based models, econometric models and least-cost optimisation models, all of which are useful in adding perspective to possible futures and in making decisions. In this case, the availability of an existing model of the latter type, in which an abstraction of the structure of the South African energy system across economic sectors had been developed, determined the tool for the study. ERC's TIMES model for South Africa lies at the centre of a modelling framework called SATIM which is documented online (ERC, 2013) but is discussed briefly below.

1 THE TIMES PLATFORM AND ERC's SATIM MODEL

The Energy Research Centre at the University of Cape Town has developed, and maintains, an energy modelling framework, now in its third generation, called SATIM built around the TIMES platform. TIMES has been developed by ETSAP, one of the International Energy Agency's implementing agencies, and is a successor to MARKAL. TIMES is a partial equilibrium linear optimisation model capable of representing the whole energy system of a country, including its economic costs and its emissions, and is thus particularly useful for infrastructure planning or for modelling potential mitigation policies.

The model is capable of solving for a variety of constraints including emissions constraints, by sector, for the whole economy, or cumulatively over a period, and can be used to identify the more complex consequences of mitigation actions or infrastructure decisions. The South African TIMES model structure is contained in a database, and constructed via a user interface called ANSWER, which provides a framework for both structuring the model and scenarios, and also for interpreting results. ANSWER compiles the model data into a set of tables, which is then interpreted by

TIMES (GAMS code), and GAMS compiles a set of linear equations, which are then solved by a linear solver such as CPLEX.

The approach is fundamentally sectoral, as would be the case with other bottom-up models, even simple spreadsheet models like the International Atomic Energy Agency's MAED, and so the analysis of the structure of energy demand from economic sectors is a fundamental building block of the modelling process. SATIM as a framework therefore includes a number of sub-models that pre-process inputs for the TIMES model itself, for instance a vehicle parc model that profiles base year vehicle technologies in detail for application to the Transport Sector Module.

2 PROJECTING THE DEMAND FOR ENERGY IN THE MODEL

The projection of future demand for energy is critical to the outcome of the model and therefore, although this is documented elsewhere in detail (ERC, 2013), this aspect is discussed briefly below. In SATIM, useful energy is an exogenous input disaggregated by an energy carrier, for each demand sector. Final energy demand is determined endogenously using the assumed efficiencies of the least cost demand-side technology mix selected by the model. The two supply sectors and primary energy sources must meet the sum of these demands. The supply sectors do not, therefore, have their own demand projections and the model optimiser will select supply-side technologies to meet the demand for final energy at least cost.

For most sectors and sub-sectors, it makes sense to reflect the demand for useful energy by means of a demand for an energy service like lighting or passenger transport, as reflected by an appropriate indicator. Thus, for the transport sector, which is of primary interest to this study, future useful energy demand is determined as a projection of demand for passenger travel in units of passenger.km and the demand for freight travel in units of ton.km. In the case of freight, this is assumed to be directly proportional to economic growth moderated by an elasticity. This is determined for the sub-sectors of light, medium and heavy commercial vehicles. The relationship between passenger.km and economic growth in SATIM is more indirect and deploys a sub-model called the 'time-budget' model which assesses the demand for passenger travel on the following basis:

- Passenger demand for road and rail is modelled for three income groups representing low-, medium- and high-income households. This was done because growth of private transport at the expense of low speed public modes is strongly related to household income. The low-income group includes households with income of up to R19 000 per annum; the middle-income group includes households with an income between R19 000 and R76 800; and the high-income group the remainder. All amounts are in 2007 Rands.

- Motorisation (car ownership) per capita for each of the income groups was estimated for the base year using survey data (DoT, 2005). Private vehicle motorisation for the reference case in SATIM is thus driven by mobility of the populace between income groups and is thus essentially GDP/capita proportional.

- Assumptions around ratios of public and private transport, average speed, and travel time budget have been made for each of the income groups and used to calculate their net demand for passenger travel. Due to the sparseness of activity data the model is calibrated to match the vehicle parc model for only two modes, private and public, for the base year of 2006.

- Time-use and travel surveys from numerous cities and countries throughout the world suggest that the travel time budget is on average approximately 1.1 hours per person per day across the spectrum of per capita income (Schafer & Victor, 2000) and this principle is applied in SATIM.

- Mobility not met by private transport was assumed to be met by public transport (for example minibus, bus, Metrorail, BRT or rapid train) and distributed between modes according to anticipated investment in infrastructure supporting each of the modes.

From our projections of demand for freight and passenger travel, a projection for vehicle-km is calculated for each demand (say private passenger car travel or heavy commercial freight) by means of an assumed occupancy (passenger/veh for passenger vehicles) or load factor (tons/veh for freight). These vehicle-km demands are the exogenous inputs to TIMES and the technologies that are loaded to meet these energy services, therefore have efficiencies attributed to them that are expressed in terms of energy used per km (MJ/km) which are in turn are used to calculate final energy endogenously by the model.

3 The evolution of new technology within the vehicle parc

Clearly though, vehicles remain in the system for some time, which affects the rate of evolution of the vehicle parc. The rate of scrappage thus determines the rate at which technology evolves in the system as much as sales. In TIMES this is modelled in a relatively unsophisticated way for new technology, the model being originally designed around large power plants with relatively predictable lifetimes. In SATIM new road vehicles are assigned a lifetime of 12 years except for buses (18 years) and motorcycles (10 years) at which time they disappear from the model. The persistence of existing stock from the base year over the projection is input as a lookup table and follows a Weibull distribution (Merven, Stone, Hughes, & Cohen, 2012).

4 Setting up SATIM for hydrogen – structural assumptions and constraints

Prior to this study SATIM did not include hydrogen. It was necessary to add the hydrogen energy chain from supply through to end-use to SATIM in a likely configuration, given current data. While more detailed TIMES models exist, SATIM has organically become fairly complex during its development and certainly the potential exists to engage in very detailed disaggregation of technologies and myriad scenarios. It was important for this short study to limit the scope and place appropriate constraints on the application of SATIM to the problem. In this regard, the set-up of the model was characterised as follows:

- While a number of operational demonstration FCV passenger cars, SUVs and buses are extant, a freight solution seems far in the future. Hydrogen-fuelled demand-side technologies were therefore only made available in the model to meet the passenger travel energy service as follows:
  - Hydrogen Fuel Cell Passenger Cars
  - Hydrogen Fuel Cell Sport Utility Vehicles (SUV)
  - Hydrogen Fuel Cell Buses
  - Hydrogen Fuel Cell Minibus/Midibus Taxis

Other hydrogen-fuelled options that appear to be further from commercialisation, for instance hydrogen-fuelled internal combustion engine vehicles or plug-in hydrogen fuel cell electric hybrids were not included in the model. A follow-on study should investigate the development status of light commercial fuel cell vehicles.

- It did not make sense to include all sectors in model runs because this would slow down processing. The model included the following sectors and their technologies: on the demand-side: transport; on the supply-side: primary energy, liquid fuels (natural gas and hydrogen), and electricity supply by the power sector. The projected electricity demand for all sectors, is however, an exogenous input to the model so we can optimise a full generation mix for various scenarios. This is of interest for this project because the current rationale of a hydrogen economy is to reach a point where hydrogen is produced using water electrolysis supplied by renewable power so that the whole energy pathway is carbon neutral. The effect of input assumptions on the least-cost generation mix for South Africa is therefore an important output of this project.

- There are a number of hydrogen production technologies of which three (those most likely to be practical for mass production) were included in generic form in the model as follows: steam methane reforming (SMR), coal gasification, and water electrolysis. The cost and efficiency data input for these technologies assumes large-scale centralised production and not medium-scale or distributed production which was not modelled for this study.

- Learning rates were not assumed for investment costs and fixed O&M for hydrogen production technologies in the model. The UKSHEC model (Dodds & McDowall, 2012), however, assumes significant cost reductions for hydrogen production technologies in the region of 0.5–1.5% per annum but much higher (greater than 5%) for small distributed SMR and electrolysis till 2025, although the latter does not drop below large-scale production costs by 2050. In contrast, an in-depth hydrogen production cost data between 1940–2007 for the three production processes above (Schootsa, Feriolia, Kramer, & van der Zwaan, 2008) found 'only limited learning at best in a couple of cases, and no cost reductions can be detected for the overall hydrogen production process'. Observed cost reductions seemed to relate to improvements in general engineering construction techniques rather than to 'learning by doing' on the processes themselves. For this project, therefore, hydrogen production costs were assumed static for the modelling horizon, although this should be reviewed for any follow-up studies, if possible in collaboration with HySA.

- The coal gasification hydrogen production costs assumed were international but could potentially be lower in South Africa, given the large-scale coal gasification operation run by SASOL. Determining possible local costs was outside the scope of this project, but a future more in-depth study should include SASOL and other stakeholders in the synthetic fuel industry, so that this potential local advantage in kick-starting a hydrogen economy can be taken into account.

· SATIM does not currently profile distribution and fuel dispensing costs, which would include the cost of pipelines, pumps, tanks and compressor stations, or liquefaction in the case of gases, although this is a longer term goal. These are therefore not included in most of the model runs for this study. Given, however, that these costs are high for hydrogen, a sensitivity analysis was undertaken on selected scenarios using an assumed variable O&M cost of R65 (2010) per GJ (Gupta, 2009) for hydrogen to reflect the costs of hydrogen distribution and dispensing. These costs are quite uncertain and depend on the distribution technology, and so the competing liquid fuels were modelled as having costs a third to a half of this number, to examine the sensitivity to relative costs.

5 KEY ASSUMPTIONS – DEMAND-SIDE

TIMES is a least-cost optimisation platform so cheaper technologies will tend to dominate. Real world consumer decisions with regard to passenger cars are not, however, based solely or sometimes even partially on cost. This reality has to be taken into account by imposing bounds on the rate of penetration of technologies. Given the cost reductions assumed for fuel cell vehicles and no bounds on penetration rate, they debut in the market near 2050 in an unconstrained model, but then displace all other technologies. An emerging hydrogen economy is likely to start earlier and proceed more gradually and bounds were imposed on all vehicle technologies to reflect this.

A reference case with which to compare our scenarios of alternative cases of possible outcomes for a hydrogen economy was required and it was decided that this would exclude fuel cell vehicles altogether. Bounds on market penetration were, however, imposed on the competing electric technologies of hybrid electric vehicles (HEV) and battery electric vehicles (BEV). An alternative case was developed that included fuel cell vehicles but also imposed bounds on their market penetration such that their numbers grow steadily from a small initial sales share, but attain significant market share by 2050 as shown below. The model optimiser can select a market share based on least-cost between the indicated upper and lower limits.

Table 29: Bounds on vehicle technology market penetration rates – reference case

Technology	Type	Base Year - 2006	Upper Bounds				Lower Bounds			
			U2010	U2023	U2030	U2050	L2010	L2023	L2030	L2050
Diesel ICE	SUV	40%	42%	57%	65%	90%	40%	20%	10%	5%
Gasoline ICE	SUV	60%	62%	76%	85%	100%	58%	46%	40%	25%
Diesel HEV	SUV	0%	0%	16%	25%	50%	0%	3%	5%	9%
Gasoline HEV	SUV	0%	0%	16%	25%	50%	0%	5%	8%	20%
Hydrogen FCV	**SUV**	**0%**	**0%**	**0%**	**0%**	**0%**	**0%**	**0%**	**0%**	**0%**
Diesel ICE	CAR	8%	10%	23%	30%	40%	8%	9%	10%	12%
Gasoline ICE	CAR	92%	100%	100%	100%	50%	61%	34%	20%	10%
Diesel HEV	CAR	0%	0%	20%	30%	80%	0%	1%	2%	4%
Gasoline HEV	CAR	0%	10%	23%	30%	80%	0%	4%	6%	25%
BEV	CAR	0%	10%	23%	30%	50%	0%	0%	4%	10%
Gas ICE	CAR	0%	10%	23%	30%	100%	0%	0%	0%	0%
Hydrogen FCV	**CAR**	**0%**	**0%**	**0%**	**0%**	**0%**	**0%**	**0%**	**0%**	**0%**

HEV: Hybrid Electric Vehicle
BEV: Battery Electric Vehicle
FCV: Fuel Cell Vehicle

Hybrid Vehicles are expected to gain significant market share because of range advantage combined with fuel economy, particularly in the high income segment and thus in the reference case they are constrained to at least 30% share of sales in the SUV and car markets by 2050, with BEV

Table 30: Bounds on vehicle technology market penetration rates – alternative case

Technology	Type	Base Year - 2006	Upper Bounds				Lower Bounds			
			U2010	U2023	U2030	U2050	L2010	L2023	L2030	L2050
Diesel ICE	SUV	40%	42%	57%	65%	90%	40%	20%	10%	5%
Gasoline ICE	SUV	60%	62%	76%	85%	100%	58%	46%	40%	25%
Diesel HEV	SUV	0%	0%	16%	25%	50%	0%	3%	5%	9%
Gasoline HEV	SUV	0%	0%	16%	25%	50%	0%	5%	8%	20%
Hydrogen FCV	SUV	0%	0%	0%	0%	0%	0%	0%	0%	0%
Diesel ICE	CAR	8%	10%	23%	30%	40%	8%	9%	10%	12%
Gasoline ICE	CAR	92%	100%	100%	100%	50%	61%	34%	20%	10%
Diesel HEV	CAR	0%	0%	20%	30%	80%	0%	1%	2%	4%
Gasoline HEV	CAR	0%	10%	23%	30%	80%	0%	4%	6%	25%
BEV	CAR	0%	10%	23%	30%	50%	0%	0%	4%	10%
Gas ICE	CAR	0%	10%	23%	30%	100%	0%	0%	0%	0%
Hydrogen FCV	CAR	0%	0%	0%	0%	0%	0%	0%	0%	0%

constrained to a minimum of 10% in the car market.

For the alternative case, FCV market penetration of new vehicle sales by 2050 is constrained to a minimum of 30% for cars, 35% for SUV where CO_2 regulations should drive faster uptake, and 60% for buses, which, being captive fleets and more cost driven, should take up commercially competitive fuel cell technology and hydrogen fuelling at a far higher relative rate if it were available. A few scenarios were run with no bounds on market penetration to illustrate the hypothetical outcomes if least-cost was the only technology decision and if hydrogen distribution infrastructure could be implemented without practical delays.

A critical assumption is for the investment costs of new technologies. The assumptions for future passenger car technology are shown below and have been modified from the original SATIM assumptions (Merven, Stone, Hughes, & Cohen, 2012), taking into account the premium of each technology relative to gasoline ICE project in the UK-SHEC assumptions (McDowall & Dodds, 2012). The rate of change in HEV costs was adjusted by current prices for HEV in South Africa (very high), and the US (surprisingly low) assuming that local prices would converge on US prices by 2020.

We assume thus that conventional gasoline ICE technology costs remain constant in real terms and competing technologies converge over the study period with the relative premium on HEV shrinking to 8% by 2050, and the premium for BEV and FCV shrinking to 20% and 15% respectively by 2050.

	Investment Costs (R000/vehicle)					
Technology	2006	2020	2030	2040	2050	Ratio to Gasoline ICE in 2050
Diesel ICE	129.7	127.2	124.7	122.1	119.6	1.06
Diesel HEV	297.1	146.4	136.6	132.9	129.2	1.14
Gasoline ICE	112.8	112.8	112.8	112.8	112.8	1.00
Gasoline HEV	258.3	129.8	123.7	122.8	121.8	1.08
BEV	296.2	256.0	215.8	175.6	135.4	1.20
CNG ICE	124.1	124.1	124.1	124.1	124.1	1.10
Hydrogen FCV	843.5	168.2	153.5	141.7	130.1	1.15

Table 31: Assumed investment costs for future passenger car technologies

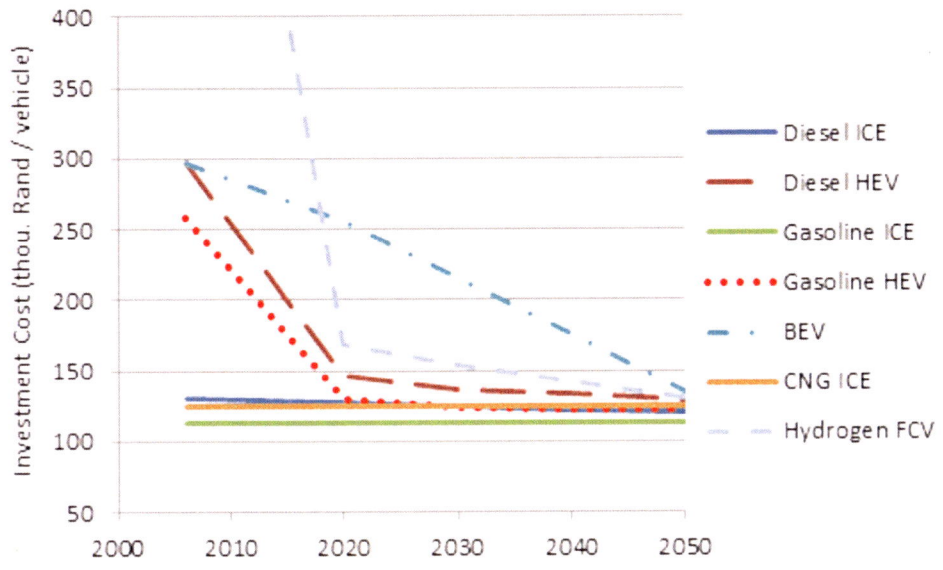

Figure 12: Assumed investment costs for future passenger car technologies

Some attempt was made to deal with the inherent uncertainty of these assumptions by means of a sensitivity analysis that inflated FCV costs in 2050 by 10%, 20% and 40%, scaling the curve accordingly. All costs shown are in 2010 Rands.

6 KEY ASSUMPTIONS – SUPPLY-SIDE

The three hydrogen production technologies that currently dominate world hydrogen production were set up in the model to supply hydrogen to the energy chain. These technologies transform other primary and secondary sources into hydrogen and have the following basic characteristics:

Technology Description	Lifetime	Efficiency[1]	Utilisation factor	Start Year	Lead Time (years)	Investment Cost[1]	Fixed O&M Cost[1]	CO₂
Units	years	(sum OUT/IN)	Annual			(mill. (2010) ZAR/PJa)	(mill. (2010) ZAR/PJa)	(kton/PJ)
Steam methane reforming (SMR)	40	80%	0.90	2020	4	73.84	2.46	88.6[2]
Coal gasification	40	65%	0.90	2020	4	254.34	12.31	180.8[3]
Water electrolysis	40	75%	0.90	2020	4	254.34	12.31	0.0

Table 32: Key assumptions for hydrogen production technologies in the base model

[1] Source - Dodds & McDowall (2012) for year 2000 without learning – so possibly conservative
[2] Source - Gupta (2009)
[3] Source - Simbeck & Chang (2002)

At this time, the model is limited to large centralised production and medium-sized and small distributed plants on the dispensing site were not investigated. The earliest start time for plant construction is 2020 with a construction lead time of four years. The earliest that hydrogen is available in the model is thus 2024. Stem methane reforming (SMR) and coal gasification have steam and electricity inputs in addition to the feedstock for which assumptions are presented below in Table 33.

Steam Methane Reforming (SMR)		Coal Gasification		Water Electrolysis	
Natural gas	94%	Coal	70%	Electricity	100%
Steam	0%	Steam	24%		
Electricity	6%	Electricity	6%		
SUM	100%	SUM	100%	SUM	100%

Table 33: Energy input proportions assumed for hydrogen production technologies

The SMR process includes exothermic oxidation of CO which supplies heat for steam production internally and, as such, the efficiency of SMR can be expressed as follows (Gupta, 2009):

$$\eta = \frac{E_{H2} + E_{steam.4.8\,MPa}}{E_{NG} + electricity + E_{steam.2.6\,MPa}}$$

The typical proportions of this steam balance were not found for this study, though, and thus steam inputs and outputs are assumed zero for now, and the natural gas feedstock is assumed to provide energy for process steam inclusive in the plant efficiency in Table 32.

A levelised cost[3] analysis of these costs outside of the model showed that fuel price has a large influence on production cost and that cheap coal favours coal gasification in spite of its inefficiencies and high capital costs. This is similar to the case of coal to liquids (CTL) production of liquid fuels in South Africa. The assumed price projections of coal and natural gas in the model thus have a large influence on which technology will be cheapest in the model over the time horizon.

The coal price was updated according to the recent MYPD3 application (ESKOM, 2013). The price of imported or domestically produced gas is indexed to the price of oil and thus rises as the oil price rises. The oil price is assumed to rise from around US$80/bbl in 2010 to just over US$150/bbl in 2040 as per the US EIA AEO Early Release Overview (EIA, 2013). The gas

---

3. The levelised cost of production by a supply technology is the production cost per unit of energy, taking into account the costs over the whole lifetime of the plant including capital costs, taxes, finance costs, operation and maintenance costs, and fuels costs taking plant efficiency and availability into account. Future costs are discounted by an assumed discount rate. Typically in SATIM a real discount rate of 8% is assumed.

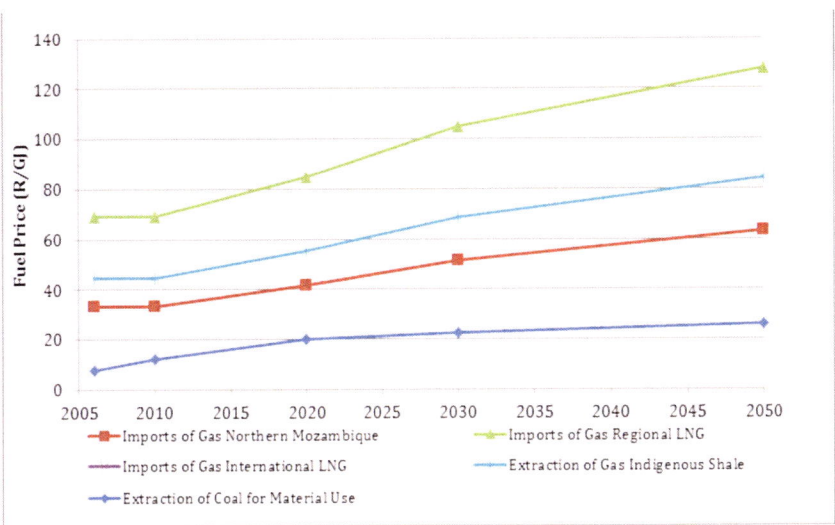

Figure 13: Assumed coal and natural gas feedstock prices (excluding infrastructure) for hydrogen production technologies All costs shown above are in 2010 rands

price seen by power plants includes supply infrastructure costs assuming a 90% utilisation factor. The detailed assumptions for the gas infrastructure costs and the oil indices for the different gas supply options are given in Table 34 and Table 35 respectively.

Gas Project		Ibhubezi/Shale	Northern Mozambique	LNG Terminal	FSRU
Range	km	500	2,500		
ODC	$m	100	501	110	41
EPC	$m	1,037[1]	4,517[1]	845[2]	315[3]
Total overnight cost	$m	1,137	5,017	955	356
Lead time	years	4	4	4	2
Capacity (size of minimum investment)	PJ/a	45.4	45.4	172.8	48
Model input overnight cost for infrastructure	2010 R/GJ	141.8	708.9	40.9	54.9

Table 34: Gas infrastructure cost assumption

1 Assuming $70k/km/inch of pipeline, and a 22-inch pipeline with compression
2 Assuming a 1xLNGC berth (3.6 mtons) with 2x165000 m3 tanks
3 Based on the 400m$ quoted in the media for a 1mton FSRU unit for the Petro SA GTL plant

The resulting hydrogen production feedstock prices for the model over its time horizon are shown below. Note the relative cheapness of coal and the divergence of coal and gas prices resulting from the linking of gas and oil prices.

Source of Gas		Ibhubezi	Domestic Shale	Northern Mozambique	LNG
Crude oil index	$/Mbtu/ $/bbl	0.10	0.08	0.06	0.12
Gas price at 100$/bbl	$/Mbtu	10.0	8.0	6.0	12.0

Table 35: Gas cost assumptions

7 Investment costs for renewable & nuclear technologies in the power sector

The version of SATIM used to investigate a hydrogen economy for this project is a 3-sector model and includes the transport, upstream, and power sectors. The latter is required because both transport and upstream sectors use electricity, and a hydrogen economy based on water electrolysis has significant implications for the power sector and its generation mix, particularly carbon neutral technologies. A hydrogen economy is an enormous risk, effort and expense to undertake and then have higher net emissions than before. Therefore, the assumptions around renewable and nuclear costs are of interest to this study. Investment costs for coal and gas, biomass and hydro technologies are derived from the IRP2010 in SATIM. The investment costs for Nuclear and Renewable Technologies have been updated to reflect current experience in the Renewable Energy IPP Programme.

8 Nuclear

The initial assumption for the overnight cost of nuclear plants in the IRP2010 was around US$3,500/kW. After stakeholder consultation this figure was adjusted upwards by 40% and an overnight cost of $5,000/kW was used in the IRP2010, despite the fact that Eskom had been quoted in the region of $6,000/kW by Areva in the 2008 bidding process. More recent publications of the cost estimates put the overnight cost in the region of $7,000/kW (Harris et al 2012). An overnight cost of $7,000/kW has been adopted for this study for scenarios where we attempt to drive an outcome of carbon-neutral hydrogen production via water electrolysis supplied by renewable power. The reference case assumption is $5,000/kW. Other parameters for nuclear remain at the IRP2010 values.

9 Renewable technologies

The recent REIPPP Windows One and Two have helped to uncover what some of the renewable technologies would actually cost in South Africa. Table 36 shows the project costs and total capacity for the second Window of the REIPPP of 2012 (DOE 2012). Given this data we can estimate what this means in terms of the overnight costs in 2010 rands, which we can use in the SATIM model.

Table 36: REIPPP window 2 cost data on renewable technologies

		Wind	PV	CSP
Total project cost	mR(2012)	10,897	12,048	4,483
Capacity	MW	563	417	50
Project cost	2012 R/kW	19,355	28,892	89,660
Project cost[1]	2010 R/kW	16,592	24,768	76,861
Lead time	years	2	1	3
IDC[2]		0.12	0.08	0.17
Overnight cost	2010 R/kW	14,772	22,933	65,766
Overnight cost[3]	2010 $/kW	1,996	3,099	8,887

Costs for these technologies are still coming down and are projected to continue to do so as the global installed capacity rapidly increases from a relatively low base. As per the modelling in the IRP2010, we assume that South Africa is a price taker on technology costs and that the learning would continue to depend on what happens globally rather than locally and therefore investment cost reductions due to technology learning are specified exogenously.

Three scenarios which reflect global technology learning were considered in this analysis, namely an 'optimistic', 'conservative', and 'pessimistic' cost reduction scenario. The 'optimistic' annual cost reductions for PV and CSP are based on the IRP2010 and are shown on the left-hand side of Figure 14 (page 364). For wind, the 'conservative' cost reduction scenario was based on IRP2010 as shown in Figure 15 (page 364). In all three cases, the annual cost improvement for the 'conservative' case is half that of the 'optimistic' case, and the 'pessimistic' case is half the 'conservative' case. The resulting impact on the overnight cost, using a 2012 value that is based on the REIPPP Window 2 (as per Table 36), is shown on the right hand side of the figures. For PV we see that the 'optimistic' case tracks the assumptions in the International Energy Agency's Energy Technology Perspectives (IEA-ETP 2012) quite well. For CSP with three hours of storage we see a similar pattern. The Sunshot and NREL figures fall somewhere in between our 'conservative' and 'pessimistic' cases for the CSP with six hours of storage, and similarly for CSP with 12 hours of storage. For the New Power Plan scenario we have assumed the conservative, mid-range, learning curves. Later we run sensitivity analyses that assume more optimistic learning curves.

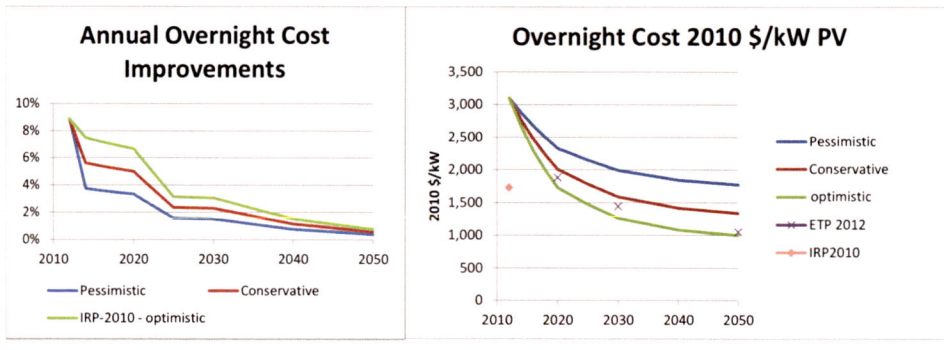

Figure 14: Annual cost reductions for PV

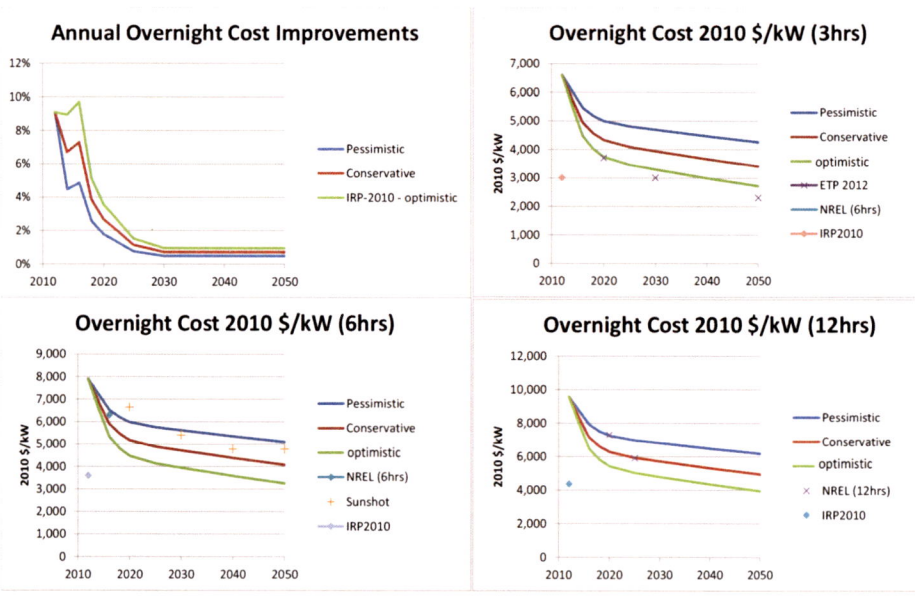

Figure 15: Annual cost reductions for CSP

10 SCENARIOS

The set-up of the model allowed the examination of the following aspects of a future hydrogen economy:

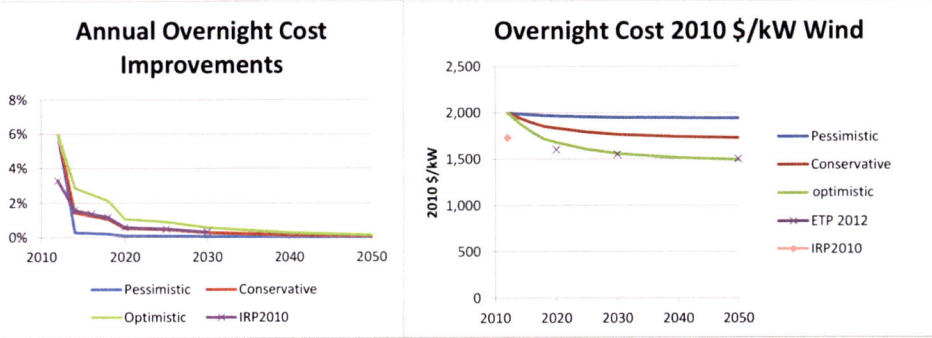

Figure 16: Annual Cost reductions for wind

- Market penetration of fuel cell vehicles;
- The impact of fuel cell vehicle roll out and hydrogen production on CO2 emissions;
- The switch to less carbon-intensive hydrogen production technologies using policy instruments like a carbon tax or carbon cap;
- The impact on a power generation mix of increasing demand for water electrolysis driven by a carbon tax or carbon cap; and
- The sensitivity of shares of technology competing with and within the hydrogen energy chain to cost uncertainty.

The scenario matrix that was adopted to achieve this is presented below.

There are three main scenarios as follows which have a number of 'cases' which investigate sensitivity to uncertainty, CO_2 taxes and a CO_2 cap:

- **REF** – Reference Scenario – Hydrogen FCV vehicles are constrained to zero.
- **NOCON** – Unconstrained Scenario – The model optimiser is unconstrained in the market share of competing vehicle technologies.
- **ALT** – Alternative Scenario – Hydrogen FCVs and other vehicle technologies are constrained to a 'realistic' growth rate between the roll-out of hydrogen in 2024 and 2050.

Scenario Name	Vehicle Market Share	FCV Costs	H2 Dist/Disp Costs[1]	H2 Prod. Techs	Renewable Learning	Nuclear Costs	CO2 Tax (R/ton)	CO2 Cap[2]
REF	**FCV Zero**	**Base**	-		**conservative**	-	-	-
NOCON	**No constraints**	**Base**			**conservative**			
NOCON1	No constraints	Base	double		conservative	-	-	-
NOCON2	No constraints	Base	triple		conservative	-	-	-
NOCON3	No constraints	+10%			conservative	-	-	-
NOCON4	No constraints	+20%			conservative			
NOCON5	No constraints	+40%			conservative			
NOCON6	No constraints	Base		+30%[2]Inv. Cost	conservative			
CAP1	No constraints	Base			conservative	High		Sector
CAP2	No constraints	+20%			conservative	High		Sector
CAP2	No constraints	Base	triple		conservative	High		Sector
ALT	**Bounds**	**Base**	-		**conservative**		-	-
ALT1	Bounds	Base	-		optimistic		-	-
ALT2	Bounds	Base	-	COG to ELE	optimistic			
ALT3	Bounds	Base	-	SMR to ELE	optimistic			
CO2A	Bounds	Base	-		conservative	High	R 2 000	-
CO2B	Bounds	Base	-		conservative	High	R 600	-
CO2C	Bounds	Base	-		optimistic	High	R 400	
CO2D	Bounds	Base	-		conservative	High	R 500	
CO2E	Bounds	Base	-		conservative	High	R 2 300	

Table 37: Modelling scenario matrix

1: Relative to conventional liquid fuels and gas distribution & dispensing costs.

2: 'Sector' means separate caps for transport, upstream & power sectors. 'System' means one cap for all.

APPENDIX B

HYDROGEN PRODUCTION COST DATA PUBLISHED BY THE NATIONAL ACADEMY OF SCIENCES

Scale	Primary Energy Source						
	Natural Gas	Coal	Nuclear	Biomass	Grid-Based Electricity (from any source)	Wind	Photovoltaics
Central station	CN: CS NG-C FN: CS NG-F CY: CS NG-C-Seq FY: CS NG-F-Seq	CN: CS Coal-C FN: CS Coal-F CY: CS Coal-C-Seq FY: CS Coal-F-Seq	F: CS Nu-F				
Midsize				CN: MS Bio-C FN: MS Bio-F CY: MS Bio-C-Seq FY: MS Bio-F-Seq			
Distributed	C: Dist NG-C F: Dist NG-F				C: Dist Elec-C F: Dist Elec-F	C: Dist WT-Gr Elec-C F: Dist WT-Gr Elec-F	C: Dist PV-Gr Elec-C F: Dist PV-Gr Elec-F

NOTES: C = current technology; Y = sequestration (hydrocarbon feedstock in central station and midsize plants only); F = future technology; N = no sequestration. The abbreviations for the hydrogen supply

SOURCE: (National Academy of Sciences, 2004)

Table 38: Hydrogen supply chain pathways examined by National Academy of Sciences study

Pathway	CS NG-C	CS NG-F	CS NG-C-Seq	CS NG-F-Seq	CS Coal-C	CS Coal-F	CS Coal-C-Seq	CS Coal-F-Seq	CS Nu-F
Production costs, $/kg H2									
Variable costs									
Feed	0.75	0.71	0.79	0.73	0.21	0.19	0.21	0.19	0.20
Electricity	0.03	0.03	0.08	0.06	0.11	0.04	0.17	0.08	
Decommission fund									
Decommission Fund									0.00
Non-fuel O&M, 1%/yr of capital	0.01	0.01	0.02	0.01	0.03	0.02	0.03	0.02	0.06
Total variable costs	*0.79*	*0.74*	*0.89*	*0.80*	*0.35*	*0.25*	*0.41*	*0.30*	*0.32*
Fixed costs, 5%/yr of capital	0.06	0.04	0.08	0.05	0.15	0.11	0.15	0.11	0.31
Capital charges, 18%/yr of capital	0.18	0.13	0.25	0.17	0.46	0.35	0.47	0.36	1
Total Production Costs	*1.03*	*0.92*	*1.22*	*1.02*	*0.96*	*0.71*	*1.03*	*0.77*	*1.63*
Distribution costs, $/kg H2									
Variable costs									
Labor	0	0	0	0	0	0	0	0	0
Fuel	0	0	0	0	0	0	0	0	0
Electricity	0.01	0.01	0.01	0.01	0.01	0.01	0.01	0.01	0.01
Non-fuel O&M, 1%/yr of capital	0.02	0.01	0.02	0.01	0.02	0.01	0.02	0.01	0.01
Total variable costs	*0.03*	*0.02*	*0.03*	*0.02*	*0.03*	*0.02*	*0.03*	*0.02*	*0.02*
Fixed costs, 5%/yr of capital	0.09	0.07	0.09	0.07	0.09	0.07	0.09	0.07	0.07
Capital charges, 18%/yr of capital	0.29	0.21	0.29	0.21	0.29	0.21	0.29	0.21	0.21
Total Distribution Costs	*0.42*	*0.31*	*0.42*	*0.31*	*0.42*	*0.31*	*0.42*	*0.31*	*0.31*
Dispensing costs, $/kg H2									
Variable costs									
Electricity	0.14	0.11	0.14	0.11	0.14	0.11	0.14	0.11	0.11
Non-fuel O&M, 1%/yr of capital	0.09	0.06	0.09	0.06	0.09	0.06	0.09	0.06	0.06
Total variable costs	*0.23*	*0.17*	*0.23*	*0.17*	*0.23*	*0.17*	*0.23*	*0.17*	*0.17*
Fixed costs, 5%/yr of capital	0.05	0.04	0.05	0.04	0.05	0.04	0.05	0.04	0.04
Capital charges, 18%/yr of capital	0.25	0.18	0.25	0.18	0.25	0.18	0.25	0.18	0.18
Total Dispensing Costs	*0.54*	*0.39*	*0.54*	*0.39*	*0.54*	*0.39*	*0.54*	*0.39*	*0.39*
H2 costs, $/kg									
Production	1.03	0.92	1.22	1.02	0.96	0.71	1.03	0.77	1.63
Distribution	0.42	0.31	0.42	0.31	0.42	0.31	0.42	0.31	0.31
Dispensing	0.54	0.39	0.54	0.39	0.54	0.39	0.54	0.39	0.39
CO_2 disposal	0	0	0.09	0.08	0	0	0.16	0.15	
Carbon tax	0.13	0.12	0.02	0.02	0.26	0.23	0.04	0.03	
Total H₂ Costs	*2.11*	*1.73*	*2.28*	*1.82*	*2.17*	*1.63*	*2.19*	*1.64*	*2.33*
Carbon dioxide vented to atmosphere									
kg carbon/kg H_2	2.51	2.39	0.42	0.35	5.12	4.56	0.82	0.60	
Direct use	2.45	2.34	0.26	0.24	4.90	4.50	0.49	0.45	
Indirect use	0.06	0.05	0.16	0.12	0.21	0.07	0.33	0.15	
kg CO_2/kg H_2	9.22	8.75	1.53	1.30	18.76	16.73	3.00	2.21	
Direct use	8.99	8.56	0.95	0.88	17.98	16.49	1.80	1.65	
Indirect use	0.23	0.18	0.58	0.42	0.77	0.25	1.20	0.56	
Carbon charge ($/kg)	0.13	0.12	0.02	0.02	0.26	0.23	0.04	0.03	
Carbon dioxide sequestered									
Carbon dioxide (kg CO_2/kg H_2)			8.56	7.91			16.19	14.84	
Carbon			2.34	2.16			4.41	4.05	

SOURCE: (National Academy of Sciences, 2004)

Table 39: Cost estimates for large plant centralised production of hydrogen

Pathway	Dist NG-C	Dist NG-F	Dist Elec-C	Dist Elec-F	Dis NGASE	Dis WI Ele-C	Dis WT El-F	Di PV El-C	Di PV El-F	Di WI Gr-C	Di WT Gr F	Di PV-Gr-C	Di PV-Gr-F
Capital investment, MM	1.85	0.96	2.54	0.57	1.3	6.86	0.89	9.94	1.43	2.75	0.59	2.74	0.59
Production costs $/kg H_2													
Variable costs													
Road tax or (subsidy)													
Gas Station mark-up					1.34								
Feed	1.37	1.17											
Electricity	0.15	0.12	384	3.31	0.4	3.29	1.9	17.48	4.64	3.67	2.75	6.57	3.58
Non-fuel O&M, %/yr of capital	0.12	0.06	0.16	0.04	0.08	0.44	0.06	0.63	0.09	0.17	0.04	0.17	0.04
Total variable costs	*1.64*	*1.35*	*4*	*3.35*	*1.83*	*3.73*	*1.95*	*18.11*	*4.73*	*385*	*2.78*	*6.74*	*3.61*
Fixed costs, %/yr of capital	0.23	0.12	0.32	0.07	0.16	0.87	0.11	1.26	0.18	0.35	0.07	0.35	0.07
Capital charges	1.64	0.85	2.26	0.51	1.15	6.09	0.79	8.82	1.27	2.44	0.52	2.43	0.52
Total Production Costs	3.51	2.33	6.58	3.93	3.15	10.69	2.86	28.19	6.18	6.64	3.38	9.52	4.21
Carbon Tax	0.17	0.14	0.24	0.21	0.18					0.17	0.12	0.19	0.17
TOTAL H_2 Costs	*3.68*	*2.47*	*6.82*	*4.13*	*3.32*	*10.69*	*2.86*	*28.19*	*6.18*	*6.81*	*3.5*	*9.71*	*4.37*
Carbon Dioxide Vented to Atmosphere													
KG Carbon/KG H_2	3.31	2.82	4.79	4.13	355					3.35	2.48	3.83	3.3
Direct Use	3.11	2.67			3.05								
Indirect Use	0.19	0.15	4.79	4.13	0.5					3.35	2.48	3.83	3.3
KG CO_2/KG H_2	12.13	10.34	15.13	15.13	13.03					12.28	9.08	14.04	12.11
Direct Use	11.42	9.79			11.19								
Indirect Use	0.71	0.55	15.13	15.13	1.85					12.28	9.08	14.04	12.11

SOURCE: (National Academy of Sciences, 2004)

Table 40: Cost Estimates for distributed plant production of hydrogen

APPENDIX C

SELECTED HYDROGEN ECONOMY ENERGY MODEL DATA PUBLISHED BY THE UK SHEC PROJECT

	Investment Cost (INVCOST) in millions of year 2000 GBP per billion vkm									
Year	Petrol ICE NH	Diesel ICE NH	Petrol HEV	Diesel HEV	Petrol PHEV	Diesel PHEV	FCV NH	FCHV	FC PHEV	BEV
2000	702	722								
2005	702	722								
2010	702	722	933	952	1577	1597	4048	3605	3662	2219
2015	690	728	762	781	1042	1062	1473	1396	1521	1255
2020	697	734	738	757	876	896	1039	1021	1073	1030
2025	703	741	738	757	860	880	1003	986	1027	998
2030	710	753	738	762	853	877	966	950	989	966
2035	710	753	734	757	840	865	929	915	952	934
2040	710	753	729	753	827	852	892	879	915	902
2045	710	753	725	748	815	839	856	844	877	869
2050	710	753	720	744	802	827	819	809	840	837

Source: (McDowall & Dodds, 2012)

Table 41: Investment cost data for energy models in UKSHEC II

	Vehicle efficiency data for models (bvkm/PJ)									
Year	Petrol ICE NH	Diesel ICE NH	Petrol HEV	Diesel HEV	Petrol PHEV	Diesel PHEV	FCV NH	FCHV	FC PHEV	BEV
2000	0.33	0.40								
2005	0.34	0.38								
2010	0.40	0.41	0.61	0.65	0.88	0.93	0.70	0.78	1.04	1.45
2015	0.45	0.54	0.73	0.77	1.04	1.10	0.80	0.89	1.18	1.57
2020	0.45	0.54	0.75	0.80	1.08	1.14	0.83	0.93	1.22	1.62
2025	0.45	0.54	0.78	0.82	1.12	1.18	0.87	0.97	1.27	1.68
2030	0.45	0.54	0.80	0.85	1.16	1.22	0.91	1.01	1.32	1.75
2035	0.46	0.55	0.82	0.87	1.18	1.25	0.94	1.05	1.36	1.79
2040	0.47	0.56	0.84	0.89	1.21	1.28	0.98	1.09	1.42	1.84
2045	0.48	0.58	0.86	0.91	1.24	1.31	1.02	1.14	1.47	1.89
2050	0.48	0.58	0.86	0.91	1.24	1.31	1.02	1.14	1.47	1.89

Note: Data for hydrogen internal combustion vehicles, CNG, LPG and flex-fuel E85 vehicles not shown, for brevity
Source: (McDowall & Dodds, 2012)

Table 42: Vehicle technology efficiency data for energy models in UKSHEC II

Table 43: Vehicle technology efficiency data for energy models in UKSHEC II

Year	Fixed O&M Costs in £/bvkm in year 2000 GBP									
	Petrol ICE NH	Diesel ICE NH	Petrol HEV	Diesel HEV	Petrol PHEV	Diesel PHEV	FCV NH	FCHV	FC PHEV	BEV
2000	84	85	98	99						
2005	84	85	98	99						
2010	84	85	98	99	136	128	272	246	249	165
2015	83	86	88	89	104	97	121	116	124	108
2020	84	86	86	87	94	87	95	94	97	95
2025	84	86	86	87	93	86	93	92	95	93
2030	85	87	86	88	93	86	91	90	92	91
2035	85	87	86	87	92	85	89	88	90	89
2040	85	87	86	87	91	84	87	86	88	87
2045	85	87	85	87	91	84	84	84	86	85
2050	85	87	85	87	90	83	82	82	84	83

Source: (McDowall & Dodds, 2012)

Table 44: Recommended TIAM-UCL investment cost and energy efficiency data for hydrogen production technologies for the period 2000–2050, for use in energy systems models

Technology	Size	First year	Investment costs			Energy efficiency		
			2000	2025	2050	2000	2025	2050
Coal gasification	Large	2000	31	27	24	65%	65%	65%
IGCC generation (after coal gasification)	Large	2020	217*	217*	217*	60%	60%	60%
SMR	Large	2000	9	7	5	80%	85%	85%
SMR	Medium	2000	21	17	14	75%	80%	80%
SMR	Small	2020	77	17	14	65%	80%	80%
Biomass gasification	Large	2020	26	26	26	50%	50%	50%
Biomass gasification	Medium	2020	52	34	34	50%	50%	50%
Biomass gasification	Small	2020	77	43	43	50%	50%	50%
Biomass oil pyrolysis	Medium	2020	52	34	34	50%	50%	50%
Waste gasification	Medium	2020	52	34	34	50%	50%	50%
Nuclear	Large	2030		69	69		75%	75%
Electrolysis	Medium	2000	31	17	17	75%	85%	90%
Electrolysis	Small	2000	112	27	17	75%	85%	90%

Note: Investment costs have units $(2005) GJ^{-1} y^{-1}, except for those denoted by * that have units $m(2005) GW^{-1}

Source: (Dodds & McDowall, 2012)

Table 45: TIAM-UCL hydrogen fuelling station best-estimate costs (US$(2005) GJ-1 y-1)

Year	LH2-LH2 tanker	LH2-GH2 tanker	GH2-GH2 pipeline (1)	GH2-GH2 pipeline (2)	GH2-GH2 on-site
2000	20	31	25	39	59
2010	18	28	23	34	50
2020	16	25	20	29	43
2030	15	23	18	26	37
2040	13	20	16	22	31
2050	12	18	15	20	27
Source:	(Dodds,	PE;	McDowall,	W;,	2012	b)

Appendix D

Market share of new technologies and the evolution of the vehicle parc in the model

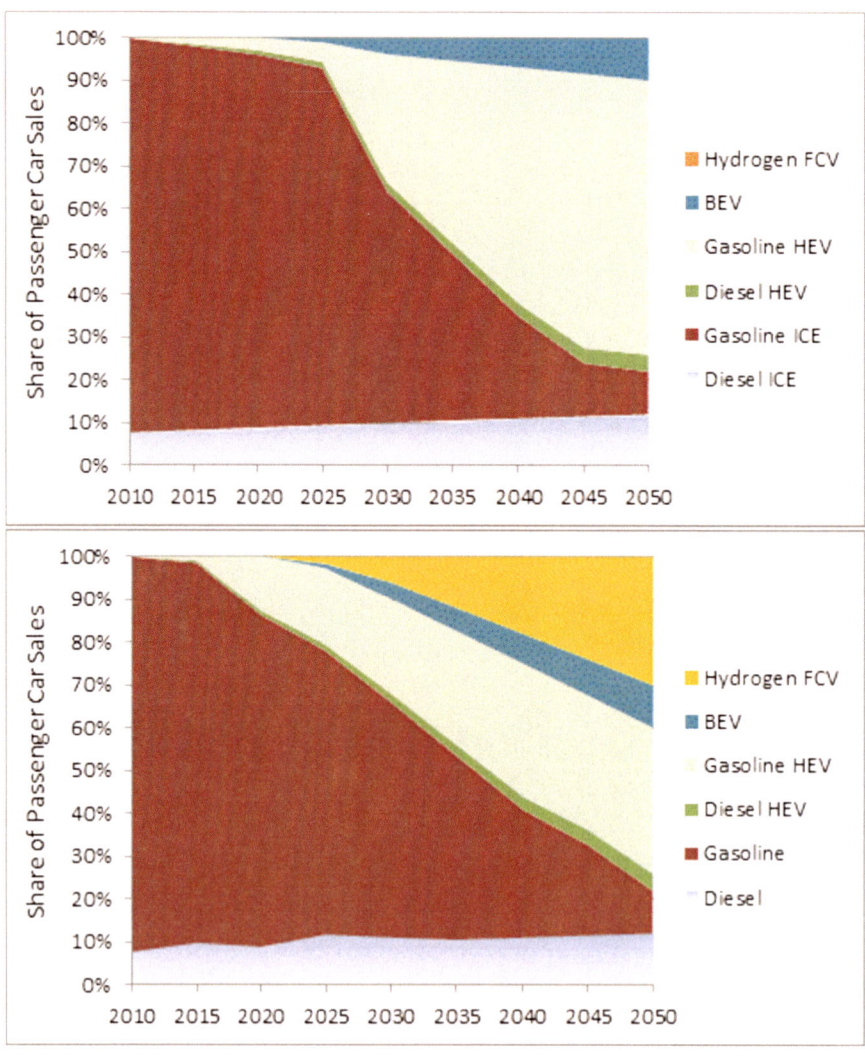

Figure 17: Share of total passenger car sales for the reference scenario (top) and alternative scenario (bottom)

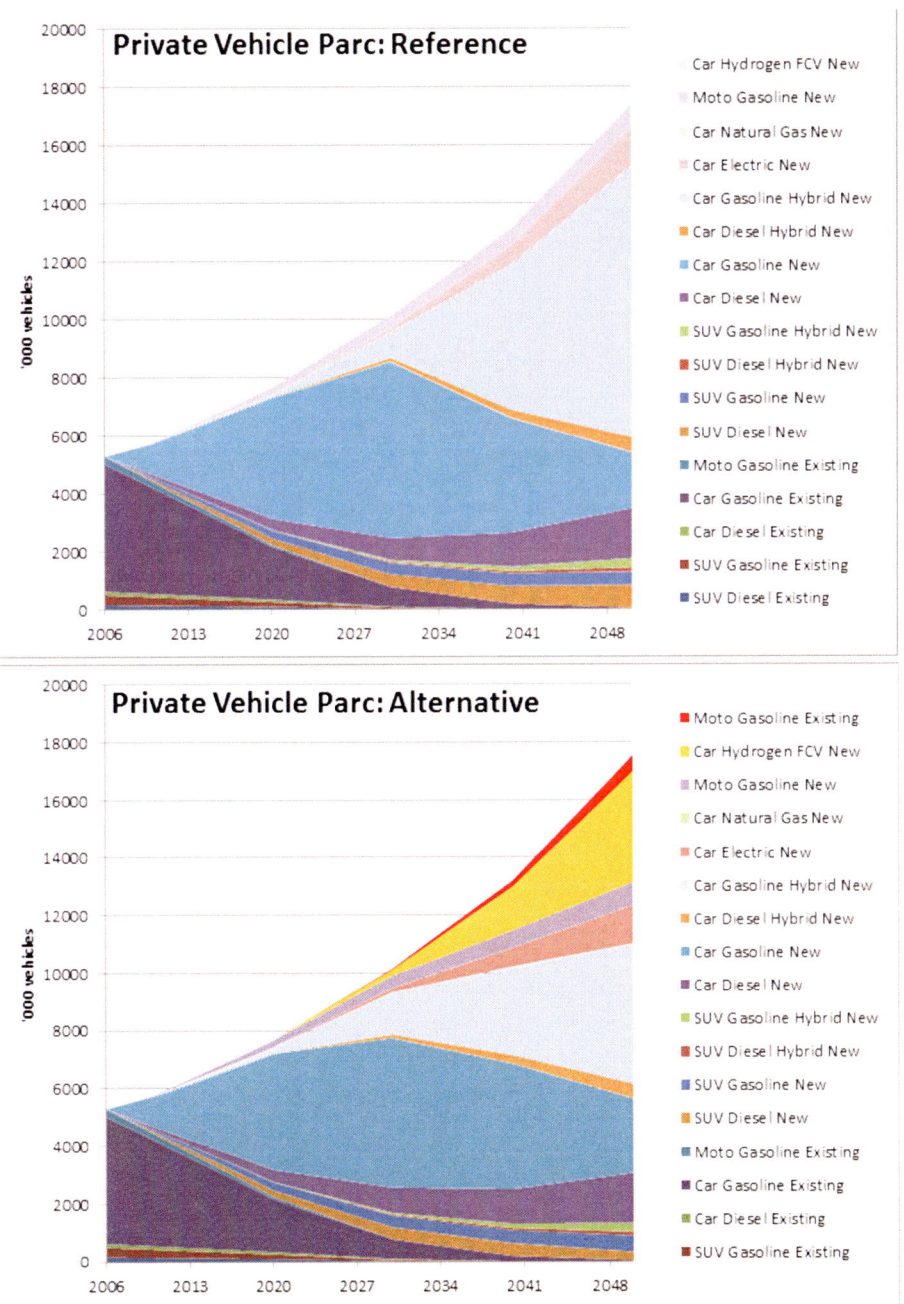

Figure 18: Evolution of the private vehicle parc for the reference scenario (top) and alternative scenario (bottom)

CHAPTER 6
APPENDIX A1

DISCOURSES OF GEOPOLITICS

A tabulation of the discourses of geopolitics is presented in Table 1 (after Ó Tuathail, 1998). 'Imperialist geopolitical' discourse evolved from imperialist strategy and racist white supremacist thought from the late 1800s, with key protagonists, the British Empire and the German state. This discourse was overtaken after World War II when a different world order emerged and 'Cold War' geopolitics predominated. The key protagonists in this period were the United States and the Soviet Union. The Cold War ended in the late 1980s when the Soviet Union fragmented, leaving the United States as the sole superpower. This heralded the New World Order geopolitical discourse, led by the US and by transnational liberalism and neoliberalism. 'Environmental geopolitics' is the most recent discourse to emerge. It has introduced the concept of sustainable development and focuses on questions of resource depletion and the political struggles over environmental change. In the last two decades scholars have begun to question the assumptions of those working in the discipline of geopolitics termed, 'critical geopolitics'.

It is important to take cognisance of the larger geopolitical context when setting the strategy for a policy. Economic advantages today are not necessarily in the hands of those with direct or territorial access to the mineral resource. The benefits emerge from negotiated access, and strategic collaboration. There is now distinction drawn between post-Cold War geopolitics and post-modern geopolitics (now termed critical geopolitics). What is becoming clear is that '[s]overeignty, therefore, operates through a number of spatial modalities: territorial, spatial-interactional, and place-based' (Agnew, 2009:21).

The New World Order has geopolitical and geo-economic dimensions. 'The geo-economic dimension of the New World Order involves the doctrine of transnational liberalism or neoliberalism. The fundamental principal of this doctrine is economic liberty for the powerful; that is, that an economy must be free from the social and political "impediments", "fetters" and "restrictions" placed upon it by states trying to regulate in the name of the

Discourse	Key intellectuals	Dominant lexicon
Imperialist geopolitics	Alfred Mahan	Sea Power
	Friedrich Ratzel	Lebensraum
	Halford Mackinder	Landpower / Heartland
	Karl Haushofer	Landpower / Heartland
	Nicholas Spykman	Rimlands
Cold War geopolitics	George Kennan	Containment
	Soviet & Western political & military leaders	First/Second/Third World countries as satellites & dominoes
		Western vs Eastern bloc
New World Order geopolitics	Mikahil Gorbachev	New political thinking
	Francis Fukuyama	The end of history
	Edward Luttwak	Statist geo-economics
	George Bush	US led New World Order
	Leaders of G7, IMF, WTO	Transnational liberalism / neoliberalism
	Strategic planners in the Pentagon & NATO	Rogue states, nuclear outlaws & terrorists
	Samuel Huntington	Clash of civilisations
Environmental geopolitics	World Commission on Environment & Development	Sustainable development
	Al Gore	Strategic environmental initiative
	Robert Kaplan	Coming anarchy
	Thomas Homer-Dixon	Environmental scarcity
	Michael Renner	Environmental security

public interest. These "impediments" – which include national economic regulations, social programmes and class compromises (i.e. national bargaining agreements between employers and trade unions, assuming these are allowed) – are considered barriers to the free flow of trade and capital…' (Routledge, 1998, p. 252). The nature of geopolitical discourse revolved around nation states as key players during the Cold War. This is no longer possible in quite the same way in view of the emergence of powerful transnational corporations. Consequently, the economy, defence, politics and information as the foundation for geopolitical positioning includes new actors, not only nation states.

'Environmental geopolitics', the most recent discourse to emerge, has introduced the concept of sustainable development and a focus on questions of resource depletion and environmental change.

Increasingly, the discourse highlights the need to recapture governments from global corporations. 'All over the world, national, provincial, and local governments have become the pawns of global corporations and the Corporate Agenda. Coalitions of popular movements and organisations, utilising tactics adapted to the political context at hand, need to challenge this domination. People need to reassert the right to use governments to regulate corporations and markets in the public interest' (Brecher & Costello, 1994, p. 301). Further, 'Globalisation by transnational corporations continues apace and the international financial markets are beginning to flex their muscles in ways that suggest that economic sovereignty for large powers other than the Germans and Americans is a historical matter. ...Meanwhile, at the other end of the geographical scale peasants continue to be dispossessed from lands in many villages around the world, feeding massive urban slums in the growing megacities of the "South"' (Dalby, 1998:307).

It is important to note as Karlin (2009) states: 'The first point [Smil, 2008] ... makes is that the basis of today's industrial system was formed a long time ago and that improvements since then paled in significance. "The most important concatenation of these fundamental advances took place between 1867 and 1914", when engineers realized electricity generation, steam and water turbines, internal combustion engines, inexpensive steel, aluminium, explosives, synthetic fertilizers, electronic components, thus laying the "technical foundations of the twentieth century" [much like men like Marx, Bismarck and Garibaldi laid its ideological foundations]. A second Golden Age occurred in the 1930s and 1940s, which saw the introduction of gas turbines, nuclear fission, electronic computing, semiconductors, key plastics, insecticides and herbicides'. Since then, the technological base has continued to evolve, becoming more pervasive, reflected in, for example, the growing role of social media. Such a technological base needs many reliable and uninterrupted supplies of energy for its existence, hence the prominence of discussions on the security of supply in relation to energy in the geopolitical discourse.

INDEX